RIVER QUALITY

Dynamics and Restoration

Edited by

Antonius Laenen
U.S. Geological Survey
Water Resources Division
Portland, Oregon

David A. Dunnette
Portland State University
Environmental Studies and Resources Program
Portland, Oregon

LEWIS PUBLISHERS

Boca Raton   New York

Contact Editor:	Joel Stein
Project Editor:	Carole Sweatman
Marketing Manager:	Greg Daurelle
Direct Marketing Manager:	Arline Massey
Typesetter:	Pamela Morrell
Cover design:	Dawn Boyd
PrePress:	Greg Cuciak
Manufacturing:	Sheri Schwartz

Library of Congress Cataloging-in-Publication Data

River quality : dynamics and restoration / edited by Antonius Laenen,
    David A. Dunnette.
        p.   cm.
        Includes bibliographical references and index.
        ISBN 1-56670-138-4 (alk. paper)
        1. Water quality—Oregon—Willamette River Watershed—Congresses.
    2. Water quality—Poland—Vistula River Watershed—Congresses.
    3. Water quality management—Oregon—Willamette River Watershed–
    –Congresses.   4. Water quality management—Poland—Vistula River
    Watershed—Congresses.   5. Water quality—United States—Congresses.
    6. Water quality—Poland—Congresses.   I. Laenen, Antonius,
    II. Dunnette, David A., 1939–
    TD224.O7R58   1996
    363.73′94—dc20

96-22308
CIP

© 1997 by CRC Press, Inc.
Lewis Publishers is an imprint of CRC Press

No claim to original U.S. Government works
International Standard Book Number 1-56670-138-4
Library of Congress Card Number 96-22308
Printed in the United States of America 1 2 3 4 5 6 7 8 9 0
Printed on acid-free paper

Contents

Restoration: Water-Quality Monitoring and Assessment

Restoration: Water-Quality Management

Acknowledgments

Proceedings of the International River Quality Symposium, March 22–24, 1994 in Portland, Oregon.

ORGANIZING COMMITTEE

Teresa Bogacka
Institute of Meteorology and Water Management
Gdańsk, Poland

Elżbieta Niemirycz
Institute of Meteorology and Water Management
Gdańsk, Poland

David Dunnette
Portland State University
Portland, Oregon

SPONSORSHIP

The following organizations and institutions contributed to the support of this symposium:

United States

- Portland State University
  Environmental Sciences and Resources Program
  Center for Public Health Studies
  Portland, Oregon
- U.S. Geological Survey
  Water Resources Division
  Portland, Oregon
- Oregon Department of Environmental Quality
  Portland, Oregon
- Metropolitan Service District (METRO)
  Portland, Oregon
- Unified Sewerage Agency
  Hillsboro, Oregon
- Environmental Protection Agency
  Portland, Oregon

Poland

- Ministry of Environmental Protection, Natural Resources, and Forestry
  Warsaw
- Institute of Meteorology and Water Management
  Warsaw
- Technical University of Gdańsk
  Gdańsk

This symposium was supported by a grant from the Maria Skłodowska-Curie Joint Fund of the United States-Poland Joint Commission.

Foreword

The idea for the symposia originated in Gdańsk, Poland in the winter of 1992. It was recognized that the Willamette River system of Oregon and the Vistula River system of Poland had been intensively studied and could both serve as instructive models for an international gathering of river-quality scientists and specialists. The Willamette River is an internationally cited case of successful water-quality restoration, and the Vistula River is a river under stress. In addition, the cities of Portland and Gdańsk possess many common qualities. Gdańsk, near the mouth of the Vistula River on the Baltic Sea, is the location for intensive scientific water-quality activities involving numerous agencies and institutions, is famous for its cultural and historical assets, and is the birthplace of democracy in Central and Eastern Europe. Portland, located at the mouth of the Willamette River, is also a national center of scientific water-quality assessment and management activity and is well known for its cultural and recreational assets and its dedication to environmental quality.

The main goal for this symposia was the promotion of international cooperation and understanding of river-quality assessment and management. A fundamental goal alongside of the main goal was to identify and explore opportunities for international partnerships and cooperation in research, education, and information-transfer between individuals and agencies in the U.S., Poland, and elsewhere. Intrinsic to these goals was the need for exchanging information and sharing perspectives on river quality. Although the primary focus of these symposia was the Willamette and Vistula River systems, it was recognized from the outset that other river systems should be encompassed in order to provide a sufficiently broad perspective.

These unique symposia consisted of two separate meetings: the first was held in Portland, Oregon, March 21–25 and the second in Gdańsk, Poland, June 13–17. They were supported by a grant from the Maria Sklodowska-Curie Joint Fund of the United States-Poland Joint Commission. The symposium proceedings of the meeting in Gdańsk, Poland will be reported in a later publication, *International River Quality,* edited by G. A. Best, T. Bogacka, and E. Neimrycz.

On March 23, 1994, after the International River Quality Symposium, participants attended an open meeting to discuss opportunities for continuing and enhancing international partnerships and cooperation in research, education, and information exchange. Attendees from the U.S. and Poland expressed their willingness and the willingness of their respective organizations for a continued partnership. Specifically emphasized was the importance of international exchange efforts such as education and technology transfer, especially related to wastewater treatment, wetland protection, and environmental inventories. Participants from academia suggested broader participation in Ph.D. research at institutions in Poland and the U.S., and participation in parallel research projects. Managers suggested an exchange of information particularly related to analytical methodologies and data storage/retrieval systems. All participants at the meeting felt that the opportunities for international cooperation in this river-quality area, and particularly rivers located in Central and Eastern Europe, were currently very high and that this partnership was developing within a unique window of opportunity.

As a result of these symposia and associated meetings, a proposal was sent from David Dunnette (Portland State University) and Elżbieta Niemirycz (Institute of Meteorology and Water Management, Warsaw) to the Maria Skłodowska-Curie Joint Fund II Commercialization Project on July 20, 1994. The project proposed was to continue the initial partnership through a continued exchange of ideas and personal exchanges between the two countries. Specific to the new proposal was the organization of several workshops, lectures, and technical exchanges.

Antonius Laenen

Dedication

This publication is dedicated to the memory of Professor David Arthur Dunnette, Ph.D. David's life was dedicated to conservation, environmental protection, environmental health, and ecology. He had passionate convictions about these issues, and he spoke, taught, worked, and lived a life that was consistent with these beliefs. To this end, he worked to develop a partnership between water-quality scientists and managers in the U.S. and Poland. One of the first steps in this partnership was to create a steering committee whose work initiated the Willamette-Vistula River Symposia. Other aspects of this partnership continue to evolve. David Dunnette died of complications of a viral illness on April 29, 1995. He was 56.

Prof. David Dunnette with the two other symposia organizers, Dr. Elżbieta Niemirycz (left) and Dr. Teresa Bogacka (right).

For the past ten years, Professor Dunnette taught Environmental Health in the Department of Public Health Education, and in the Environmental Sciences Doctoral Program. He was Director of the Center of Public Health Studies at Portland State University, Portland, Oregon. At the University, he developed a master's program in Public Health and helped develop the curriculum for a new undergraduate program in Environmental Studies. Prior to his university career, from 1973 to 1983, David worked as a senior scientist for the Oregon Department of Environmental Quality. In this capacity, he first conceived and developed water-quality indices that could be used in the management of water resources. His departmental responsibilities also included the assessment of many difficult problems of toxicology.

From 1991 to 1992, David was a Fulbright Scholar at the Gdańsk Technical University in Gdańsk, Poland. While there, he provided advice in the areas of transportation, hazardous chemicals in soil and water, and potential health effects due to water pollution. In Poland, David gained the respect of his colleagues and became instrumental in obtaining grant funds to begin development of a scientific exchange program between the two countries.

Scientists with the full scope of knowledge and the insightful perspective possessed by Dr. Dunnette are rare. He bridged the gap between human and ecological health, between scientist and manager, and between international boundaries. In his teaching, he included aspects of mathematics, physics, chemistry, ecology, and human understanding. David brought a well-rounded intellectual approach to environmental issues.

Dr. Dunnette is a well-published scientific author. The contents of his work transcend pure science — voicing concerns about human impacts on the environment. David strived to show how scientific information might best be used to provide better health for mankind while maintaining an ecological balance. The following selected bibliography is included for those interested in the same pursuits.

Selected Bibliography

Dunnette, D. A., Kowalik, P., Kreehniak, J., and Makinia, J., Perspectives on hazardous emission and public health in Poland. In *Hazardous Waste and Public Health: International Congress on the Health Effects of Hazardous Waste*, J. Andrews, et al., Eds., Princeton Scientific Publishing Co., Princeton, NJ, 178–181, 1995.

Dunnette, D. A., Assessing global river water quality I, Overview and data collection. In *The Science of Global Change — The Impact of Human Activities on the Environment*, D. A. Dunnette and R. J. O'Brien, Eds., American Chemical Society, Washington, D.C., 1992.

Dunnette, D. A., Assessing global river water quality II, Case study using mechanistic approaches. In *The Science of Global Change — The Impact of Human Activities on the Environment*, D. A. Dunnette and R. J. O'Brien, Eds., American Chemical Society, Washington, D.C., 1992.

Dunnette, D. A., Toxic chemicals in the environment. A perspective on the value of the risk assessment process. Batlyckie Forum Ecologiczne, Gdańsk, Poland, 1991.

Dunnette, D. A., Mancy, K. H., and Cynoweth, D. P., Use of a kinetic bioassay procedure to estimate sulfate and cysteine in sediment. *Water Res.* (England) 24, 1395–1400, 1990.

Dunnette, D. A., Assessing risks and preventing disease from environmental chemicals. *J. Com. Health* 14, 169–186, 1989.

Dunnette, D. A., Cooking while the house is burning. An environmental scientist's view of health education needs in the 1990's. *Health Educ.* 20, 4–8, 1989.

Dunnette, D. A., The origin of hydrogen sulfide in freshwater sediment. In *Biogenic Sulfur in the Environment*, E. Saltzman and W. Cooper, Eds., American Chemical Society, Washington, D.C., 1988.

Dunnette, D. A., Silent spring revisited, *AURA* (Poland), 1988.

Dunnette, D. A., Health risks from naturally occurring mercury in Oregon. *Ore. J. Environ. Health,* October, 1–4, 1988.

Dunnette, D. A., Environmental health chemistry — an emerging discipline. *Ore. Health Perspect.* 1, 2, 1986.

Dunnette, D. A., Mancy, K. H., and Cynoweth, D. P., The source of hydrogen sulfide in anoxic sediment. *Water Res.* (England) 19, 875–884, 1985.

Dunnette, D. A., Effect of an industrial ammonia discharge on the dissolved oxygen regime of the Willamette River, Oregon. *Water Res.* (England) 17, 997–1007, 1983.

Dunnette, D. A., Sampling optimization using a water quality index. *J. Water Pollut. Control Fed.* 52, 2807–2811, 1980.

Dunnette, D. A., A geographically variable water quality index used in Oregon. *J. Water Pollut. Control Fed.* 51, 53–58, 1979.

Dunnette, D. A., et al., The energy cost of environmental control. In: *The Energy Conservation Papers,* R. Williams, Ed., The Ford Foundation Energy Policy Project, Ballinger Press, 1976.

Preface

An appropriate preface to this publication would be to prompt the reader to think about river systems as they were conceived by the organizers of the International River Quality Symposium held in Portland, Oregon and presented at that meeting. The importance of understanding the physical, chemical, and biological dynamics of a river system and the realization that the restoration of these systems is imperative to our collective well-being is underscored in the following remarks. Dave Dunnette's introductory remarks have been lost, so I have taken the liberty of quoting from one of his publications. The remarks by Teresa Bogacka and Elżbieta Niemirycz were obtained directly from their personal notes.

The following is an excerpt from "Assessing global river water quality" by Dave Dunnette in *The Science of Global Change* (D.A. Dunnette and R.J. O'Brien):

> Rivers are of interest in a global context because, although the characteristics of rivers reflect predominantly local and regional conditions, they nevertheless are an indicator of the health of the total environment. This is because rivers, like the atmosphere and oceans, are integrative and reflect the conditions within their boundaries. Like the winds, rivers are powerful agents of the internationalization of environmental problems. Rainfall into a river drainage basin absorbs gaseous and particulate material, is supplemented by urban and agricultural runoff, melting snows, lakes, swamps, and groundwater, and discharges from municipal and industrial activity. Rainwater flows down land slopes and river beds, dissolving and transporting various chemical and geochemical species and receiving tributaries from adjacent subbasins including substances transported across political boundaries.

The following are notes from welcoming remarks given on Tuesday, March 22, 1994 at the symposium in Portland, Oregon:

> The symposium will focus on the following areas: (1) water quality assessments mainly for two rivers; the Willamette of Oregon and the Vistula of Poland, (2) chemical, biological, and physical dynamics of river systems, and (3) water quality management and control. The symposium was also designed to promote international cooperation and activity related to river quality improvement and water management.

> Taking the efforts to restore the water ecosystems is of great importance to the Polish society, which has the right to live in an unpolluted environment and consume their necessary quantity of clean water. It is also important to the societies of neighboring countries, because they are recipients of our devastated water and the repercussions to the flora and fauna.

Teresa Bogacka

> The Willamette and Vistula Rivers are about 12,000 km apart. It is a great distance. But taking into consideration the fact that the groundwaters are flowing all around the globe, we cannot exclude the possibility that these two rivers have at least some small influence on each other. Despite this distance, scientists, engineers, and ecologists (in other words, people who are studying rivers) have met together in Portland and will meet in Gdańsk.

> The rapid worldwide change of conditions in the past decades, caused by growing demands for water and increasing environmental deterioration, has resulted in the awareness that the disturbed equilibrium between exploitation of natural resources and environment should be restored. To find the ways of restoration of the aquatic environment should be one of the aims of this international meeting.

Elżbieta Niemirycz

The Editor

Antonius Laenen has recently retired from the U.S. Geological Survey (USGS) but continues to work with the USGS on a part-time basis as scientist emeritus. Prior to his retirement, he was Chief of the Hydraulic and Hydrologic Applications Section of the Portland, Oregon District Office of the USGS. Since retirement, he has continued his research with the Willamette National Water-Quality Assessment Program and has provided guidance for the Environmental Quality Section of the Oregon District.

Mr. Laenen graduated in 1967 from the University of Florida, Gainesville, with a B.S. degree in civil engineering. He has worked for the Water Resources Division of the USGS for 29 years in the state of Oregon. In his capacity as a hydrologist for the USGS, Mr. Laenen has spent extensive time collecting and analyzing streamflow and water-quality data and supervising others in these tasks. He has designed water-quality programs and has worked with river basin committees and water-resource managers to evaluate concerns and remediation.

Mr. Laenen is currently Secretary of the Board of Registration of the American Institute of Hydrology and is President of the Oregon Section. He, and others in the Oregon Section, were largely responsible for a water-quality category being added to the professional membership of the institution.

Mr. Laenen has presented over 20 papers at national and international meetings and has published more than 30 papers in his career. He was co-editor of the *1992 Proceedings of the American Institute of Hydrology: Interdisciplinary Approaches in Hydrology and Hydrogeology.* His current research is the definition of flows with implications for chemical interactions in the hyporheic zone of river systems.

Contributors

P. J. Ashton, Water Quality Information Systems, Division of Water Technology, CSIR, Pretoria, South Africa

Patricia A. Benner, Department of Fisheries and Wildlife, Oregon State University, Corvallis, Oregon

Gerald A. Best, Clyde River Purification Board, East Kilbride, Glasgow, Scotland

Marek Biziuk, Department of Analytical Chemistry, Technical University of Gdańsk, Gdańsk, Poland

Teresa Bogacka, Institute of Meteorology and Water Management, Gdańsk, Poland

Romuald Ceglarski, Institute of Meteorology and Water Management, Gdańsk, Poland

Russell N. Clayshulte, Denver Regional Council of Governments, Denver, Colorado

Jacek Czerwiński, Department of Analytical Chemistry, Technical University of Gdańsk, Gdańsk, Poland

Curtis L. DeGasperi, Tetra Tech, Inc., Redmond, Washington

S. Tad Deshler, Tetra Tech, Inc., Redmond, Washington

Andrezej Dobrowolski, Department of Hydrology, Institute of Meteorology and Water Management, Warsaw, Poland

Jan R. Dojlido, Institute of Meteorology and Water Management, Warsaw, Poland

Joseph L. Domagalski, U.S. Geological Survey, Water Resources Division, Sacramento, California

Doug Drake, Oregon Department of Environment Quality, Portland, Oregon

Steven G. Ellis, Tetra Tech, Inc., Redmond, Washington

William Fish, Department of Environmental Science and Engineering, Oregon Graduate Institute of Science & Technology, Portland, Oregon

Eugene Foster, Oregon Department of Environmental Quality, Portland, Oregon

Gregory J. Fuhrer, U.S. Geological Survey, Water Resources Division, Portland, Oregon

Marshall W. Gannett, U.S. Geological Survey, Water Resources Division, Portland, Oregon

Barbara Głowacka, Department of Hydrology, Institute of Meteorology and Water Management, Warsaw, Poland

Derek C. Godwin, Watershed Management Extension Agent, Oregon State University Extension Service, Gold Beach, Oregon

Gordon E. Grant, U.S.D.A. Forest Service, Pacific Northwest Research Station, Corvallis, Oregon

Marek J. Gromiec, Department of Water Management, Institute of Meteorology and Water Management, Warsaw, Poland

Desmond Hammerton, Clyde River Purification Board, Bothwell, Glasgow, Scotland

Alan Herlihy, Department of Fisheries and Wildlife, Oregon State University, Corvallis, Oregon

Elżbieta Heybowicz, Institute of Meteorology and Water Management, Gdańsk, Poland

Stephen R. Hinkle, U.S. Geological Survey, Water Resources Division, Portland, Oregon

Margaret A. House, Middlesex University, Enfield, United Kingdom

Waldemar Jarosiński, Institute of Meteorology and Water Management, Warsaw, Poland

Wesley M. Jarrell, Department of Environmental Science and Engineering, Oregon Graduate Institute of Science and Technology, Portland, Oregon

Philip R. Kaufmann, Department of Fisheries and Wildlife, Oregon State University, Corvallis, Oregon

Valerie J. Kelly, U.S. Geological Survey, Water Resources Division, Portland, Oregon

Tarang Khangaonkar, ENSR Consulting and Engineering, Redmond, Washington

Irena Kulik-Kuziemska, Technical University of Gdańsk, Gdańsk, Poland

Antonius Laenen, U.S. Geological Survey, Water Resources Division, Portland, Oregon

Judith Li, Department of Fisheries and Wildlife, Oregon State University, Corvallis, Oregon

Zbigniev Makowski, Institute of Meteorology and Water Management, Gdańsk, Poland

Cheryl L. Martin, Oregon Graduate Institute of Science & Technology, Beaverton, Oregon

Tim D. Mayer, Water Resources, U.S. Fish and Wildlife Service, Portland, Oregon

Stuart W. McKenzie, U.S. Geological Survey, Water Resources Division, Portland, Oregon

Adam Mierzwinski, State Inspectorate of Environmental Protection, Warsaw, Poland

Richard Miller, Taxon Aquatic Monitoring Service, Corvallis, Oregon

J. Ronald Miner, Bioresource Engineering Department, Oregon State University, Corvallis, Oregon

P. Morawiec, Institute of Meteorology and Water Management, Gdańsk, Poland

Neil Mullane, Oregon Department of Environmental Quality, Portland, Oregon

Jacek Namiesńik, Department of Analytical Chemistry, Tech. University of Gdańsk, Gdańsk, Poland

Elżbieta Niemirycz, Institute of Meteorology and Water Management, Gdańsk, Poland

Bogdan Ozga-Zielinski, Warsaw University of Technology, Institute of Environmental Engineering Systems, Warsaw, Poland

Gregory Pettit, Oregon Department of Environmental Quality, Portland, Oregon

Lou Reynolds, Wisconsin Department of Natural Resources, Bureau of Fish Management, Madison, Wisconsin

Joseph F. Rinella, U.S. Geological Survey, Water Resources Division, Portland, Oregon

John C. Risley, U.S. Geological Survey, Portland, Oregon

E. George Robison, Oregon Department of Forestry, Salem, Oregon.

William C. Romanelli, Bureau of Environmental Services, Portland, Oregon

Faith E. Ruffing, Sun Mountain Reflections, Portland, Oregon

Jacek Rulewski, Department of Chemistry, University of Gdańsk, Gdańsk, Poland

Jerzy Rybiński, Institute of Meteorology and Water Management, Gdańsk, Poland

Pat Sandra, Department of Organic Chemistry, University of Ghent, Ghent, Belgium

James R. Sedell, USDA Forest Service, Pacific Northwest Research Station, Corvallis, Oregon

Kenneth A. Skach, U.S. Geological Survey, Water Resources Division, Portland, Oregon

Maciej Smoręda, Department of Hydrology, Institute of Meteorology and Water Management, Warsaw, Poland

William J. Sobolewski, Consulting Engineer, Vancouver, Washington

Mary Lou Soscia, Columbia River Inter-Tribal Fish Commission, Portland, Oregon

Timothy D. Steele, HSI GeoTrans (A Tetra Tech Co.), Golden, Colorado

Karin Sundblad, Department of Water and Environmental Studies, Linkoping University, Linkoping, Sweden

Wojciech Szczepański, Institute of Meteorology and Water Management, Warsaw, Poland

Regina Taylor, Institute of Meteorology and Water Management, Gdańsk, Poland

Andrzej Toderski, Department of Water and Environmental Studies, Linkoping University, Linkoping, Sweden

Henk R. Van Vliet, Institute for Water Quality Studies, Department of Water Affairs and Forestry, Pretoria, South Africa

Dennis G. Woodward, U.S. Geological Survey, Water Resources Division, Albuquerque, New Mexico

Jan Zielinski, Institute of Meteorology and Water Management, Warsaw, Poland

Conversion Factors
[SI = International System of units, a modernized metric system of measurement]

Multiply	By	To obtain
Length		
micrometer (μm)	0.00003937	inch (in)
millimeter (mm)	0.03937	inch
centimeter (cm)	0.3937	inch
decimeter (dm)	0.3281	foot (ft)
meter (m)	3.281	foot
hectometer (hm)	0.06214	mile (mi)
kilometer (km)	0.6214	mile
Area		
square kilometer (km^2)	0.3861	square mile (mi^2)
Volume		
cubic decimeter (dm^3)	0.03532	cubic foot (ft^3)
cubic meter (m^3)	35.31	cubic foot
cubic hectometer (hm^3)	810.8	acre-foot (ac-ft)
microliter (μL)	0.000001057	quart (qt)
milliliter (mL)	0.001057	quart
liter (L)	1.057	quart
liter	0.2642	gallon (gal)
liter	0.03531	cubic foot (ft^3)
Mass		
microgram (μg)	0.00000003527	ounce (oz avoirdupois)
milligram (mg)	0.00003527	ounce
gram (g)	0.03527	ounce
kilogram (kg)	2.205	pound (lb)
metric ton	2205.00	pound
Density		
grams per cubic centimeter (g/cm^3)	0.5780	ounces per cubic inch (oz/in^3)
Temperature		
degree Celsius (°C)	°F = 1.8 (Temperature °C) + 32	degree Fahrenheit (°F)
Concentration in Water		
micrograms per liter (μg/L)	1	parts per billion (ppb)
milligrams per liter (mg/L)	1	parts per million (ppm)
Concentration in Sediment and Tissue		
micrograms per gram (μg/g)	1	parts per million

Introduction

1 The Willamette and Vistula Rivers: Contrast and Comparison

Antonius Laenen and Jan R. Dojlido

The Willamette and Vistula Rivers are both low-gradient streams located in temperate latitudes. At first glance, they seem to have more dissimilar characteristics than similar ones. If, however, we regard the two rivers as being in different stages of restoration, then some good water-quality contrasts and comparisons can be made between them. Both basins have been subjected to stress from agriculture, industry, and urbanization, allowing insights into the physical, chemical, and biological dynamics in river ecosystems. The Vistula River Basin is much larger and dryer than the Willamette River Basin and contains industry and agriculture characteristic of a self-supportive country. The Willamette River Basin is one-sixth the size of the Vistula River Basin and has less diverse agricultural and industrial activities. The Willamette River, however, is an internationally cited example of a restored river; the restoration process of the Vistula River has just begun. Much can be learned from the histories of each basin and from the data collected and interpreted within those basins.

Readers are encouraged to consider the entire physical, chemical, and biological dynamics of a river system and to ask the following questions:

1. What constitutes the restoration of a river system?
2. Was the Willamette River totally restored? Does it need further or continued restoration?
3. Is the Vistula River responding to cleanup measures and new management controls? How far in the restoration process has it come and will the current measures be enough to bring ecological restoration to the river?

HISTORIES

WILLAMETTE RIVER BASIN

The geologic structure and rocks of the Willamette River Basin (Figure 1) are generally much younger than their counterparts in Poland, although both areas have been affected by Pleistocene glaciation. The Coast Range, which forms the western part of the basin, consists of marine sedimentary rocks deposited 2 to 60 million years ago, during the Tertiary period. The Cascade Range, which forms the eastern part of the basin, consists primarily of andesitic volcanic deposits of Tertiary and Quaternary age, 0.01 to 14 million years old. Even to this day, volcanism continues in the Cascade Range. Mt. Hood last erupted in about 1800, just before Lewis and Clark journeyed to the area in 1805, Mt. St. Helens, just 80 km north of Portland (but outside of the basin), last erupted in 1982. During Pleistocene times, large glacial outburst floods that originated in the upper Columbia River Basin entered the Willamette Valley and deposited hundreds of meters of sediment at lower elevations. Later, in the fifteenth and sixteenth centuries during the "Little Ice Age," many of the mountain glaciers advanced and then retreated creating numerous cirque and moraine lakes in the upland watersheds of the Cascade Range.

0-56670-138-4/97/$0.00+$.50

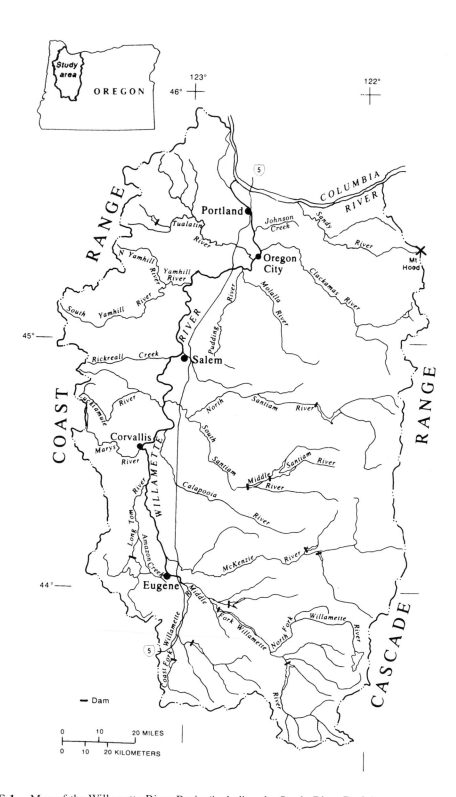

FIGURE 1. Map of the Willamette River Basin (including the Sandy River Basin).

It is generally assumed that prior to settlement by non-Indians, the Willamette River Basin remained in a natural state; this, however, is not true. The Indians commonly burned the lowlands so that berries, nuts, grass, and other edibles could be harvested. Reducing the forests and the undergrowth probably enhanced hunting as well. The practice of burning probably began in about the fourteenth to fifteenth century, when the Indians that dwelled along the Columbia River began to populate the Willamette Valley. Fires were not a frequent natural phenomenon in the valley, because lightning is not common.

The first major anthropogenic changes to the basin began in the early 1840s concurrent with the settlement of the area by non-Indians. Wagon trains carrying settlers left the economically depressed Missouri and Ohio River Valleys to travel the Oregon Trail. Their destination was the Willamette Valley and the Puget Sound area where the fabled "milk and honey" could be found. The lowlands were rapidly cleared for farming by these pioneers. By the time Oregon City became the state capital in 1845, the Willamette Valley had a population of about 1300, and mills for sawing wood and grinding grain were established along the main stem and tributaries. By 1860, the valley population had increased to almost 40,000. In 1870, the U.S. Army Corps of Engineers began clearing snags and dredging the main stem channel for navigation upstream to Eugene. The Corps cleaned out the channels and protected banks, and farmers cut trees that once grew back to the river banks in order to establish farms in the fertile alluvium of the Valley.

Logging became an important industry by 1900; about 300,000 people were living in the Willamette Valley by this time. In the early days of logging, the easiest way to harvest and transport timber was to clear-cut an area close to a major stream. It was common practice to dam a stream with logs, fill the pond behind with more logs, then blow up the "splash" dam with dynamite to wash all the logs downstream for further transportation or milling. Although transport practices like the use of "splash" dams have not been used for more than half a century, they were extremely destructive to stream channel environments at the time. By 1920, many pulp and paper mills had been established in the Willamette River Basin, and water quality began to decline rapidly.

By 1950, the Willamette River had become an open sewer, transporting refuse from paper mills, canneries, slaughterhouses, and communities. The bacteria count rose, oxygen levels dropped to near zero in some reaches, and fish died. Rafts of sludge covered the water surface, and blood from slaughterhouses turned the river red. The Willamette River was the most filthy waterway in the Northwest and one of the most polluted in the U.S. Legislation was passed in 1967 to control the effluents causing the pollution. Pulp and paper mills — once the worst polluters — installed chemical recovery and secondary waste-treatment facilities. All incorporated communities in the basin installed secondary waste-treatment facilities, and some communities installed tertiary treatment. By 1970, the deteriorating chemical and bacteriological quality of the Willamette River was significantly reversed.

VISTULA RIVER BASIN

In north-central Europe, landforms reflect the effects of continental glaciation during the Pleistocene epoch, which ended about 10,000 years ago. The most notable features created by these glacial advances and retreats are the thousands of lakes and hillocks that are located in northern and central Poland (Figure 2). The Carpathian Mountain Range and adjoining Beskidy Mountains that form the southern boundary of Poland consist primarily of rocks of Mesozoic age (65 to 230 million years old) that were uplifted from 22 to 100 million years ago. Volcanic activity last occurred in the eastern Carpathians about 65 million years ago. At the base of these mountains are the areas of Upper and Lower Silesia, which consist of flat lying Mesozoic and Tertiary sediments that were deposited on folded metamorphic and sedimentary rocks of Carboniferous age (280 to 345 million years). These rocks contain abundant deposits of salt, coal, and sulfur.

Human cultural history in the Vistula River Basin greatly predates that of the Willamette River Basin. Since the early Stone Age, the Vistula River has served both as a trade route and a means of expansion to both the north and the south. Initially, raw materials and flint tools were sent

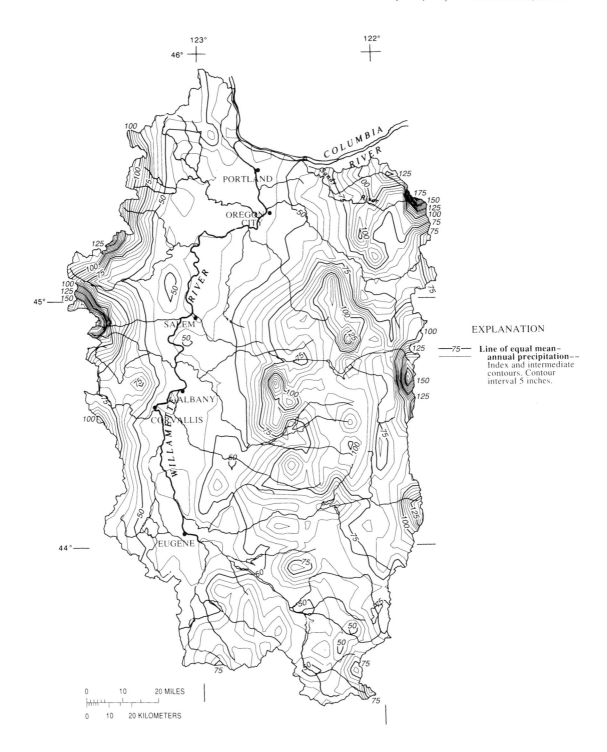

FIGURE 3.   Mean annual precipitation in the Willamette Basin, Oregon, 1961 to 1990.

    The upper Willamette River is called the Middle Fork Willamette River. The river originates
in a mountain pass at an elevation of about 2340 m and, after flowing for 196 km, reaches an
elevation of 135 m at the confluence with the Coast Fork Willamette River near Eugene. The main
stem of the Willamette River is formed by this confluence and flows another 301 km to the Columbia
River. The main stem can be divided into four distinct reaches. The upper reach from Eugene to

Corvallis is 110 km in length and is characterized by a meandering and braided channel with many islands and sloughs; the river is shallow, and the bed is composed almost entirely of cobbles and gravel. The middle reach from Corvallis to the confluence of the Yamhill River is 103 km in length and is characterized by a meandering channel deeply incised into the valley floor. The next reach from the Yamhill River to Oregon City is a 46 km long natural pool impounded behind the 15 m high Willamette Falls. Hydraulically, the deep, slow-moving pool can be characterized as a reservoir. The pool is a depositional area for small gravel- to silt-size material and provides a source of gravel, which is regularly mined from the streambed. The lower 42 km of river below the falls is affected by tides, and, during spring and early summer, by backwater from the Columbia River.

The 11 major reservoirs in the Willamette River Basin (Figure 1) have a combined usable capacity of nearly 2300 hm^3. None of the reservoirs are on the main stem of the river. The reservoirs are designed for multipurpose use, but their primary, legally designated function is the maintenance of a minimum navigable depth on the main stem of the Willamette River during the summer. The required minimum flow for navigation is 170 m^3/s at Salem. Most of the flow in the Willamette River from November to March is a result of frequent winter rainstorms and spring snowmelt. About 35% of the annual runoff is from snowmelt that flows either directly to the stream or indirectly through the groundwater system. Reservoir regulation affects low flows in the Willamette River and has curtailed the duration of low flow, which prior to regulation typically occurred from mid-July to mid-October. Since regulation, low flow now occurs from mid-July to mid-August. Increased flows from mid-August to mid-October are utilized to facilitate the mitigation of anadromous fish.

The Willamette River Basin contains Oregon's three largest cities: Portland, Eugene, and Salem. Approximately 1.91 million people, representing 69% of the state's population (1990 census), live in the basin. The basin supports an economy based on timber, agriculture, manufacturing, and tourism. Timber (forest harvesting, forest products, and pulp-paper manufacture) is still the largest industry in Oregon, with an estimated economic value of $8.5 billion (U.S. dollars) annually. Approximately 70% of the basin area is forest, with 60% of these forests being publicly owned (mostly federal). Timber will continue to be a leading industry in the state for at least another decade. Agriculture is the second-leading industry in Oregon (est. $8 billion); the state leads the nation in the production of Christmas trees, grass seed, hazelnuts, peppermint, raspberries, black-berries, and is a major producer of hops, strawberries, prunes, plums, onions, cauliflower, apples, pears, and nursery products. Tourism is Oregon's third-largest industry. Skiing and water sports such as fishing, wind surfing, and whitewater rafting are some of the state's most popular attractions.

Portland (metropolitan population 0.84 million in Oregon, 1.2 million including population in Vancouver, Washington, and vicinity) is the largest population center in the Willamette River Basin. The Port of Portland (Figure 4) on the Willamette River, located 176 km from the Pacific Ocean, is Oregon's largest and most diversified port. It exports a larger volume (Portland is third in tonnage of all the West Coast ports) of goods than any other West Coast port. In 1993, imports and exports totaled $12 billion, with Japan (first) and Canada (second) as the major sources and destinations. The port consists of five marine terminals served by three major rail lines and two interstate highway systems. Shipping on the Willamette River, however, extends only 27 km upstream from the mouth. Grain, forest products, and containerized goods are trucked, carried by rail, or barged to and from destinations. The Portland ship repair yard handles more than 45% of all commercial repair work done on the West Coast. In addition, Portland has become an area of growth in high-tech industries, such as electronics manufacturers, with an estimated combined economic value of $4.5 billion reported in 1993.

Vistula River Basin

The Vistula River Basin (Figure 2) covers an area of 194,400 km^2, including 25,700 km^2 outside of Poland. The basin occupies 64% of Poland as defined by the current political boundaries. The basin is bounded on the south by the Carpathian Mountain Range and on the north, east, and west

FIGURE 4.   Photograph of the port facilities in Portland. Mt. Hood is shown in the background.

by low-elevation divides. About 100 km south of Cracow, the mountains of Rysy in Poland, and Gerlach in Slovak, rise to elevations of 2499 m and 2655 m, respectively. In contrast, to the east, the divide between the Vistula and Dnepr River Basins becomes nearly indistinguishable, and the two river systems are connected by the Dnepr-Bug Canal. To the west, the basin divide also becomes hard to define. Near the city of Bydgoszcz, a canal connects the Vistula River with the Noteć River, a major tributary to the Oder River (Odra in Polish). At this location, the drainage divide is only about 10 km from the main Vistula River channel. To the north, low hills and many small lakes create a highly irregular drainage boundary. Basin elevations range from 2600 m in the Carpathiam Mountains to sea level at the Baltic Sea; the average elevation of the Vistula River Basin is 270 m. The basin consists of 40% mountains or hills and 60% lowlands.

The Polish climate ranges from oceanic to continental, and varies not only geographically from north to south and east to west, but seasonally depending on storms spawned in the Baltic and North Seas. Weather conditions are subject to marked changes. Winter is relatively short, lasting only from December through February; however, winter conditions can be severe and in many northern regions temperatures drop to –30°C. Most years, but not every year, an ice sheet forms on the main stem of the Vistula River from about mid-January to late February. Summer is generally hot with temperatures often exceeding 30°C. Late summer and fall are usually dry. Mean annual precipitation (Figure 5) is about 700 mm in the north part of the basin, 525 mm in the central part, and 1100 mm at the crest of the Carpathian Mountains. Precipitation is sparse and the water resource is severely limited. After evapotranspiration and human consumption is considered, only 175 mm of the average precipitation of 620 mm that falls on the basin flows into the Baltic Sea. The period of high water usually occurs in early spring and is caused by snowmelt.

The Vistula River originates from two small rivers: the Biała Wiselka and the Czarna Wiselka. From this union to the confluence with the Przemsza River, the Vistula caries the name Mała Wisła (Small Vistula). The river from Przemsza to the Baltic Sea is named Wisła in Polish and Vistula in English. At a distance of 1063.3 m, the headwaters of the Vistula River originate in the Carpathian Mountains at en elevation of about 1100 m. The Upper Vistula River course begins at the confluence of the Przemsza River at river kilometer (RK) 941.3 and ends at the confluence of the San River at RK 661.6. The middle course of the Vistula River extends from the San River to the confluence with the Narew River at RK 390.8. In its middle and lower courses, the river has a low gradient of about 1 m/km. In its middle course from Cracow to Warsaw, the river has many braided channels,

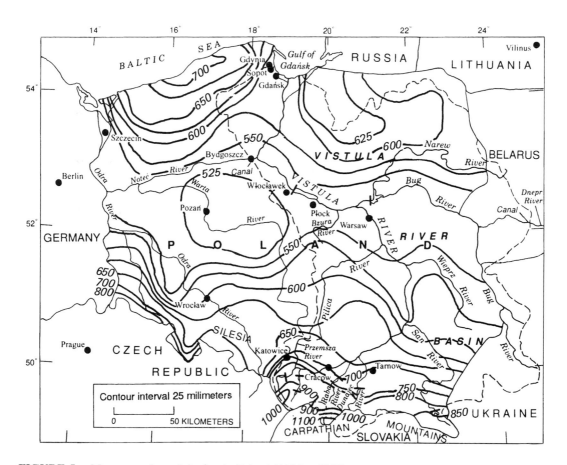

FIGURE 5. Mean annual precipitation in Poland (1966 to 1990).

permanent and temporary islands, and a richly vegetated flood plain. All along the river there are longitudinal flood-protection dikes to prevent the constant problem of bank erosion. From RK 54.7 to the mouth (RK 0.0) a delta has formed, and the river flows to the Baltic Sea in three primary channels. The delta land, bounded by the former Vistula channel and the Nogat channel, is diked and farmed. The former main channel of the Vistula River connects Gdańsk to the river and the Baltic Sea.

The Vistula River is essentially unregulated and therefore has a large variation in flow throughout any given year, and from wet years to dry years. There are eight significantly large reservoirs in the Vistula River Basin (Figure 2), with a combined capacity of about 1500 hm³. The reservoirs are primarily used to supply water, but two of the dams generate power (hydroelectric power production in Poland represents only 3% of the total power production). Reservoir regulation has little impact on low flows in the river. In fact, reservoir regulation, along with deforestation and agriculture, have resulted in an increase of the annual fluctuation of discharge and have contributed to stream erosion. Pollution and eutrophication are continual problems in many of the reservoirs. Vistula River water is used to cool many power plant operations and industrial processes. As a result, thermal pollution of the river also exists.

In Poland, the Vistula River Basin is inhabited by 22.8 million people (1990 census) — about 61% of the country's population. The average population density of the basin is 122 inhabitants per km², but it is much higher in the Upper Vistula River Basin in the vicinity of Cracow, where there are 400 inhabitants per km². Three large cities are located along the Vistula water course:

Cracow in the upper course, Warsaw in the middle course, and Gdańsk in the lower course. The Vistula River Basin supports an economy based on industry, mining, and agriculture. Most industry and mining are located in the upper part of the basin, upstream from Cracow in the industrial area of Upper Silesia. Light and heavy industry of almost every description is represented in Poland. Oil refineries, steel mills, chemical plants, textile manufacturers, tanneries, paper mills, automobile and tractor assembly plants, and animal and crop processing plants are just a few of the larger industries. Coal mining is the primary mining activity in the area (130 to 150 million tons annually), although sulfur and other minerals also are mined. Most mining is underground, and land subsidence has become a problem in some areas. Hard coal (anthracite) and brown coal (coke) are both used for power production and industry within the country. Agriculture is found in almost all parts of the basin, but the central part is the most productive region. Primary crops include potatoes, sugar beets, wheat, oats, apples, and such specialty crops as strawberries, black currants, and mushrooms. Cattle and swine are also raised in sufficient numbers for exports of beef and pork. In Poland, for both heavy industry and agriculture, the estimated economic value is several tens of billions of U.S. dollars. In 1995, exports ($14.0 billion) exceeded imports ($13.5 billion), with Germany and Russia being the leading countries for both import and export goods.

Warsaw is the capital and cultural center of Poland. During World War II, 700,000 citizens lost their lives when the city was nearly totally destroyed. Now, above the slow-flowing Vistula River, a city center has been rebuilt from the rubble in a faithful reconstruction of the original. Warsaw is the largest population center in Poland with about 1.7 million inhabitants, many of whom dwell in homes and apartments outside the older city center. The drinking water in Warsaw is of poor quality and taste. Drinking water comes from two sources: the Vistula River which requires much filtration, and a reservoir on the Narew River which contains high concentrations of humic acids.

The Port of Gdańsk is situated near the Baltic Sea on the Gulf of Gdańsk and on one of the distributary channels of the Vistula River delta. From the tri-city of Gdańsk, Sopot, and Gydnia (population about 0.9 million), goods are distributed to and from ports worldwide. Gdańsk is the largest and most diversified port on the Baltic Sea and is one of Europe's largest harbor cities. Produce and containerized goods are trucked and carried by rail to and from destinations throughout Poland and central Europe. The Gdańsk shipyards are famous for being the cradle of the Solidarity revolution and as the builders of world-class ships. Although they now only operate at about 50% of their production capacity, the shipyards are nonetheless a significant asset to the Polish earning power.

A significant geomorphic feature of the Vistula River Basin is the large delta created by the river in the vicinity of Gdańsk. Floods and siltation continue to change the lower course of the river in this area, and hundreds of kilometers of levees and canals have been constructed to protect and drain the adjacent lowlands. At the beginning of the twentieth century, engineers cut a new main channel straight through to the Gulf of Gdańsk to help drain the flooded lands and prevent future flooding. The former natural channel of the Vistula River called the Motława Vistula still flows through the city of Gdańsk (Figure 6).

Cracow is the third largest city in Poland (population 0.8 million) and the former capitol of Poland (from the middle ages to the end of the sixteenth century). It is situated on the Vistula River just downstream from the industrial area of Upper Silesia. Cracow once used the Vistula River for drinking water, but it can no longer do so because of pollution from toxic chemicals. The highest pollution levels are found in the river reach between the Przemsza River and the Dunajec River. Cracow must now import water a great distance from the Dobczyce reservoir on the Raba River.

CONTRASTS AND COMPARISONS

Low precipitation, intensive industrialization, and a relatively large human population combine to impose a great environmental stress on the Vistula River Basin. In contrast, the Willamette River has the benefit of a large quantity of precipitation and a smaller population, so it experiences much less environmental stress. In each situation, however, both basins are the most stressed at times

FIGURE 6. Photograph taken from St. Mary's Church in Gdańsk showing the Motława Vistula (no longer the main course of the Vistula River). The Gulf of Gdańsk is in the far background.

when river flow is low and pollutant contributions are high. In the Willamette River Basin, total precipitation is abundant on an annual basis but is not evenly distributed throughout the year. Summertime deficiencies occur in many of the tributary basins that do not have the benefit of a reservoir. The Vistula River Basin is much more water limited, with very little reserve available for flow augmentation, and the limited amount of water has quality problems. The Vistula River Basin is six times larger than the Willamette River Basin (Table 1); however, because rainfall is 2.6 times greater in the Willamette River Basin and because evapotranspiration and water use in the Willamette River Basin are much less, the mean annual discharges for each basin are nearly the same. Population density in the Vistula River Basin (122/km^2) is almost twice that of the Willamette (71.5/km^2), and 30% of sewage effluent into the Vistula remains untreated. Point source pollution is a major problem in the Vistula River system; the volume of untreated sewage and the wastes created by heavy industrial production are currently larger than on the Willamette River. Much of the pollution in the Willamette River Basin is from nonpoint sources. In the Willamette River Basin, there is more fertilizer, herbicide, and pesticide applied to agricultural cropland and urban greenspace than in the Vistula River Basin.

Environmental concerns are great in both the Willamette and Vistula River Basins. In the Willamette River Basin, shrinking forest reserves and related ecological stresses have necessitated that timber-management practices consider the ecological integrity of the entire watershed. The listing of various species of animals for protection by the U.S. Government, such as the northern spotted owl, the marbled murrelet, and various stocks of anadromous fish, has had a profound effect on the management of both forest and water resources. Now that Poland is no longer part of the Eastern Bloc, there are major concerns by the current government about the environment, and water use and resource abuse are being more carefully scrutinized. Industrial pollution in the upper river introduces toxic compounds that are harmful to all species. Pollution in the lower river causes eutrophication that endangers existing ecosystems in both the river and the Baltic Sea. Water pollution also is closely related to air pollution. About 80% of Poland's energy is obtained from burning coal; as a result, the emission of sulfur dioxide compounds has reached levels of very high

TABLE 1
Physiographic and Demographic Basin Characteristic
Comparisons Between the Willamette and Vistula Rivers

Basin characteristic	Willamette River	Vistula River
Drainage area	31,100 km²	194,400 km²
Main channel length	497 km	1063 km
Latitude range	43°21'N 45°38'N	48°55'N 54°48'N
Mean basin elevation	556 m	270 m
Mean annual discharge	900 m³/s	1090 m³/s
Mean annual precipitation	1600 mm	620 mm
Total discharge as untreated sewage effluent	0%	30%
Population (1990 census)	1,910,000	22,800,000
Forested area	70%	28%[a]
Agricultural area	24%	60%[a]
Urbanized area	5%	10%[a]
Landscape protected area	3.5%	8.8%[a]

Note: km², square kilometers; m, meters, m³/s, cubic meters per second; mm, millimeters.

[a] Percentages apply to all of Poland and are probably similar to what can be expected in the Vistula River Basin.

risk to human and ecological health. Poland has been fortunate in one respect: There are no nuclear power plants located on Polish soil or in any Polish river basin, and no nuclear storage has ever occurred within Polish boundaries. The Polish use the white-tailed eagle and the osprey as their environmental indicators of uncontaminated land. Both bird species are threatened with extinction in Europe, and these birds are found only in large tracts of forest that have clean water nearby.

Water availability is a concern in both river basins. In the Willamette River Basin, agriculture is feeling the stress. Urban growth, recreational uses, and the need to accommodate endangered species have all put pressures on the limited water resource of the Willamette River. In the Vistula River Basin, concerns of water pollution caused by upstream sources are great because the river is used to supply drinking water for many communities. There are more than 30 major surface-water intakes within the Vistula River Basin; several are on the polluted main stem which serves large cities, including Warsaw and Plock. Industry is the primary water user in the basin, using three times more water than either municipalities or agriculture. In central and southern Poland, the shortage of water during periods of low rain has become critical. More than 100 large towns, thousands of smaller communities, and nearly all agricultural activities are affected by temporary shortages of water. The main causes of water shortages are water that is too polluted to use, losses of water in old, faulty distribution systems, and poor conservation practices.

The Willamette and Vistula River Basins are blessed with beautiful scenery, and tourism is a major industry for both. The Willamette River Basin contains many rugged mountains, lush forests, and cascading rivers, whereas the Vistula contains more pastoral scenery and a much longer legacy of human cultural history. Tourism requires that basin ecology be kept in good health and that streams and forests be aesthetically pleasing. Recreation creates increasing competition for water use. Both basins contain fish and wildlife habitats that are major tourist attractions, but these resources require that water be allocated for their benefit.

WATER-QUALITY DATA COLLECTION

Industry, agriculture, and urbanization each place demands on a river basin. Water-quality data are collected to monitor and understand the effects of these contaminant sources. All data referred to in this section are from the U.S. Geological Survey's WATSTORE files, the U.S. Environmental Protection Agency STORET files, and the water-quality files of the Polish State Inspectorate of Environmental Protection.

WILLAMETTE RIVER BASIN

In the Willamette River Basin, the largest industries are pulp and paper mills, located primarily on the lower McKenzie River and main stem of the Willamette River. Seven large mills exist on the main stem of the river which together discharge about 100 hm^3 of effluent annually (1993). Effluent is monitored by each individual mill and reported to the Oregon Department of Environmental Quality (ODEQ). In turn, downstream from these effluents the river is monitored routinely by ODEQ and other agencies for various characteristics and chemical constituents, such as dissolved oxygen (DO), pH, suspended solids, temperature, total organic carbon (TOC), and turbidity. Much less frequently (approximately once every 3 years) samples for analysis of toxic compounds are collected from water, streambed sediment, and fish tissue for analysis.

Agriculture consumes the most water in the Willamette River Basin, using 194 hm^3 of ground-water and 315 hm^3 of surfacewater annually (1985 to 1990), primarily for irrigation. Grass seed is the largest crop grown in the Willamette Valley, and the use of pesticides is high. In the valley, about 2.3 million kg of pesticides were applied in 1987 (Oregon State University Extension Service), and 0.21 million metric tons of fertilizers were applied in 1993 (National Agriculture Statistic Service). Various agencies regularly monitor for DO, pH, specific conductance, nutrients, and bacteria at selected times and locations. Sampling for trace elements and organic compounds has begun only recently and is less frequent than for other water-quality constituents.

Municipalities are the next largest water consumer in the basin, using 50 hm^3 groundwater and 360 hm^3 of surfacewater annually (1985 to 1990 average). There are 21 major sewage treatment plants that collectively discharge 186 hm^3 annually into the Willamette River system. Untreated spills occur several days each year (mostly in the Portland Harbor), because Portland is the only municipality on the river that uses a combined storm and sewage collection system. Fecal coliform and streptococci bacteria have been used in the past to indicate human pollution, but, recently, enterococci or *Escherichia coli* are being used as indicators. Frequent sampling for phosphorous is also done to monitor urban effluent. Individual municipalities are responsible for collecting and analyzing their own stormwater-quality samples. These samples are usually analyzed for major ions and cations, nutrients, trace elements, and organic compounds.

VISTULA RIVER BASIN

Industry is by far the largest consumer of water in the Vistula River Basin, using 4040 hm^3 annually from streams. Industries are located along the length of the river but are especially concentrated in the upper basin. This situation is unfortunate because the concentration of industries in the upper basin creates a high degree of contamination for the entire length of the river. The primary point sources of industrial pollution are chemical plants, nitrogen plants, pulp and paper mills, petro-chemical plants, sulfur plants, foundries, wood distillation factories, and natrium plants.

Industrial water consumption and the accompanying pollution cannot be easily separated from mining, agriculture, and municipal uses. Approximately 70 hm^3 of saline water is pumped from mines annually into Vistula streams. The processing of sugar from sugarbeets in many small plants in the central and southeast part of the basin also creates significant pollution from November to February in many of the smaller streams in the basin. Agricultural water withdrawal from streams in the Vistula River Basin is only about 750 hm^3 annually.

Power production cannot be separated from industrial or municipal water use. The largest plants utilize large open cooling systems. Although they are equipped with cooling towers, these and other plants lack any temperature moderating devices and thus pollute the Vistula River with warm water. Because of the extensive use of the Vistula River and because the river water is too polluted for certain industrial uses, many industries are looking for other water sources. In Poland as in the U.S., individual industries are required to monitor and report their effluent quantity and quality to a regulatory agency.

Municipalities use 1070 hm^3 of water from streams in the Vistula River Basin. A population of 8.7 million, about 38% of the basin population, uses surface-water supplies. The remaining population uses groundwater, which is consumed more conservatively. Many surface-water supplies are polluted and require considerable treatment. Only 70% of the sewage effluent into Vistula River Basin streams is treated and, of that, only 37% is treated biologically. Both the untreated and treated sewage effluent, in combination with agricultural and urban nonpoint pollution, provides a source of nutrients that causes many of the Vistula River Basin streams and the main stem to become eutrophic where water is pooled behind impoundments. In addition, Vistula River Basin waters are very saline because of industrial effluent; chloride concentrations reach nearly 2000 mg/L in the upper reach of the main stem. Only a small part of the effluent from basin industries is desalinized.

Water-quality monitoring in Poland is organized by the State Inspectorate of Environmental Protection. The assessment of water quality is done by the Ministry Council, which has divided Polish streams into four classes of quality with respect to water use. In the Vistula River Basin, there are 540 measuring locations where 20 to 30 water quality parameter values are determined at a frequency of 12 to 24 times per year. In addition, measurements are made twice a week at 60 locations on the main stem of the Vistula River and major tributaries, and analyses are done for 50 water quality characteristics (including heavy metals, pesticides, and PCBs).

COMPARISON OF WATER-QUALITY CHARACTERISTICS

Water-quality characteristics of the Willamette and Vistula Rivers are compared in Tables 2 and 3. The Willamette River can be regarded as a river that has been restored but that still has some background pollution that is becoming greater with increasing urban, industrial, and agricultural loads. The Vistula River can be viewed as a river that is in serious need of restoration and that may be at or above the level of pollution that existed in the Willamette River in about 1950.

Dissolved oxygen (DO) is a good indicator of the overall ecological health of a river. Although other indicators also signify general river health, an adequate supply of oxygen is essential for animal life. For many species of fish, DO levels below 6 mg/L for any length of time can be lethal. DO levels in both the lower Willamette and lower Vistula Rivers are marginal at times. When the Willamette River was highly polluted prior to 1970, DO concentrations of 0.0 mg/L often were measured.

The value of pH indicates the acidity or alkalinity of water. For the Willamette River the pH is nearly neutral (~7.0) most of the time, but for the Vistula River the water is alkaline (pH ~8.8), especially in the lower reach. Both the geology and discharge of saline water from mining contribute to this higher than normal pH.

Total dissolved solids (TDS), like specific conductance, indicates the total concentration of chemical constituents that are dissolved in the water. The TDS value, together with the values of other chemical constituents (nitrate, phosphorous, sulfate, and chloride) provides the relative proportion of these constituents to the overall dissolved component. For the Vistula River, the TDS is very high, especially in the upper basin, and can be explained by the high concentrations of sulfate (SO_4) and chloride (Cl). The high sulfate and chloride concentrations are a result of the mining and burning of coal. For the Willamette River, the TDS is relatively low and is more typical of a natural stream. Suspended solids indicate the component of suspended material in a sample and are only an approximate indicator of sediment transport and water clarity. For the Willamette and the Vistula Rivers, suspended solid concentrations are small compared with the dissolved solid concentrations.

Nitrate and phosphorous are indicators of agricultural pollution; however, nitrate can also indicate sewage contamination, and phosphate can reflect the use of detergents. Naturally occurring sources of both these chemical constituents exist. Neither the Willamette nor the Vistula River

TABLE 2
Water-Quality Characteristic Comparisons Between the Lower Willamette (1980 to 1990) and Lower Vistula (1994) Rivers. Samples Collected at the Willamette River at Portland, OR, and the Vistula River at Gdańsk

Water-quality characteristics	Willamette River	Vistula River
Basin drainage area	28,800 km^2	194,400 km^2
Agricultural surface-water use	315 hm^3	750 hm^3
Industrial surface-water use	207 hm^3	4040 hm^3
Municipal surface-water use	360 hm^3	1070 hm^3
Dissolved oxygen	11.0 (5.8–13.8) mg/L	(6.7–14.5) mg/L
pH	7.4 (6.6–7.9) std. units	(7.8–8.9) std. units
Total dissolved solids	56 (45–71) mg/L	(340–920) mg/L
Suspended solids (total)	7 (4–14) mg/L	(1–70) mg/L
Nitrate (total)	0.51 (0.12–1.25) mg/L as N	1.5 (0.1–5.0) mg/L as N
Phosphorous (total)	0.08 (0.04–0.21) mg/L as P	0.2 (0.06–0.45) mg/L as P
Ammonia (NH$_4$)	0.08 (<0.02–0.26) mg/L as N	(0.1–1.4) mg/L as N
Sulfate (SO$_4$)	3.7 (1.7–6.1) mg/L	(50–90) mg/L
Chloride (Cl)	4.1 (2.7–6.3) mg/L	(50–300) mg/L
BOD (5 day)	1.6 (0.6–3.3) mg/L	(1.9–9.2) mg/L
Coliform, fecal	162 (3–2400) cols./100 mL	(2500–1,000,000) cols./100 mL
Copper (Cu)	(1.3–8.0) μg/L	(1–5) μg/L
Lead (Pb)	(<0.5–1.0) μg/L	(n.d.–40) μg/L
Zinc (Zn)	(1.1–10) μg/L	(1–59) μg/L
Chromium (Cr)	(<0.5–1.1) μg/L	(3–30) μg/L
DDT	<0.001 μg/L	(n.d.)
Atrazine	(0.005–0.18) μg/L	—
Diazinon	(<0.002–0.009) μg/L	—

Note: Values reported as median (range); km^2, square kilometer; hm^3, cubic hectometer; mg/L, milligram per liter; μg/L, microgram per liter; cols./100 mL, colonies per 100 milliliters; n.d., not detected.

water at the sampling locations indicated in Tables 2 and 3 contain unusually large amounts of these constituents. Smaller streams in agricultural or urban areas of both basins are likely to have greater values than shown for the main stems.

The pesticides DDT, atrazine, and diazinon are indicators of agricultural pollution. DDT has been banned as an agricultural pesticide in both the U.S. and Poland since the early 1970s. Even though it is no longer applied, DDT remains in the environment. The disparity in DDT values shown in Tables 2 and 3 is probably a result of application and time differences. Atrazine and diazinon have both been detected in the lower Willamette River. These pesticides are used in significant quantities in the Willamette River Basin and are detected at high concentrations in many of the smaller agricultural basins. These constituents are not sampled for in Polish rivers because they are not widely used.

Fecal coliform bacteria and 5-day biochemical oxygen demand (BOD$_5$) can be indicators of sewage pollution and the efficiency of sewage treatment facilities. They are also, however, indicators of other animal wastes; livestock (and even wild animal) wastes contribute to bacteria that affect these tests. Because of the large amount of untreated sewage that is released into the Vistula River, fecal coliform bacteria populations are enormous. Five-day BOD measurements on the Vistula River are high and confirm a very high bacteria level. On the Willamette River, fecal coliform bacteria populations are higher in the Portland area because of occasional sewage spills and livestock activity on tributary streams.

TABLE 3
Water-Quality Characteristic Comparisons Between the Upper Willamette (1980 to 1990) and Upper Vistula (1994) Rivers. Samples Collected at the Middle Fork Willamette River at Jasper (near Eugene), OR and the Vistula River at Cracow

Water-quality characteristics	Willamette River	Vistula River
Basin drainage area	3470 km²	7524 km²
Dissolved oxygen	11.0 (9.0–14.0) mg/L	(4.6–13.5) mg/L
pH	7.4 (6.4–8.6) std. units	(7.4–8.2) std. units
Total dissolved solids	—	(550–3900) mg/L
Suspended solids (total)	—	(10–100) mg/L
Nitrate (total)	0.02 (<0.02–0.18) mg/L as N	1.5 (0.5–2.7) mg/L as N
Ammonia (NH$_4$)	0.04 (<0.02–0.07) mg/L as N	(0.6–4.8) mg/L as N
Phosphorous (total)	0.32 (0.027–0.18) mg/L as P	0.6 (0.4–1.2) mg/L as P
Sulfate (SO$_4$)	6.0 (0.3–11) mg/L	(50–280) mg/L
Chloride (Cl)	8.0 (0.5–14) mg/L	(250–1900) mg/L
Coliform, fecal	28 (<2–230) cols./100 mL	(2500–250,000) cols./100 mL
BOD (five day)	1.1 (0.1–2.6) mg/L	(5.2–12.2) mg/L
Copper (Cu)	<0.5 µg/L[a]	(2–650) µg/L
Lead (Pb)	<0.5 µg/L[a]	(8–60) µg/L
Zinc (Zn)	<0.5 µg/L[a]	(12–3500) µg/L
Chromium (Cr)	<0.5 µg/L[a]	(3–200) µg/L
DDT	<0.001 µg/L[a]	(0.004–0.16) µg/L

Note: Values reported as median (range); km², square kilometer; hm³, cubic hectometer; mg/L, milligram per liter; µg/L, microgram per liter; cols./100 mL, colonies per 100 milliliters; n.d. is not detected.

[a] Data collected at the Willamette River at Corvallis which should be representative of the more upstream site.

Copper, lead, zinc, and chromium are indicators of industrial and urban pollution. Based on these indicators, the Vistula River has a much greater level of industrial pollution than the Willamette River.

SUMMARY

There are major differences between the basins. The Willamette River Basin (including the Sandy River) has a drainage area of 31,100 km², whereas the Vistula River Basin is 194,400 km². According to the 1990 census of both countries, the population of the Willamette River Basin is 1.9 million people, whereas the population of the Vistula River Basin is 22.8 million people. The mean annual precipitation for the Willamette River Basin is 1600 mm, but only 620 mm for the Vistula River Basin.

Both basins have environmental stresses placed on them from human activities. Low precipitation, intensive industrialization, and a relatively large human population combine to impose a great environmental stress on the Vistula River Basin. In contrast, the Willamette has the benefit of a large quantity of precipitation and a smaller population, so it experiences much less environmental stress. In the Willamette River Basin, total precipitation is abundant on an annual basis but is not evenly distributed throughout the year. Summertime deficiencies occur in many of the Willamette River tributary basins that do not have the benefit of reservoir flow augmentation. The Vistula River Basin is much more water limited, with very little reserve available for flow augmentation, and the limited amount of water has quality problems. Point sources of pollution are major problems in the Vistula River Basin; the volume of untreated sewage and the wastes created by heavy industrial production are currently larger than on the Willamette River Basin. Much of the pollution in the Willamette River Basin is from nonpoint sources. In the Willamette River Basin, more fertilizer, herbicides, and pesticides are applied to agricultural cropland and urban greenspaces than in the Vistula River Basin.

Measurements made on the Willamette and Vistula Rivers indicate some of the water quality problems that exist. DO is a good indicator of overall ecological health. DO levels in both the lower Willamette and lower Vistula Rivers are marginal at times. The Willamette River is much improved from the 1950s when DO levels reached zero, but vigilance and improvement are still necessary. In the Vistula River Basin, only 70% of the sewage effluent into streams is treated and, of that, only 37% is treated biologically. The untreated and treated sewage effluent, in combination with agricultural and urban nonpoint pollution, provides a source of nutrients that causes many of the Vistula River Basin streams and the main stem to become eutrophic where water is pooled behind impoundments. The Vistula River also has a high level of industrial pollution, which is reflected in measurable quantities of trace elements in water-quality samples. For the Vistula River, the water is alkaline (pH ~8.8), especially in the lower reach. Both the geologic characteristics of the basin and the discharge of saline water from mining contribute to this higher than normal pH. For the Vistula River, the TDS load is very high, especially in the upper basin, and can be explained by the high concentrations of sulfate and chloride. The high sulfate and chloride concentrations are a result of the mining and burning of coal. Agriculture is the primary water user in the Willamette River Basin. Several million kilograms of pesticides and about one-quarter million metric tons of fertilizers are applied each year, which is reflected in raised levels of nutrients and measurable amounts of atrazine and diazinon from water samples collected in the basin.

After reading this introduction to the Willamette and Vistula Rivers, one can now begin to read the following papers to obtain more details about the two river systems, to understand river dynamics, and to determine effective monitoring techniques and restoration processes.

REFERENCES*

Keisling, Phil. *Oregon Blue Book,* Secretary of State Office, Portland OR, 1995–1996, 459.
Encyclopedia of Polish Industry. London, Sterning Publication, 1993, 304.
Horn, A. and Bozena, P., Eds., Poland — Insight Guides. Singapore, APA Publications Ltd., 1992, 316.

* These references have been used many times throughout the chapter regarding historic and geographic facts and demographic statistics.

the baseline for evaluating or judging the condition of other fluvial systems. Yet these systems have likely undergone many anthropogenic changes prior to a person's lifetime.

This chapter discusses some of the anthropogenic modifications made to the river system since the advent of Euro-American influence, and synthesizes material from the authors' own research, and published material on the upper Willamette River. The objective is to broaden awareness of cumulative impacts of landscape modifications on river systems for future assessment of a river system's ecological health. The Willamette River case study provides an opportunity to examine cultural expectations and assumptions regarding river landscape appearance and river behavior.

Historical data such as found about the Willamette River can also broaden the base of understanding about river systems elsewhere and their ecological functions and processes. Such information is especially valuable in the context of land-use planning and other resource policy arenas, for restoration activities, and other resource-related actions or decisions. It can help identify remnants of the river system that merit protection as significant natural areas.

BACKGROUND

The Willamette Basin in Western Oregon covers approximately 11,500 mi^2 (29,800 km^2) (Tetra Tech, Inc, 1993). The Willamette Valley, through which the main stem Willamette River flows, comprises about 3500 mi^2 (9100 km^2). The Willamette River main stem begins where the Willamette Coast and Middle Forks merge. It flows north from near Eugene to Portland, a river distance of about 187 miles (300 km), where it enters the Columbia River. This chapter will focus on the 58-mile (93 km) section of river from the McKenzie River confluence near Eugene (RM 175) to Albany (RM 117) (see general map of Willamette River basin in publication introduction). Much of the Willamette Valley is in agricultural use. Most of the valley's major cities are located along the river, and the valley population is growing rapidly.

WILLAMETTE RIVER CHANNEL LOSS

The Willamette River has experienced extensive channel loss since Euro-American settlement in the 1800s. A comparison of the 1850s township original survey plats, the maps produced by the General Land Office, and late 1960s U.S. Geological Survey topographic maps show a loss of approximately 60 to 70% of channel length between the McKenzie River confluence and Harrisburg (Sedell and Froggatt, 1984). Over time, this river section has been altered from a multiple channel to a simplified and often single channel system (Figure 1).

The river between Harrisburg and Albany has experienced significant channel loss as well (Figure 2). A comparison of the 1850s township original survey plats and late 1960s U.S. Geological Survey topographic maps at a scale of 1:24,000 shows about a 40% channel length loss for this section of river. The 42 miles (68 km) between Harrisburg and Albany had more than 120 miles (195 km) of channel in the 1850s, but only about 75 miles (120 km) in 1970. The combined channel losses along the two sections of the Willamette River between Eugene and Albany were approximately 45 to 50% of the original total channel distance, or from about 190 miles (305 km) to 100 miles (168 km). What is especially significant is that much of this loss was of secondary channels, the areas where there was a large proportion of the habitat for juvenile fish, and there were connections between the riparian vegetation and the water. The character of these secondary channels, including depth and vegetative cover, were significantly different from today's main channels.

The total distance of secondary channels in the 1850s between Harrisburg and Albany was probably somewhat greater than mapped in the 1850s plats because some of the watered "slough" crossings within the bottomlands — at least between Peoria and Corvallis — recorded in the survey field notes were not incorporated into the map drawings. Dry sloughs and swales noted by the surveyors may also have functioned as winter channels and water courses.

Willamette River

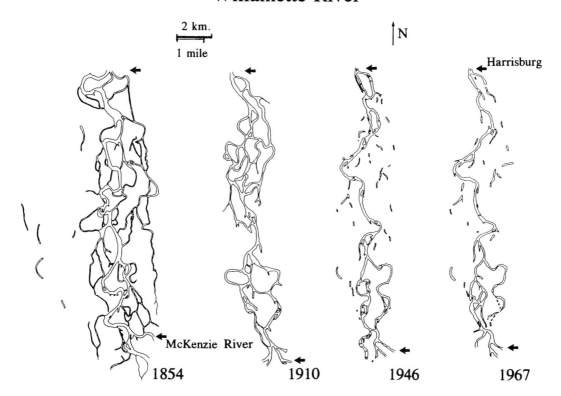

FIGURE 1.   Willamette River channel simplification over time between 1854 and 1967. This 14-mile (23 km) section of river is between the McKenzie River confluence (RM 175) just downriver of Eugene to Harrisburg (RM 161). (From Sedell, J. R. and Froggatt, J. L., *Verh. Internat. Verein. Theor. Ange. Limnol.,* 22, 1828, 1984. With permission.)

In 1875 the U.S. Army Corps of Engineers described the Willamette River channel upriver of Corvallis as being:

> …cut up into so many useless sloughs, and at each liable to undergo very marked and frequent changes… New channels are not infrequently cut out and old ones have been nearly left bare; the latter, after being closed for years are again reopened…The river above Corvallis is very tortuous, at places very narrow, and runs at many localities with great rapidity (Report of the Chief of Engineers, U.S. Army, 1875).

Though the Willamette above Corvallis was reported to be troublesome for boat travel, the most difficult stretch of river for "light-draught steamers" appeared to have been above Harrisburg.

> Captain Miller, one of the oldest and most experienced pilots in shoal waters of the same nature as the Willamette, has stated that he has never run the same channel for two consecutive years between Harrisburg and Eugene (Report of the Chief of Engineers, U.S. Army, 1875).

Large downed trees and piles of wood contributed to the formation and movement of Willamette River channels (Sedell and Froggatt, 1984). The wood obstructed and diverted channel courses, and contributed to the dynamics of the river that were reported historically.

Several factors have probably contributed to Willamette River channel simplification. Anthropogenic modification of river system processes such as deposition and scour, channel downcutting,

Willamette River

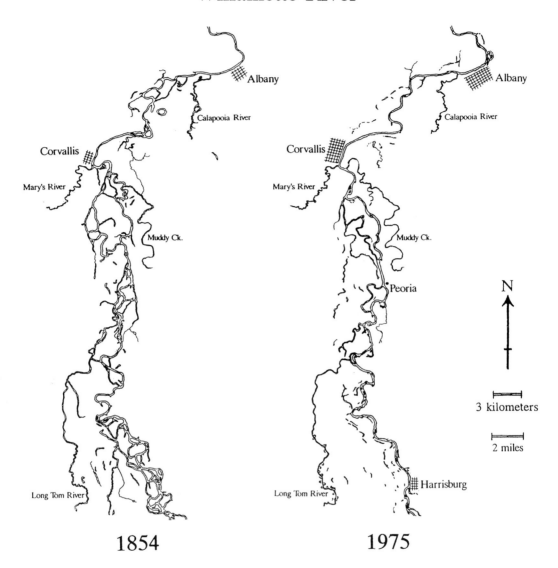

FIGURE 2. Willamette River channel loss between Harrisburg (RM 161) and Albany (RM 117) (70 km section), from the 1850s to the 1970s. These maps are a compilation of the original survey plats and U.S. Geological Survey topographic maps.

lateral migration, and flooding may have contributed to channel separation and the river's ability to create replacement channels. Dredge spoil at times were disposed of at the heads and mouths of small side-channels within the sections of river that were dredged. There have also been specific efforts to eliminate channels.

Improving the channel for navigational purposes was one activity that simplified the channel system, though boat travel on the Willamette varied depending on the section of river and the water level. The first of the federal projects to improve the navigability of the Willamette above Oregon City to Eugene was adopted in 1870 (Report of the Chief of Engineers, U.S. Army, 1881). Over several years, a number of secondary river channels were cut off through efforts by the U.S. Army

Corps of Engineers to confine the water into fewer channels, primarily for navigational purposes. This was accomplished in several ways. One technique was to close unwanted channels by filling their mouths or entrances with drift.

> Many useless sloughs were closed by cutting trees across them, and damming them with floating drift, in order to increase the volume of water in the main channel…[and by] filling up as far as possible the heads of all useless sloughs (Report of the Chief of Engineers, U.S. Army, 1875).

The Corps constructed "closing dams" or "cut-off dams" that were built across the heads of sloughs (Figures 3 and 4). They were generally made of wood pilings and timbers along with other material such as rock or brush. To further deepen the flow, "shoal bars" were "scraped," and "water-contracting low dams" (Figures 5 and 6) were built in the channel to sluice the gravel bed and deepen the flow.

The use of the river-bottom lands for agricultural purposes also led to gradual side-channel elimination (Figure 7). The clearing and farming of the wooded bottomland happened over time, and mostly occurred after 1900 (Towle, 1982). The level, fertile alluvial lands on floodplain islands were more accessible for farming once secondary and seasonal channels were eliminated or detached from the larger channels.

Channel loss has resulted in the loss of land-river interface, and its components. Natural structural features such as exposed tree roots, overhanging branches, and undercut bank provided

FIGURE 3.   Snag boat *Mathloma* at the 1000 feet-long cut-off dam at Davis Chute, about 3 km downriver of Eugene, in about 1899 (Report of the Chief of Engineers, U.S. Army, 1899, doc. #3908).

FIGURE 4. Dam at Head of Lambert's Slough, an attempt to cut off this large side channel (1899). This dam was downriver of Salem, at about river mile 71. (From Report of the Chief of Engineers, 1899, doc. #3908.)

a variety of protective habitat for aquatic organisms, including both juvenile and adult fish. With the loss of channel, there were fewer opportunities for ecological interactions between the land and the river. Terrestrial inputs such as leaves, insects, and large wood from the riparian area all contributed to the river food chain and habitat. Though there is little historical data on water temperature for verification, the riparian tree canopy must have shaded the many secondary channels providing cooler water refuges for fish. With the loss of channel, there were fewer opportunities for ecological interactions between the land and the river.

There may have also been an increase in the average width of primary channels in some sections of river, in which channel loss may have played a role. A limited study of channel width was done on the section of river above Corvallis. A comparison was made of the width of the present primary channel in the 77-mile (11 km) section upriver of Corvallis, and the two primary channels present in the mid-1800s along the same river section. The average width of these "East" and "West" Channels in 1852 was about 280 ft (85 m), and in 1990 the average width of the single main channel was approximately 440 ft (134 m). This approximation was based on 1990 aerial photographs (1 inch:400 feet scale), and on a calculation that converted the historical survey field data channel crossings to

FIGURE 5.   Skinners Bar dam, 1530 feet long, built in 1898–1899 just downriver of Eugene. Part of the dam cut off the head of a side channel, and the remainder "extended to direct water into the old channel." (From Report of the Chief of Engineers, 1899, doc. #3908.)

estimated channel widths (General Land Office, 1852). The present-day channel above Corvallis is roughly 50% wider than the 1852 channel as measured by the surveyors. The comparison was limited in scope, and is worth further study. At Corvallis where the river was single-channeled, both in 1852 and today, the channel width has changed little. A combined loss of channel and an increase in river width reduces the degree to which the riparian vegetation has influence on the active channel.

FLOODPLAIN VEGETATION

At the time of the federal land surveys of the Willamette Valley townships in the mid-1800s, the "low bottom" lands within the floodplain along the Willamette River and its tributary corridors were predominately covered by forest (General Land Office, 1852–1853; Habeck, 1961; Johannessen et al., 1970; Towle, 1974). Today, most of this bottomland woodland has been eliminated. In the 1800s much of the rest of the valley was in "prairie" and "oak openings" communities that were maintained by aboriginal population through annual burning (Habeck, 1961; Hussey, 1967; Johannessen et al., 1970; Towle, 1974; Boyd, 1986). The original government survey field notes for Benton County reported that:

> The bottoms along the Willamette are heavily timbered with fir, maple, ash, and a dense undergrowth of vine maple, hazel, and briers…there are numerous sloughs which would make the township almost impossible to survey in the winter (General Land Office, T13S R4W, 1852).

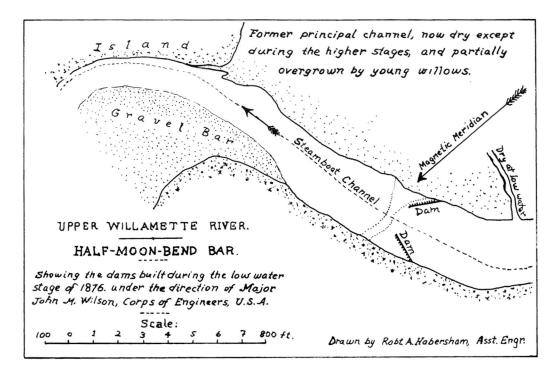

FIGURE 6. Facsimile copy of an 1876 U.S. Army Corps of Engineers map of two wing dams in the channel at Half-Moon Bar, about the present river mile 127. The dams directed current to mid-channel. (From Report of the Chief of Engineers, 1876, doc. #1676.)

FIGURE 7. Farming on Kiger Island just upriver of Corvallis (ca. 1910), probably on a former prairie area within the bottomland forest. (From Benton County Historical Museum #985-120, 903. With permission.)

The vegetation on Grand Island created by the East and West Channels upriver of Corvallis "…has some splendid Groves of Fir, but the principal proportion of the Timber is Ash, Maple & Balm of Gilead" (General Land Office, T12S R5W, 1852).

The U. S. Army Corps of Engineers also reported that the frequently flooded areas were largely timbered, and were connected with the channel through flood events.

> The adjacent country [between Eugene and Harrisburg] is level…the river bottom is from one to two miles in width…The timber, consisting of cottonwood, or Balm of Gilead, maple, ash, alder, and willows, is dense, and…is traversed by sloughs and bayous, large and small, and in times of floods is covered by swiftly-running water to a depth of from 5 to 10 feet (Report of the Chief of Engineers, U.S. Army, 1875).

These wooded floodplains contributed in many ways to the river system. The woodland community was habitat for wildlife. The trees slowed river flood waters, and not only protected the land from erosion, but also captured and recycled the water-borne sediment and nutrients. Fallen leaves contributed carbon and other nutrients to the floodplain and river.

> Before the Willamette bottomlands were cleared and placed under cultivation the heavy timber and underbrush with which they were covered retarded the water during periods of their overflow… (Report of the Chief of Engineers, 1891).

Both the original government township survey field notes and donation land claim survey field notes were used for this paper as sources of data to characterize the mid-1800s section of river corridor upriver of Corvallis (Figure 8). In addition to enlarging the data set by supplying landscape features such as streams, swales, and vegetation, the donation land claim survey field notes sometimes included site descriptions much like the township field notes, such as for Samuel Gage's claim west of the Willamette.

> That portion of this claim lying east of slough is rich bottomland, mostly timbered with balm, ash & maple, subject to inundation from one to six feet deep; Balance [of] claim flat prairie (General Land Office, Claim 57, T12S R5W, 1853).

Prairie communities, often with scattered "oak openings," came fairly close to the river when higher bench lands bordered the channels. However, even when the prairie grew close to the river as it did at the site of the town of Corvallis, there appears to have been a significant number of trees such as oak, fir, and maple near to the river bank, because surveyors nearly always had trees with which to mark the river meander survey posts.

The corridor of timbered lands associated with the Willamette River varied in width. The presence and type of woodland community on the islands and adjacent lands were influenced by the presence of surface and subsurface water. The river-bottom forests probably were buffered from Native American valley burning by the physical separation of the higher bench-ground from the lower bottomland areas, and surface water and wet soil wherever it was present during the drier seasons.

The surveyors' notes regarding conifer tree types generally only stated "Fir." Though recent authors interpreted the references to "fir" to be Douglas fir (*Pseudotsuga menziesii*), it is as likely that the bottomland conifers were very often grand fir (*Abies grandis*) (Hibbs, 1995). Grand fir is tolerant of seasonally wet sites, and young grand fir plants are more shade-tolerant than Douglas fir (Burns and Honkala, 1990). In a few cases, bearing trees in the Corvallis study area were noted more specifically to be "White Fir" and "Red Fir." White fir is a name that has sometimes been used for grand fir (Ross, 1991).

Though the floodplain islands and bottomlands were predominantly in timber, there were a number of pockets of "prairie" located within the bottomland forests. Surveyors noted "fern" in a few of them in the section of river mapped for this paper. These timber breaks reported by the

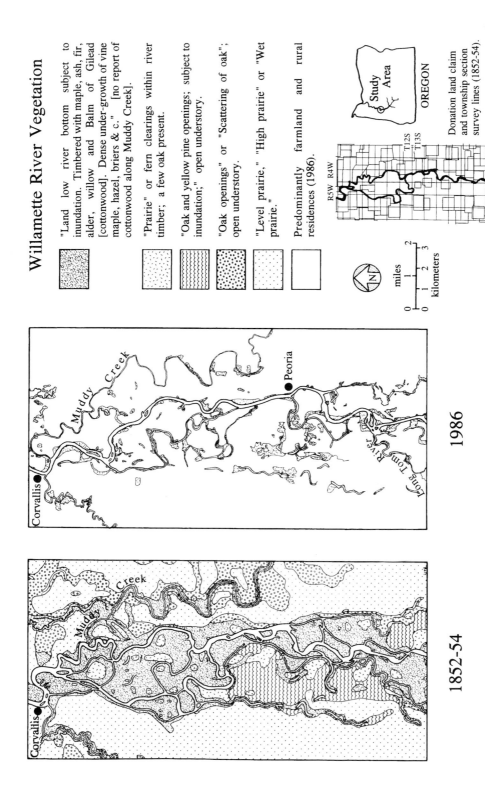

Willamette River Vegetation

"Land low river bottom subject to inundation. Timbered with maple, ash, fir, alder, willow and Balm of Gilead [cottonwood]. Dense under-growth of vine maple, hazel, briers & c." [no report of cottonwood along Muddy Creek].

"Prairie" or fern clearings within river timber; a few oak present.

"Oak and yellow pine openings; subject to inundation;" open understory.

"Oak openings" or "Scattering of oak"; open understory.

"Level prairie," "High prairie" or "Wet prairie."

Predominantly farmland and rural residences (1986).

OREGON

Study Area

Donation land claim and township section survey lines (1852-54).

R5W R4W

T12S
T13S

miles
0   1   2   3
kilometers

N

1986

1852-54

FIGURE 8. A comparison of the extent of 1852 and 1986 woodland vegetation within the Willamette River corridor in townships 12 and 13 south (1852–54 map based on original township and donation land claim field survey notes).

surveyors varied in size from about 200 ft (62 m) to one over one-half mile (770 m) in diameter. Boyd (1986) cited 1800s information that documents valley burning to maintain prairie, hunt wildlife, and collect food plants. He suggests that Native American groups used fire to clear forest understory for deer and elk, and that burning was used to increase berry species and bracken fern harvests. Though fire was reportedly used by the Willamette Valley Kalapuyan Indians until the 1840s to manage and harvest both plant and animals food resources, there is no discussion in the historical literature on burning examined by Boyd (1986) of bottomland openings created by Native American burning.

The original survey field data was collected at a time when there was very little impact on the river landscape by Euro-Americans with the exception of trappers. Beaver activity in the Willamette Valley influenced the character of the historical bottomland vegetation community. However, beaver trapping in the valley significantly reduced the number of beavers by the 1830s (Hussey, 1967). It would be interesting to determine if the vegetation, especially the "dense under growth" often observed by surveyors in the 1850s, was less patchy than two decades earlier when there was a larger beaver population.

The study area also had "oak and pine openings" and "scattering oak and yellow pine" (*Quercus garryana* and *Pinus ponderosa*). This vegetation sometimes grew on lands located between the low forested bottoms of cottonwood, maple and fir, and the higher prairie. These areas were described by the surveyors as "subject to inundation," but were within a higher floodplain flooded less frequently and not directly influenced by the river-connected groundwater. From the description of these areas, it is probable that Native American burning influenced this vegetation community.

A comparison of the timbered land along the Willamette between the 1850s and the present shows that most of the bottomland forest has been replaced by farmland. Remnant bottomland woodlands are growing, in many instances, where the ground is broken or otherwise difficult to farm. Residential lots along the Willamette River also have reduced vegetation on and near the bank. These timber losses have continued to occur, even recently. Frenkel et al. (1983) compared vegetative cover in Linn and Benton Counties within the Willamette River Greenway, a land-use zone that includes a degree of protection for vegetation. From 1972 to 1981 there was a loss of 294 hectares, or more than 12% of riparian vegetation in the Greenway. Most of this loss of natural and semi-natural vegetation was because of conversion to agricultural land use, and sand and gravel extractive activities (Frenkel et al., 1983).

IN-CHANNEL LARGE WOOD

Sedell and Froggatt (1984) estimated amounts and effects of large woody debris in the Willamette main stem. The U.S. Army Corps of Engineers reported that downed trees, called "snags," and "drift-piles" of wood were historically a common feature of the Willamette River channels. Between about 1870 and 1950, the Corps removed over 69,000 snags and overhanging trees (Figure 9) (Sedell and Froggatt, 1984).

Snags and wood jams were an important ecological component of the river channel and floodplain. Wood created and diversified habitat by adding structure that provided cover for fish, creating slackwater areas and pools in the channel, and trapping sediment, leaves, and other nutrients. The snags were "principally cottonwood, maple, ash and willow," species predominantly found in the floodplain (Report of the Chief of Engineers, U.S. Army, 1875). The source areas for the large wood included upstream reaches of the Willamette, tributaries, and nearby riparian corridors.

In 1875, the U.S. Army Corps of Engineers suggested continued annual removal of the snags, and reported that "not until the dense growth of the timber disappears from the banks…can any real and permanent good be accomplished" for navigation. A spectacular example of a wood jam was reported in 1877 by the Corps in the Centennial Slough between Corvallis and Peoria.

FIGURE 9. Snag boat *Mathloma* in operation pulling downed trees from the river. (From Report of the Chief of Engineers, U.S. Army, 1899, doc. #3908.)

Last winter's flood had brought down a large quantity of drift, which had formed a jam about 1,200 feet below the head, filling the steamboat channel and forming almost a solid raft from the point of obstruction to the head of the slough; besides this, the river for about 1,000 feet above the slough was almost closed by drift (Report of the Chief of Engineers, U.S. Army, 1877).

A U.S. Army Corps of Engineers description of channel wood near the mouth of the Willamette noted that "stumps and snags [were] from 180 to 30 feet in length, and in circumference from 4 to 19.5 feet [6 ft diameter]." The Corps reported "193 snags from 150 to 30 feet in length, the majority of them very long..." at the Booneville Slough above Corvallis (Report of the Chief of Engineers, 1875).

Other comments by the Corps regarding an 1875 flood also give a rare insight of historical sediment loads in the river during high flows.

The Willamette had risen...so high as to render it unsafe and risky to venture with boat into the channel, owing to the number of floating logs and large trees displaced from the banks. The water was so thick with mud as to render it impossible to discern the positions of snags below its surface (Report of the Chief of Engineers, 1875).

Bank erosion, a process associated with sediment loading and inherent in the Willamette system, was described in the same river survey.

The yielding nature of the [bank] soil cannot resist the action of the strong currents which strike the banks. The latter are low, composed of fine gravel overtopped with alluvial soil, and although covered with timber and thick brush to the water's edge, are easily washed out, or cut through as if they possessed the consistency of sand (Report of the Chief of Engineers, 1875).

Flood events contributed to the delivery of wood to the channel by both downstream transport and by recruitment of trees from banks. Delivery of downed wood to the Willamette River was the product of basinwide linkages that supplied not only the Willamette with material from the watershed, but that eventually transported large wood to the Columbia River and the Pacific Ocean through the Willamette corridor (Benner and Sedell, 1987). The wood, both on the ocean surface and on the ocean floor, provided cover and food for marine organism communities (Turner, 1973). Thus, the biological linkages of downed wood extended from the tributaries in the basin to the main stem Willamette, and onward to the Columbia Estuary and marine environments.

CHANNEL CONSTRAINTS

One of the earliest Willamette River revetments constructed by the U.S. Army Corps of Engineers was built in the 1890s on the outside bank of a wide bend in river across from Corvallis (Figure 10) (Report of the Chief of Engineers, 1892). There was concern that the river might cut a new course to the east and abandon the town. Today about one fourth, or 32 (51 km) of the 124 miles (200 km), of the main channel bank between Eugene and Albany has been stabilized by boulder "rip-rap" (Figure 11). On the entire main stem down to the Newberg Pool, about 18%, or 47 (76 km) of the 255 miles (410 km), of the main channel bank is revetted (U.S. Army Corps of Engineers, 1984). Most revetments have been placed on the outside of river bends, and have essentially eliminated lateral migration of the channel and the creation of new off-channel aquatic habitat (Hjort et al., 1984). This constraint has also possibly contributed to deepening (Klingeman, 1973) and widening by directing erosional forces to other parts of the channel.

Another consequence of channel constraint and change in flood regime may be the gradual loss of cottonwood stands along the Willamette River (Rood and Mahoney, 1990; Dykaar, 1995). Cottonwood seeds typically only germinate on fairly exposed soils, and the seedlings do not grow in understory shade. Lateral migration of river channels that creates new riparian zones allows for the replacement of aging cottonwood stands.

FIGURE 10. Construction of an early revetment (1890s) on the East Channel of the Willamette River just upstream of Corvallis. (From Report of the Chief of Engineers, U.S. Army, 1900, doc. #4049.)

OFF-CHANNEL AQUATIC ZONES

On occasion, the Willamette River abandoned sections of channel, which became oxbow lakes and linear ponds. Anthropogenic channel cut-offs have also created off-channel aquatic zones, many of which have a wooded margin. These off-channel aquatic areas provide habitat for organisms such as the Western pond turtle and waterfowl. Over time, however, many of these wooded remnant slough sections and swales have been eliminated, though the soils are generally hydric and still retain water in the winter. McBee Slough, across the river from the town of Peoria (RM 141) is an example of a relatively recent loss of such a remnant area (Figure 12). Today this former slough is only a depression in a large farm field.

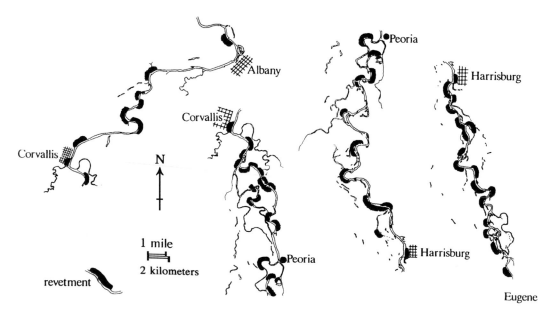

FIGURE 11. Locations of revetments, or bank-stabilizing rock, between Eugene and Albany; 24% of the main channel bank has been stabilized. (After U.S. Army Corps of Engineers, 1984.)

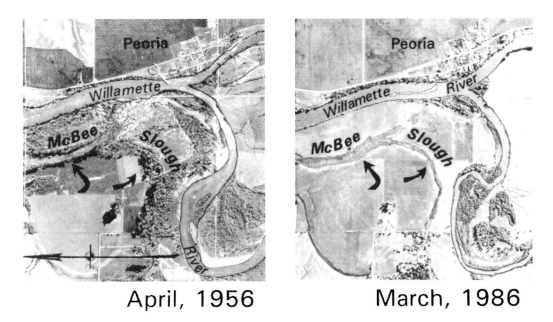

April, 1956   March, 1986

FIGURE 12.   The ongoing loss of off-channel woodland aquatic areas is illustrated at McBee Slough, across from Peoria (RM 141).

FLOODS AND FLOODPLAIN LOSS

Flooding is a process that provides opportunities for extensive ecological linkages and interactions between the active channel, the floodplain, and upstream-downstream river sections. It also creates and maintains habitat and the ecological complexity of the river system. Flood control has decreased the extent of ecological connections within the river system and altered the processes associated with frequent and significant flooding. These include nutrient exchange, sediment trapping and recycling, and the movement of large wood within the land and the river channel.

Prior to the construction of the 11 water-storage dams in the Willamette Basin beginning in the early 1940s, frequent and substantial flooding was a dominant ecological process along the main stem Willamette (Figure 13). Even with the present dam and reservoir system, large floods can still inundate considerable amounts of land along the main stem Willamette River, in part because only a portion of the basin is under dam control. At Albany and Salem, about 51 and 47%, respectively, of the basin is regulated by dams. About 76 and 65% of the basin is controlled above Springfield and Harrisburg, respectively. About 27% of the entire basin is controlled by dams (U.S. Army Corps of Engineers, 1989). However, the frequency and magnitude of these flood events has been much reduced from the pre-dam flooding patterns.

Changes to flooding patterns since Euro-American settlement of the main stem Willamette extend beyond the influence of dams, but are less easy to quantify. Channel simplification may have modified the flooding patterns. Possible channel downcutting (Klingeman, 1973) would reduce frequency of overbank flooding and decrease the area flooded. However, the cumulative loss of the valley's ability to detain and retain water, through urbanization, ditching, tributary channelization and floodplain filling, has increased the potential of larger Willamette flood peaks.

Based on accounts by fur trapping agents, the Willamette River experienced at least five major floods in the 1800s prior to the 1861 flood (Brands, 1947). The flood of 1861 was the first event commonly remembered, probably because of its size. The 1861 flood is the largest flood on record for the Willamette River (U.S. Army Corps of Engineers, 1951). The peak flow has been estimated to have been greater than a 100-year event at Salem, and about a 90-year event at Albany, based

FIGURE 13.   Willamette River at Corvallis during the 1890 flood, about a 45-year event at Albany prior to dams, with a peak flow of 291,000 cfs. (S. B. Graham, photographer, Corvallis, Benton County Historical Museum #985-119:16801. With permission.)

on the 1862 to 1942 period of record (U.S. Geological Survey, 1993). In 1881, a second major flood swept the valley. The U.S. Army Corps of Engineers reported that:

> The river experienced during the winter and spring two very prominent freshets, and three moderate ones. The one which caused the greatest damage...commenced early in January [1881], as the result of heavy snows in the Willamette Valley, followed by long continued warm rains, and reached its maximum on the 16th of January... (Report of the Chief of Engineers, 1881).

The 1881 flood was approximately a 30-year event at Albany and a 70-year event at Salem (U.S. Geological Survey, 1993). After the flood of 1890, about a 45-year and 85-year event at the Albany and Salem gauges, respectively, the U.S. Army Corps of Engineers described flooding along the stretch of the Willamette near Corvallis as follows:

> During extreme high water the whole valley, with the exception of a few isolated patches, is overflowed. The least distance between banks [of the Willamette]...within the limits of this survey is two and one-half miles distant. (Report of the Chief of Engineers, 1891.)

Because of the ecological value of flooding, it is worth comparing the frequency and magnitude of floods before and after dam construction in the basin. A large portion of land that once was within the 10-year floodplain, and flooded multiple times during a 10-year period, is now infrequently flooded. Prior to dams, a 10-year peak flow on the Willamette at Albany and Salem was similar in magnitude to today's post-dam 100-year flood event (Table 1) (U.S. Geological Survey, 1993; U.S. Army Corps of Engineers, 1982). These instantaneous peak flow analyses were based on a pre-dam period of record that was from 1862 to 1942 for Albany and 1862 to 1941 for Salem.

Based on Corps values, a 10-year flood peak flow has been reduced by 66% at Springfield, 55% at Harrisburg, 43% at Albany, and 34% at Salem since the construction of dams (U.S. Army Corps of Engineers, 1982).

The most recent major flood event on the Willamette River (the 1964 flood) can serve as a visual example of the extent to which the Willamette corridor was once frequently flooded, based on the period of record. This "rain-on-snow" storm combined a substantial watershed snowpack with warming temperatures and excessive rainfall (Figure 14).

TABLE 1
Willamette River Comparison of Pre-Dam and Regulated Flood Flows

Location	10-Year[a] (predam) discharge (ft³/s)	100-Year[b] (regulated) discharge (ft³/s)
Springfield (RM 185)	117,000 (79,100)[c]	74,000
Harrisburg (RM 161)	179,000 (no USGS data)	123,000
Albany (RM 119)	206,000 (201,000)[c]	200,000
Salem (RM 84)	288,000 (283,500)[c]	280,000

Note: A 100-year flood peak flow on the Willamette River in the regulated basin is comparable to a 10-year event prior to dams.

[a] Maximum discharge for a 10-year recurrence interval using U.S. Army Corps of Engineers average recurrence intervals for predam conditions.
[b] Maximum condition for a 100-year recurrence interval using U.S. Army Corps of Engineers average recurrence intervals for postdam conditions.
[c] U.S. Geological Survey statistical summary data, 1993, 10-year recurrence interval instantaneous peak flow.

From U.S. Army Corps of Engineers, Willamette River Basin Cumulative Frequency Curve Maximum Annual Discharge Graphs for Albany, Salem and Harrisburg gauges. U.S. Army Corps of Engineers Hydrology Section, Portland, OR, 1982.

FIGURE 14. Willamette River at Corvallis during the December 1964 flood. (From U.S. Army Corps of Engineers, Portland District.)

FIGURE 18. Artwork often portrays a society's perspective of an attractive stream landscape, though the stream or river may be highly altered and consequently in poor ecological health (artist, Jessie Swigert, *ca.* 1963).

Such culturally based perspectives are expressed and reinforced in many ways within our society. Artwork depicting what is thought to be an attractive landscape both mirrors and reinforces these perspectives (Figure 18). Language, as well, reflects and supports cultural views. For example, a Swiss-German word for the channelization and bed-bank stonework of European streams and rivers is "corrected." In some American communities, urban streams are referred to as "drainage-ways." The "improvements" that were made to the Willamette River were generally defined within a narrow set of objectives.

In addition, rivers and their tributaries have not commonly had land-use status as a resource component within the landscape. Streams and rivers typically travel through lands already designated for agricultural, forestry, urban, or other activities. As a consequence, these land uses are given precedence, and stream and river channels, banks, and floodplains have often been altered to accommodate various activities. It has not been generally recognized that a river's terrestrial corridor, including the floodplain, is a part of the river system, and is as important to the biological and ecological health of the river as is the channel (National Research Council, 1992). Recently, increased awareness of these concepts has begun to modify management and land-use strategies to give priority to the hydrology and biology of fluvial systems.

DISCUSSION

The Willamette River is an ecologically altered system. Since Euro-American settlement in the mid-1800s, the character of the river and many of its inherent features have been modified to achieve specific societal objectives. These purposeful modifications (some of which have been described in this paper) have led to a complex web of unintended secondary ecological changes to this river system. The result is that the ecological integrity of the system has been seriously compromised.

Until recently, the health of the Willamette River has been typically measured by water quality alone. Though a river is ecologically impaired when in a water quality-limited state, the health of a river system is also contingent on the river's form and function as a complex ecosystem.

Because of the multiple expectations within the society concerning the uses of resource lands within the Willamette corridor and the river water, it is not anticipated that the Willamette River system will be returned to its aboriginal state. However, the reestablishment of key elements of its antecedent state can be feasibly addressed within these societal constraints, and the ecological health of the system significantly improved.

A primary consideration is to identify and maintain the remnant areas along the Willamette corridor that retain the elements of the historical riverine landscape. Another is to reestablish the ecological relationship and linkages between the river channel and its bottomlands and terraces that once existed in abundance within this "river-floodplain" system. The National Research Council (1992) identified the conceptual framework when they wrote to "consider...the river and its floodplain as components of one ecosystem." The river system should also be provided opportunities for "self-restoration" where it shows indications of future channel changes and floodplain inundation.

Reestablishment of many of the former ecological components of the Willamette River system, though at a smaller scale, can be accomplished within the bottomland floodplain of the Willamette. The postdam remaining flooded bottomland, especially within the 10-year floodplain, is a priority area for opportunities to maintain and reestablish frequently occurring ecological linkages. Other priority sites, as suggested by Frenkel et al. (1991), are the nodes of land located at tributary confluences such as at the Santiam and Luckiamute Rivers.

Areas outside of the postdam 10-year floodplain to consider are sites where the Willamette River is likely to make avulsive changes, where there is evidence of significant lateral erosion, where there are side arms and remnant channels, or where there is still a high degree of connectivity between the river and adjoining landscape. This is especially the case along the section of river between Eugene and Harrisburg, where a larger percentage of its watershed is controlled by dams, and so the postdam 10-year floodplain covers less of the potentially restorable bottomland. However, within the 10-year floodplain along this river section there is still considerable potential to reestablish ecological interactions.

Both tributary confluence nodes and 10-year floodplain are areas within the Willamette corridor that provide significant opportunities for the reestablishment and accommodation of the river's natural disturbance processes necessary to sustain ecological integrity, including much of the channel lateral migration, avulsive channel course change, secondary channel development, and island and gravel bar formation. These areas would be of high ecological value for restoration of bottomland woodlands and for flood management. These lands provide opportunities to protect and restore linkages between the existing active channel and isolated secondary channels, surface-groundwater linkages, and river bottomland. Reestablishing areas of habitat and river process integrity are important objectives if we are to maintain any semblance of a self-sustaining river ecosystem impacted by human management.

ACKNOWLEDGMENT

Our thanks to the many individuals, museums, and agencies which have provided information, photographs, and insight necessary to develop this case study.

REFERENCES

Benner, P. A. and Sedell, J. R., Chronic reduction of large woody debris on beaches at Oregon river mouths, in *Proceedings of the Society of Wetland Scientists' Eighth Annual Meeting, May 26–29, 1987.* Mutz, K. M. and Lee L. C. Eds., Society of Wetland Scientists, Wilmington, NC, 1978, 335–341.

Berger, J. J., Restoring attributes of the Willamette River, in *Restoration of Aquatic Ecosystems*, National Research Council, National Academy Press, Washington, DC, 1992, 433–456.

Boyd, R., Strategies of Indian burning in the Willamette Valley. *Can. J. Anthropol.*, 5(1), 65–86, 1986.

Boag, P. G., *Environment and Experience*, University of California Press, Berkeley, CA, 1992, 209 pp.

Brands, M. D., Flood Runoff in the Willamette Valley, Oregon. U.S. Geological Survey, Water Supply Paper 968-A, U.S. Government Printing Office, Washington, DC, 1947, 57 pp.

Burns, R. M. and Honkala, B. H., Eds., *Silvics of North America, Vol. 1, Conifers.* Agriculture Handbook 654. U.S. Department of Agriculture, Forest Service, Washington, DC, 1990, 675 pp.

Dykaar, B.B., Personal communication, National Research Council Associate Post-doctorate, at EPA, Corvallis, OR, 1995.

Federal Emergency Management Agency, Flood Insurance Study, Benton County, OR, 1986, 32 pp.

Frenkel, R. E., Gregory, S. V., and Sedell, J. R., Presentation written for Audubon's 1991 Country in the City Symposium III, Portland, OR.

Frenkel, R. E., Wickramaratne, S. N., and Heinitz, E. F., Vegetation and land cover change in the Willamette River Greenway in Benton and Linn Counties, Oregon: 1972–1981, in *Association of Pacific Coast Geographers 1984 Yearbook*, Oregon State University Press, 1983, 63–77.

General Land Office, Original Donation Land Claim Field Notes and Plats. Townships in Benton and Linn Counties, Oregon. Bureau of Land Management Public Room, Portland, OR, 1853–1854.

General Land Office, Original Government Survey Field Notes and Plats. Townships in Benton and Linn Counties, Oregon. Bureau of Land Management Public Room, Portland, OR, 1852–1854.

Habeck, J. R., The original vegetation of the mid-Willamette Valley, Oregon, *Northwest Sci.*, 35(2), 65–67, 1961.

Haskin, L. L., *Pioneer Stories of Linn County, Oregon, from W.P.A. Interviews,* Vol. 1, Linn-Benton Genealogical Services, Albany, OR, 1984, 136 pp.

Hibbs, D. E., Personal communication, Oregon State University, Corvallis, OR, 1995.

Hjort, R. C., Hulett, P. L., LaBolle, L. D., and Li, H. W., Fish and invertebrates of revetments and other habitats in the Willamette River, Oregon, Tech. Rep. E-84-9, Vicksburg, MS, 1984, 83 pp.

Hussey, J. A., *Champoeg: Place of Transition*, Oregon Historical Society, Portland, OR, 1967, 404 pp.

Johannessen, C. L., Davenport, W. A., Millet, A., and McWilliams, S., The vegetation of the Willamette Valley, *Ann. Assoc. Am. Geographers,* 61, 286–302, 1970.

Klingeman, P. C., Indications of Streambed Degradation in the Willamette Valley. WRRI-21, Water Resources Research Institute, Oregon State University, Portland, OR, 1973, 99 pp.

National Research Council: Science, Technology, and Public Policy. *Restoration of Aquatic Ecosystems*, National Academy Press, Washington, DC, 1992, 552 pp.

Report of the Chief of Engineers, U.S. Army. Reports of the Secretary of War, in House Executive Documents, Sessions of Congress, Annual Reports, U.S. Government Printing Office, Washington, DC, (annual reports), 1875–1910.

Rood, S. B. and Mahoney, J. M., Collapse of riparian poplar forests downstream from dam in Western Prairies: probable causes and prospects for mitigation. *Environ. Manage.*, 14(4), 451–464, 1990.

Ross, C. R., *Trees to Know in Oregon*, Extension Bulletin 697, Oregon State University Extension Service and Oregon State Forestry Department, 1991, 96 pp.

Sedell, J. R. and Froggatt, J. L., Importance of streamside forests to large rivers: the isolation of the Willamette River, Oregon, U.S.A. from its floodplain by snagging and streamside forest removal. *Verh.-Internat. Verein. Theor. Ange. Limnol.* (International Association of Theoretical and Applied Limnology), 22, 1828–1834, 1984.

Tetra Tech, Willamette River Basin Water Quality Study. Component 10: Project Summary Report, 1993, Prepared for Oregon Department of Environmental Quality, Portland, OR, Tetra Tech, Inc., Redmond, WA, 1993, 167 pp.

Towle, J. C., Changing geography of the Willamette Valley Woodlands. *Oreg. Histor. Q.*, 83, 66–87, 1982.

Towle, J. C., Woodland in the Willamette Valley: a historical geography, Ph.D. dissertation, University of Oregon, Eugene, OR, 1974, 159 pp.

Turner, R. D., Wood-boring bivalves, opportunistic species in the deep sea, *Science*, 180, 1377–1379, 1973.

U.S. Army Corps of Engineers, Columbia River and Tributaries, Northwestern United States, Vol. V. House Document 531, 31st Congress, 2nd session, prepared by Portland District, U.S. Army Corps, October 1, 1948, Government Printing Office, Washington, DC, 1951.

U.S. Army Corps of Engineers, Postflood Report, December 1964, January, 1965 Flood, Portland District Office, Corps of Engineers, 1966, 237 pp.

U.S. Army Corps of Engineers, Willamette River Basin Cumulative Frequency Curve Maximum Annual Discharge Graphs for Albany, Salem, and Harrisburg gauges, U.S. Army Corps of Engineers Hydrology Section, Portland, OR, 1982.

U.S. Army Corps of Engineers, Bank Protection Work Prior to December 1984, unpublished data sheets, Portland District Office, Corps of Engineers, Portland, OR, 1984.

U.S. Army Corps of Engineers, Willamette River Basin Reservoir System Operation, Portland District Office, Corps of Engineers, Portland, OR, 1989, 43 pp.

U.S. Geological Survey, Statistical Summaries of Streamflow Data in Oregon. Vol. 2, Annual Low and High Flow, and Instantaneous Peak Flow, Open File Report 93-63, Portland, OR, 1993, 406 pp.

U.S. Geological Survey, Willamette River at Albany and Salem, Oregon, Station Numbers 14174000 and 14191000, Watstore discharge data, Portland, OR, 1995.

Wellman, R. E., Gordon, J. M., and Moffatt, R. L., Statistical Series of Streamflow Data in Oregon. Vol. 2, Annual Low and High Flow, and Instantaneous Peak Flow. Open File Report 93-63. U.S. Geological Survey, Portland, OR, 1993, 406 pp.

Williams, I. A., The drainage of farm lands in the Willamette and Tributary Valleys of Oregon. The Mineral Resources of Oregon, Oregon Bureau of Mines, 1(4), 3–81, 1914.

3 The Vistula River of Poland: Environmental Characteristics and Historical Perspective

Elżbieta Niemirycz

THE VISTULA RIVER BASIN: GENERAL INFORMATION

The Vistula River is the largest river in Poland (see map of Poland in Chapter 1 of this book) and it is also one of the 30 largest rivers in the world (Statistical Annual Report, 1992, 1993). It is situated in central Europe. The Vistula is the second largest river (after the Neva River in Russia) in the Baltic Sea Basin, in flow rank. Its source is in the mountains of southern Poland. It first flows north then eastward passing Cracow, the previous capital of Poland (Figure 1). The river then flows in a great arc passing through Warsaw in a northwestern direction. It continues northwest and then heads north-northeast to the Baltic Sea. Near Warsaw the Vistula is a typical lowland river (Figure 2). The mouth of the Vistula consists of the Vistula Delta, which is bounded by two branches: the Nogat (Figure 3) flowing into the Vistula Lagoon and the Leniwka, the main stream of the Vistula (Figure 4), flowing directly into the Baltic. These two branches are navigable at their mouths. The Vistula Delta (Żuławy) with a total area of 1740 km^2 is used for agriculture.

FIGURE 1. Photograph of the Vistula River in Cracow.

0-56670-138-4/97/$0.00+$.50
© 1997 by CRC Press, Inc.

FIGURE 2.   Photograph of the wide, slow-moving Vistula River near Warsaw.

The Vistula is 1063 km long and 300 to 1000 m wide in its middle and lower parts, with 194,424 km² of total drainage area (54% of the total area of Poland). The channel slope is gentle, from 0.10 to 0.31%. The river flow varies from year to year, as can be seen from Figure 5. It is also evident that the dry weather in central Europe in the 1980s and 1990s had a strong effect on the Vistula water flow. Variations of the river flow in a yearly cycle can also be observed (Figure 5) (Niemirycz and Taylor, 1992).The maximum spring flows (March, April) can be the cause of

FIGURE 3.   Photograph of the castle at Malbork on the Nogat, a distributary channel of the Vistula River.

FIGURE 4.   Photograph of the Leniwka, the main channel of the Vistula River near Tczew.

floods along the whole length of the Vistula, while the autumn maximum flows are severe only in the upper part (the Carpathian foothill region). The minimum flows occur in August, September, and November ("Polish golden fall") when rainfall is low. In general, during the winter season (from the second part of December to the end of February), the Vistula is frozen over. The average river flow at the mouth is 1000 m³/s; the river discharges, on average, 32 km³ of water into the Baltic Sea annually. In comparison the average flow near Warsaw, midway in the basin is 520 m³/s.

The changes in the natural environment of the last 50 years in the Vistula Basin are closely connected with historical changes, which occurred in three post-World War II periods. Historical changes also affected the economic development of the Vistula Basin. During the first period, from the 1940s to the early 1950s, reconstruction of a country destroyed by a five-year war took place. In the next period, from the 1960s to the early 1980s, government investment was unsound and inconsistent, and within the government there was contempt for environmental protection principles. Government policy in this period resulted in the degradation of the natural environment. Naturally, such a policy had serious consequences to human health. In the last period, from 1980 to present, government policy has changed from contempt to concern for the environment and for public health.

From 1945 to 1980, the population of Poland increased by about 70% (Figure 6). This growth caused a great increase in drinking water consumption. A great part of drinking water originates from the Vistula River.

At present, land use in the Vistula Basin within Poland is characterized as 48.8% arable land, 26.4% forest, 15.7% grassland, 1.4% orchards, and 8.1% other categories (Environmental Protection, 1993). Despite a significant growth of forest land after World War II, the present forest area is only 2.3 km² per 1000 inhabitants. Compare this with Canada, where there are 170.3 km²/1000 inhabitants, the U.S. has 11.7 km²/1000 inhabitants, Germany with 1.3 km²/1000 inhabitants, and France with 2.7 km²/1000 inhabitants. Deforestation has resulted in more river flow variability (Statistical Annual Report, 1992 and 1993) and the high erosion rate (16 t km-2 yr-1) (Kajak, 1992). The erosion rate is also affected by the geological structure of Poland (Figure 7).

The whole area of the Vistula Basin is populated by 22.0 million inhabitants; about 60% of this population is concentrated in urbanized agglomerations. The remaining part of the population are farmers. The variety of nutrient inputs from agriculture includes surface and underground runoff from fields, meadows, and pastures, as well as from settlements without sewerage systems (Taylor

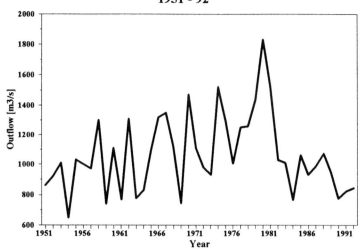

**The average annual flow during
1951 - 92**

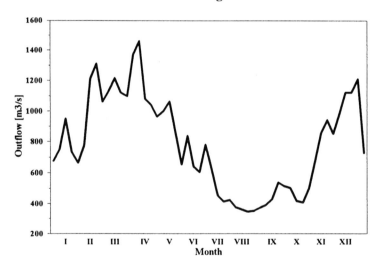

The flow during 1992

FIGURE 5.   Hydrological conditions of the Vistula River at the mouth cross section.

et al., 1992; Niermirycz and Taylor, 1993). Abatement of nutrient discharges from nonpoint sources is a very difficult problem (Niermirycz et al., 1993). In order to solve this problem, several measures should be taken, e.g., improvement of village sanitation (by providing villages with sewerage systems), creation of plant buffer zones, and construction of small treatment plants. It is also possible to reduce the nutrient inputs from agriculture by adjusting the quantity and terms of fertilization to plant and soil needs (Kajak, 1992). Surface water pollution is also caused by phosphorus introduced with waste waters from cleaning and washing (Roman, 1993). It is obvious that the sources of phosphorus are detergents based on phosphates. It has been proved that change of detergent composition can reduce phosphorus load in waste waters by up to 30%.

Manure is also a great threat to water environment (Obarska-Pempkowiak, 1994). The manure from big industrial farms, applied to soils without control causes pollution of surface and ground-waters, as well as soil degradation.

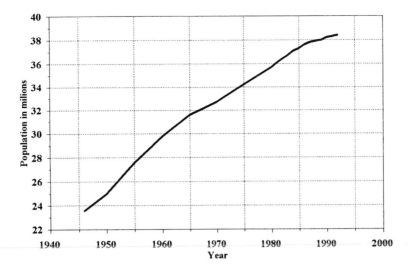

FIGURE 6.   Population of Poland.

In Poland only 52% (2297) out of 4414 large factories, were equipped with purification facilities (Korol et al., 1992). Of those factories with purification, 353 discharged their waste waters directly into surface waters while 1764 discharged their wastes through municipal sewage systems or directly into the soil. Waste waters, especially from the chemical, metallurgical, and pulp and paper industry, are characterized by high concentrations of pollutants, as well as by a high content of persistent and toxic substances. Therefore, wastes that do not undergo suitable treatment are very harmful for the water environment.

Another major problem is the salinity of the Vistula waters, which is mainly caused by coal mines situated in the upper Silesian region (Figure 2). Water from some reaches of the Vistula cannot be used for industrial purposes because the salinity of the water will corrode metallic parts in industrial machinery.

THE VISTULA IN THE HISTORY OF POLAND

The name "Wisła" appears in legends, chronicles, and various description dating as far back as the Roman times. The Vistula has been associated very deeply with Poland: with the mentality of its people, their literature, the folklore, and national emotions. The river symbol "the queen of Polish rivers" has even inspired many Polish writers and composers (Gierszewski, 1982).

The fifteenth century's changes in foreign trade encouraged the export of such goods as corn and timber. The shipping of the goods by river was easier and more cost effective than transporting them by land. The landed gentry soon saw a good opportunity to use the Vistula and its affluence for their trade purposes. During this early period, there began work to enhance the navigability of the river, to build larger watercraft, and to organize the floating of the goods to sea ports, all on a scale unknown before. In their economic considerations, the gentry not only considered the river route as the cheapest conveyance, but there was also free servile labor at hand.

To improve navigability of Polish rivers, the Seym (parliament) of the Nobility's Republic, up to the seventeenth century, from time to time, passed resolutions ordering regulation of the Vistula or its tributaries. The demand for shipbuilding timber, timber by-products such as wood-ash and potash used as dyes for fabrics was fully met in the eighteenth century by the great forests of the southeastern territories of Poland. The goods were first carried by horse and cart as far as the Bug and San Rivers and there they were loaded into boats. Taking advantage of serf labor, the supply of goods up to the given river port was organized from remote distances more than 300 km away from river transportation. The predominating feudal trade organization involved a tremendous region

SOILS ON THE UPLAND AND LOWLAND AREAS

	podsols and rustu soils created from mellow sands	
	podsols, rusty and brown soils created from poor loams	

soils created from light and middle clays and sands
  white
  brown

  brown soils created from heavy clays and silts

soils of water origin
  white
  brown

loessial soils
  white
  brown

  chemozems

bog soils
  mud soils
  peat soils
  black soils
  limestone soils
  fen soils

MOUNTAIN SOILS

  soils created from Flysch rocks
  soils from different massive formations
    skeletal soils
    rocky soils
  soils damaged by mining and industry

FIGURE 7.   Soils in the Vistula River Basin.

of Poland as the market area. This trade stimulated the building of river ports and yards for shipbuilding. Simultaneously with river trade, there came to being the specific profession of raftsman, a servile sailor, and the whole customary framing of the Vistula environment: songs, legendary tales, and raftsmen's ceremonies.

The continental blockade raised during the Napoleonic wars and the general economical crisis after them, also the commercial policy of some of the European states in the first half of the nineteenth century, led to a deep decline of importance of the Vistula for the transport of goods. After the last partition of Poland, and since the nineteenth century, each of the occupants of Polish territories, aimed at including the occupied area for their own economical system. By the nineteenth century, the railway had become a competitor to transportation on the Vistula River, since it was able to deliver the agricultural products to the whole of the occupied territory, e.g., Prussia, much faster. One only needs to mention that in the seventeenth century, the port of Gdańsk shipped about 80% of Polish corn brought along the river. This figure was still accurate until about the middle of the nineteenth century. In 1893, river transport of corn was barely 37%, and in 1912, it was only 12%.

During the interwar period (1918 to 1939) and especially in the 1930s, scientists and hydrologists often offered suggestions and plans to regulate the Vistula in order to reactivate its former economical importance. At this time there was also a plan for building a Silesia-Vistula canal to export Polish coal. Eventually, another plan was implemented. It was agreed that it would be more economical to build a railway trunk line across Bydgoszcz and Kocierzyna to Gdynia. Financial conditions of the pre-World War II Polish Government were poor and they did not invest capital in so great an undertaking as the regulation of the river. Foreign countries were also unwilling to invest in this enterprise. However, the establishment of the port of Gdynia was a significant achievement. After the construction of the Gdynia port, the Polish shipbuilding industry developed at a fast pace.

At present, in Poland, scientists and engineers are divided into either enthusiasts for Vistula River regulation as a means for restoration, increased water availability, and an economically feasible transportation alternative to rail transport and as opponents of this idea. Opponents of river regulation are afraid that significant changes in the river flow will bring changes that could devastate the natural river environment with its unique flora and fauna as well as deteriorate the self-purification process inherent in a free-flowing stream.

POLLUTION OF THE VISTULA RIVER FROM 1960 TO 1992

Water quality of the Vistula River has deteriorated during the last 30 years. Because most heavy industry is located in the drainage area of the upper Vistula, the river becomes polluted almost from its source.

The concentrations of pollutants are high, even at the mouth of the river, where they have been reduced by a long process of self purification (Table 1). Since 1972, water-quality measurements on the Vistula at its mouth have been made by the Institute of Meteorology and Water Management, Department of Water Protection in Gdańsk.

The 1960's Vistula water-quality data on the Vistula River is from the archival collection of the Provincial Inspectorate of Environmental Protection in Gdańsk.

MINERAL SUBSTANCES

The main mineral substances occurring in the Vistula water are chlorides and sulfates (Figures 8 and 9). The sources of these substances are from the geological bed of the Vistula Basin as well as from human activities. From 1960 to 1990, the concentrations of mineral components have significantly increased (Januszkiewicz et al., 1974; Niemirycz and Borkowski, 1992; and files from the archives of the Provincial Departments of Environmental Protection). However reported levels of concentrations are not yet toxic to water organisms and do not yet affect self-purification processes. It was found that the concentrations of chemical constituents are inversely proportional to water flow, and that the dilution caused by precipitation is the primary process affecting mineral concentrations in the Vistula River.

pH

The average pH values (from 6.9 to 8.5) in the Vistula water do not exceed the upper limits for class I, according to Poland's water classification scheme (Dz.U.Nr 116, 1991). These values are

TABLE 1
Mean Annual Concentrations and Variability
Coefficients in Vistula Mouth Cross Section in 1992

No.	Parameter	Unit	Mean value	Variability coefficient
1	Air temperature	[°C]	7.9	98.8
2	Water temperature	[°C]	7.6	85.8
3	Color	[mg/l]	19.7	19.1
4	Turbidity	[mg Pt/L]	6.8	24.8
5	pH	[pH]	7.7	5.9
6	Dissolved oxygen	[mg/L]	10.2	18.4
7	BOD_5	[mg/L]	5.4	32.9
8	COD-Mn	[mg/L]	7.4	28.9
9	COD-Cr	[mg/L]	25.6	36.2
10	TOC	[mg/L]	41.9	46.2
11	Ammonium nitrogen	[mg N/L]	0.62	64.9
12	Nitrite nitrogen	[mg N/L]	0.019	73.7
13	Nitrate nitrogen	[mg N/L]	1.91	59.3
14	Kjeldahl nitrogen	[mg N/L]	1.42	30.0
15	Total nitrogen	[mg N/L]	3.35	28.0
16	Phosphate phosphorus	[mg P/L]	0.15	46.6
17	Total phosphorus	[mg P/L]	0.21	43.3
18	Chlorophyll *a*	[μg/L]	21.7	104.7
19	Dissolved matter	[mg/L]	594.0	21.8
20	Suspended matter	[mg/L]	15.7	102.8
21	Chlorides	[mg/L]	155.9	38.1
22	Sulfates	[mg/L]	74.3	13.6
23	Total hardness	[mg/L]	292.8	15.0
24	Calcium	[mg/L]	86.1	13.0
25	Magnesium	[mg/L]	13.0	13.3
26	Sodium	[mg/L]	70.6	40.1
27	Potassium	[mg/L]	5.3	21.8
28	Iron	[mg/L]	0.35	75.7
29	Manganese	[mg/L]	0.11	75.5
30	Zinc	[mg/L]	0.031	81.2
31	Cadmium	[mg/L]	0.0008	94.0
32	Copper	[mg/L]	0.007	71.2
33	Lead	[mg/L]	0.005	66.4
34	Mercury	[μg/L]	0.118	143.9
35	Chromium	[μg/L]	1.500	62.3
36	Nickel	[μg/L]	3.030	52.9
37	Phenols	[mg/L]	0.006	156.1
38	Anion detergents	[mg/L]	0.082	25.2
39	PCB	[ng/L]	7.88	63.1
40	DDT	[μg/L]	0.009	101.8
41	DDD	[μg/L]	0.003	156.0
42	DDE	[μg/L]	0.005	160.0
43	Σ DDT	[μg/L]	0.017	80.8
44	DMDT	[μg/L]	0.001	0.0
45	γ-HCH	[μg/L]	0.006	40.1
46	Coli index		0.0188	136.4
47	Waterflow	[m³/s]	842.5	38.5

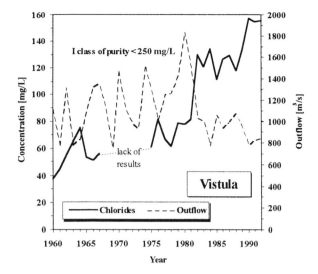

FIGURE 8. Chlorides.

typical for natural surface waters (Figure 10). It was found that due to the great buffer abilities of the Vistula waters, wastewater discharges did not significantly affect the pH in the river.

ORGANIC SUBSTANCES

Concentrations of organic compounds, expressed as BOD_5, COD-Cr, and COD-Mn, at the mouth of the Vistula River (Figures 11 to 13) are usually in the class II purity range, Class I is the best purity, and Class III the least acceptable purity, according to Polish classification. The BOD_5 values reported for the 1980s and 1990s are higher than those for the 1970s. The pollution of the Vistula River caused by organic substances does not adversely affect oxygen saturation at the mouth of the Vistula River. The dissolved oxygen (DO) content of more than 9 mg O_2/L (Figure 14) indicates that the Vistula water is well aerated. However, new data also show that the DO content has decreased from the mid-1980s; which is an alarming phenomenon.

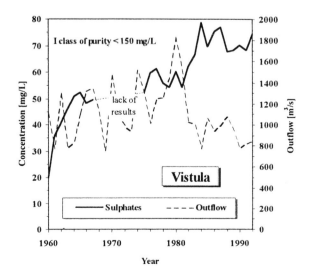

FIGURE 9. Sulfates.

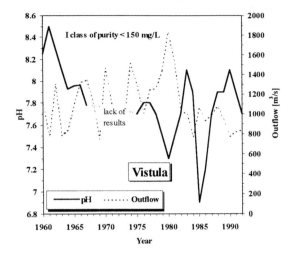

FIGURE 10. pH.

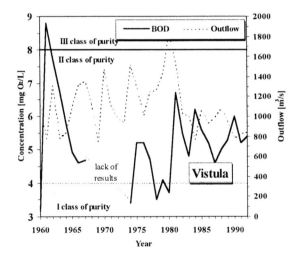

FIGURE 11. BOD₅.

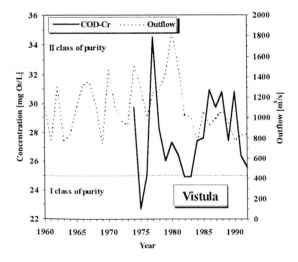

FIGURE 12. COD-Cr.

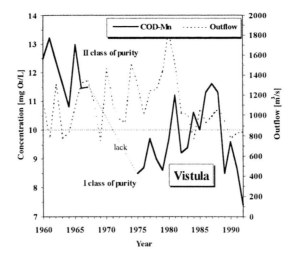

FIGURE 13. COD-Mn.

NUTRIENTS

The increased inputs of nitrogen and phosphorus from the anthropogenic sources (agriculture, industry, urbanization) into surface waters accelerate the eutrophication process. As a result, the biological equilibrium in water ecosystem is disturbed. From 1960 to 1992, a significant increase in concentration of nitrogen compounds in the Vistula River was observed (Figures 15 to 18). In the 1960s the mean concentration of ammonia nitrogen ranged from 0.2 to 0.5 mgN/dm^3. This value increased to the level 0.4 to 0.8 mg N/dm^3 in the 1980s. Within the same period, the mean concentration of nitrates increased from 0.0–1.2 mg N/dm^3 to 0.5–2.5 mg N/dm^3. The concentrations of both nitrogen forms did not exceed the upper limits for class I (according to Polish classification). Only the concentration values of nitrites were high (within the range of class III). Although the contribution of nitrites to the total nitrogen amount is small (on average less than 1%), it cannot be neglected. This form of nitrogen is very harmful to human health.

Phosphorus and its compounds have greater impact on the pollution of the Vistula River than nitrogen. The current concentrations of phosphate are significantly increased compared to observations from the early 1980s (the only exception was 1989) (Figures 19 and 20). The mean

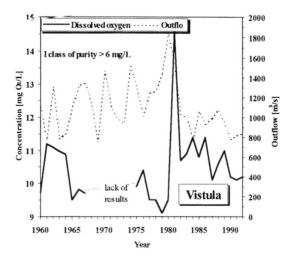

FIGURE 14. Dissolved oxygen.

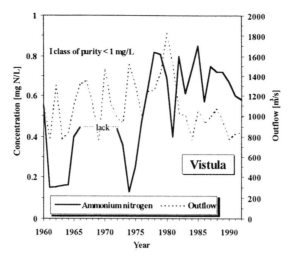

FIGURE 15.   Ammonium nitrogen.

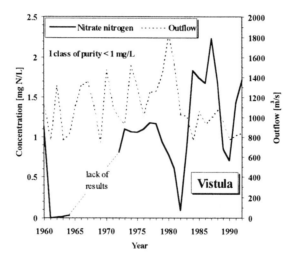

FIGURE 16.   Nitrate nitrogen.

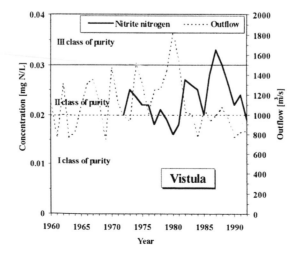

FIGURE 17.   Nitrite nitrogen.

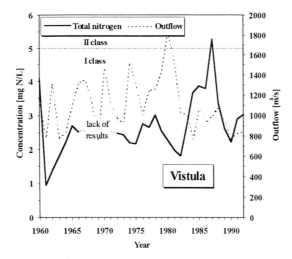

FIGURE 18. Total nitrogen.

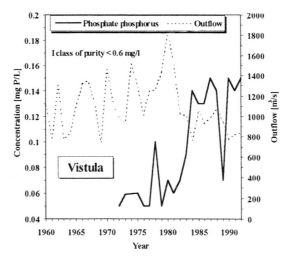

FIGURE 19. Phosphate phosphorous.

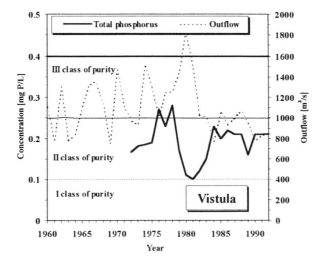

FIGURE 20. Total phosphorous.

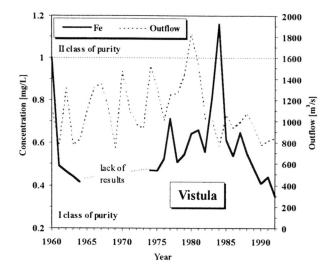

FIGURE 21. Iron.

concentrations of total phosphorus are in the class II and III range. It is obvious that hydrological conditions have a strong effect on nutrient concentrations in the Vistula water, (e.g., decrease in nutrient concentration during summer season when precipitation is low and vegetation growth is high). Despite this, the correlation between nitrogen concentration and river flow is still difficult to express mathematically.

METALS

The concentrations of metals such as iron, manganese, lead, zinc, copper and cadmium at the mouth of the Vistula River, in the most cases, do not exceed the upper limits for class I (according to the Polish classification) (Table 1; Figures 21 and 22).

CONCLUSIONS

1. The Vistula River, one of the biggest European rivers, is the source of water for 22 million Poles who inhabit its catchment area. In the past it has had a significant influence on the Polish transport and logistic system. At present it has a very logistical importance. The Vistula River flows past Cracow, the previous capital of Poland, and then past Warsaw, the present capital.

2. The water quality of the Vistula River has deteriorated during the last 50 years. Much pollution is in the upper Vistula River where heavy industry is located, but concentrations of pollutants are also high at the mouth of the river.

3. The high population density in the river basin and an insufficient number of treatment plants serving this population make the Vistula River one of the significant sources of pollutants transported into the Baltic Sea.

ACKNOWLEDGMENTS

I am obliged to Dr. Zbigniew Makowski for his help in computer data preparation.

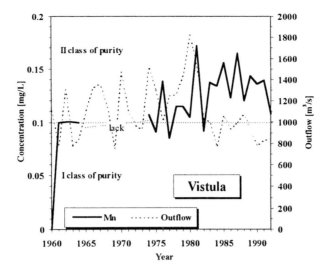

FIGURE 22. Manganese.

REFERENCES

Archival data of the Provincial Departments of Environmental Protection in Gdańsk.

Environmental Impacts of Nutrient Emission in Denmark, Nr 1, Ministry of the Environment, National Agency of Environmental Protection, 1991.

Environmental protection, Mat. GUS, Warsaw, 1993.

Dz.U. Nr 116, Regulation of the Ministry of Environmental Protection, Natural Resources, and Forestry regarding water-quality classification, November 11, 1991.

Gierszewski, S., The Vistula River in the history of Poland. Wyd. Morskie Gdańsk, 1982.

Januszkiewicz, T., Kowalewska, K., Szarejko, N., and Żygowski, B., Effect of fertilization on the Vistula water quality, Mat. IMGW, 1974.

Kajak, Z., *The River Vistula and its Floodplain Valley (Poland): Its Ecology and Importance for Conservation, River Conservation and Management,* Boon, P. J., Calow, P., and Petts, G. E., Eds., John Wiley & Sons, 1992, 35.

Korol, R., Niemirycz, E., and Szczepański, W., Water quality of main Polish rivers, Vistula and Oder in 1990, ICWS, Report 92, Netherlands, 1992.

Niemirycz, E. and Taylor, R., Land-use and nutrients to surface waters, G-1, IMGW, Mat. Oddziału Morskiego, Gdynia, 1992.

Niemirycz, E. and Taylor, R., Source division of the load of eutrophying substances. Mat. IAWQ, Cracow, 1993.

Niemirycz, E. and Borkowski, T., Riverine Input of Pollutants in 1992, Environmental conditions in the Polish zone of the Southern Baltic Sea during 1992, Maritime Branch Materials, Institute of Meteorology and Water Management, 205–207, Annuals since 1987.

Niemirycz, E., Taylor, R., and Makowski, Z., Threat of nutrients to surface waters, Mat. PIOS. Monografia, 1993.

Obarska-Pempkowiak, H., State of environmental protection in Poland, Mat. Sem. Sensors in environmental protection PG, 94–114, 1994.

Roman, M., Reduction of phosphorus load in surface waters, Mat. Symp. Intern. Assoc. on Water Quality, Cracow, 1993.

Rybiński, J., Niemirycz, E., and Makowski, Z., Pollution load, Marine Pollution (2), Studia: Materiały Ocean-ograficzne Nr 61, 25–52, 1992.

Taylor, R., Bogacka, T., Rybiński, J., Niemirycz, E., Zelechowska, A., Makowski, Z., and Korzec, E., Influence of diffusion agricultural pollution on surface water quality, Mat. MEPNRF, Warsaw, 1992.

Statistical Annual Report, Mat. GUS, Warsaw, 1992, 1993.

4 The Willamette River of Oregon: A River Restored?

Neil Mullane

The cleanup of Oregon's Willamette River was heralded as a national environmental success story in the U.S. in the early 1970s. The question now is whether the Willamette River is losing ground. To answer this question, the river's capacity to assimilate waste loads is being assessed and ecological studies are being conducted that will provide information on the biological health of the river. Preliminary study results are described. Past and future regulatory approaches are also described and the challenge of river basin management is explored in terms of protecting diverse uses of the river.

INTRODUCTION

The Gdańsk/Portland International River Quality Symposium assembled many different and complex scientific and administrative aspects of river-quality protection. The Symposium examined, among other things, river ecology, water chemistry, water-quality monitoring and modeling, and the impact of nutrients and organic chemicals on river systems. It also examined the methods, techniques, and approaches to regulate pollution sources to achieve river restoration and maintain high water quality.

The restoration efforts on one river in the U.S., the Willamette River in Oregon, involved many of these scientific elements. It also illustrates the interplay between the public's desire to protect and eventually enjoy a river resource and their use of the river as a receiving stream for treated and untreated wastewater discharges. An examination of this river cleanup will provide some insight into the necessary blend of scientific information and the formation and implementation of public policy. The regulatory approach established an essential foundation for public, industry, municipal, and state interaction throughout the restoration. Carefully considered regulatory actions were supported by scientific information describing river conditions and evaluating whether the controls implemented were having the desired effect. Follow-up monitoring and evaluation lead to increasingly higher levels of treatment until cleanup goals were met. Basin water quantity management to control massive flooding played a key role in the restoration, but left lingering questions on how it affects anadromous fish migration.

The phrase in the title of this paper, "A River Restored," suggests that the cleanup work has been completed. However, restoration is constant, not static. Time brings increasing population, expanding industries, agricultural growth, better understanding of pollution impacts, and advances in sampling techniques, all of which continually affect water quality. Therefore, to restore a river must mean constant vigilance to maintain the goals established and achieved as time brings new challenges and information on potential threats.

This chapter begins with a brief historical review of how a river was destroyed through neglect and indifference. It will describe how citizens rallied behind an initiative petition effort to force the state, the cities, and the industries to take the steps necessary to restore the river.

TABLE 1
Willamette River Basin
Beneficial Uses

Public domestic water supply
Private domestic water supply
Industrial water supply
Irrigation
Livestock watering
Anadromous fish passage
Salmonid fish rearing
Salmonid fish spawning
Resident fish and aquatic life
Wildlife and hunting
Fishing
Boating
Water contact recreation
Aesthetic quality
Hydro power
Commercial navigation and transportation

PHYSICAL BACKGROUND

The state of Oregon is located in the northwest portion of the U.S. It is the tenth largest state, with a total area of approximately 97,073 mi^2 and a population of almost 3 million (1996). The Willamette River lays completely within the state and drains approximately 12,000 mi^2 or 12% of the state (refer to Figure 4 in Chapter 1). The river is the thirteenth largest in coterminous U.S. The river valley is some 160 miles long and 75 miles wide. The basin contains most of Oregon's urban and industrial complex and is home to more than 2 million people. The basin has a diverse natural resource base supporting healthy recreational, agricultural, forestry, and fishing activities. A vast array of agricultural products including numerous vegetable and fruit crops, tree, shrub, and flower nurseries, vineyards, livestock, and grass seed are grown in the valley. Recreational opportunities abound throughout the basin. The valley flanks are heavily forested and the timber industry provides building materials for a growing economy. The Willamette Basin also contains a rapidly growing electronics industry.

The river itself supports a long list of beneficial uses (Table 1). It teems with people during the summertime and supports strong fisheries throughout the year and particularly during various salmon migrations as many people enjoy what the river has to offer. On a typical summer day you can watch waterskiers, rowers, sailors, and windsurfers moving up and down the river. Drinking water is also taken from the river by several communities. All these activities and expectations place a heavy burden on the river.

However, to truly appreciate what the river provides today, you have to travel back several decades and examine the condition of the river in the early 1900s.

HISTORIC REVIEW

At the turn of the century, the river was a commercial highway through the heart of the natural resource rich Willamette Valley. The river provided an easy and accessible mode of transportation to move people and products up and down the valley. For example, huge log rafts floated down the river to the many mills that extended along its length, and riverboats moved people (Figure 1).

FIGURE 1. Willamette River riverboats.

The river was not looked on as a natural resource, but as a commercial resource; something to be used, but not necessarily respected. It was seen as a place to discard the unwanted wastes of a rapidly growing community. It received untreated wastes from several different sources. Cities poured untreated sewage into the river throughout its length. Commercial operations such as slaughterhouses and food processors discharged untreated process water into the river. Critics of the time suggested that the municipalities and industries treated the river like an open sewer. The historical accounts would seem to support this. These discharges rapidly exhausted the river's capacity to accept and assimilate pollutants.

The river was described as ugly, filthy, and intolerable (Figure 2). Garbage was thrown down the riverbank without regard to its impact. Workmen refused to work on riverside construction projects because of the stench and fear of illness if they were to fall into the river. People could not swim or recreate in or on the river. One historical account described how people shunned the river. This was manifested in how they built their homes: facing away from the river.

Communities along the river were forced to look elsewhere for drinking water because it could no longer be taken from the river without extensive treatment. Salmonids could no longer survive in the river. In fact, in tests conducted during this time, most fish suffocated in a matter of minutes after being exposed to ambient water.

The Willamette River had once provided readily available drinking water, a vast fisheries, numerous recreational resources, and other beneficial uses. But the river could not handle the onslaught of human sewage and industrial waste products and still support these uses (Table 1). It became a river filled with the refuge of a growing human society.

INITIATIVE PETITION

The river was destroyed through neglect and indifference. Even with this obvious pollution and terrible loss of resource, no action was taken by the government, the cities, or the industries to stop the contamination. After a long period of inaction by the state legislature, the citizens of the state in 1938 rallied behind a citizen initiative petition and passed into law the Water Purification and

FIGURE 2.   Pollution problems in the lower Willamette River.

Prevention of Pollution Bill (Figure 3). The bill established the State Sanitary Authority, which in 1969 would become the Department of Environmental Quality.

The Sanitary Authority was responsible for bringing water pollution under control and restoring the river. The citizens of the state wanted action and the Sanitary Authority was to take the actions needed. It began this effort by documenting water pollution conditions in the river and establishing

FIGURE 3.   Citizen rally.

specific scientific criteria on which to evaluate river quality. Dissolved oxygen (DO), because it is essential in the support of aquatic life, became the most important criterion. Field teams were sent up and down the river to gather information on DO. In these early studies, DO levels in the Willamette, where it discharged to the Columbia River, were less than 0.5 mg/L. In the years that followed, the conditions continued to degrade, even to the point where DO was not detected during some test periods (Oregon SSA, 1964).

Sampling was also conducted for turbidity, bacteria, pH, suspended solids, biological oxygen demand, temperature, and other parameters, but DO was seen as the criterion upon which restoration success would be evaluated (State Water Resources Board, 1965).

APPROACH TO THE CONTROL

The causes and results of river quality decline were not difficult to document. The discharge of untreated sewage and wastewater from numerous cities and industries was exhausting the river's assimilative capacity. The Sanitary Authority wanted the cities and industries to install treatment facilities. The first step was to hold discussions with these sources and explain the results of the river testing and then issue administrative orders to the cities along the river to install primary treatment facilities. As these facilities were installed, the Sanitary Authority continued to sample the river to document river conditions and any changes. The sampling results showed that municipal primary treatment was not going to restore the river. The next step was to require the industries to install primary treatment. As industries installed these treatment facilities, the river was again monitored to document the changes; again the results showed little change.

The Sanitary Authority next ordered the cities to install secondary treatment and the industries were required to remove 85% of their pollutant load (Gleeson, 1972; State Water Resources Board, 1969). These controls were installed by the late 1960s, some 30 years after the initiative petition, but the effort was beginning to show results. The river's assimilative capacity was being restored. Dissolved oxygen goals were actually achieved throughout the river during low-flow conditions in 1969, 1970, and 1971. After three decades of citizens, industries, municipalities, and the state working to control pollution, the river had responded. The goals of bringing fish back into the river and providing for recreation in and on the river had been achieved. In 1972, the Willamette River cleanup became a national success story in the U.S. The goal of increasing the DO had been achieved as well as the other water standards (Oregon SSA, 1967). The most important visible sign of success was the return of the salmon and steelhead runs up the river (Gleeson, 1972). People could again recreate in and on the water. The river had been restored.

It is important to note that in 1972, when the Willamette River was responding to three decades of concerted efforts by municipalities, industries, citizens, and Oregon state government, the national Clean Water Act was passed by Congress establishing the national water quality protecting program. The Federal Clean Water Act established a national goal that waters should be fishable and swimmable by 1985. Municipal dischargers were required to install secondary treatment, and industrial dischargers were required to meet specific pollution removal efficiencies and categorical industry effluent limits. A permit program was established where dischargers were required to implement these specific levels of treatment. Permit violations were enforceable by the federal and state water pollution control agencies, as well as private citizens. The potential for citizen suit, through Section 505 of the Clean Water Act, brings considerable pressure to bare on the regulated community as well as the regulatory agencies to do the work required of each.

FLOW AUGMENTATION

An important element of river management that played a key role in the restoration was flow augmentation. Willamette River natural flows vary widely from tens of thousands of cubic feet per second in the winter and spring to less than 3000 ft³/s in the summer. The summer low flow

FIGURE 4.   General flooding in the Willamette Valley.

conditions presented a tremendous challenge to restoring river quality. The lower a river's flow the greater the waste treatment requirements. The amount of water in a river at any given time has a tremendous effect on the ability of a river to assimilate waste products. It would have been extremely difficult to achieve cleanup goals without augmentation.

At the same time that the cities and industries were improving wastewater treatment, the U.S. Army Corps of Engineers was receiving Congressional authorization to build a flood control system in the Willamette Basin (U.S. Army Corps of Engineers, 1989). The system was to protect people and property from flood damage and provide irrigation water and flows for river navigation (Figures 4 and 5). Thirteen dams were constructed in the basin (Table 2; Figure 6). Eleven were primary dams and reservoirs, and two were re-regulating dams. The reservoir system worked by storing spring rain and snowmelt flood waters for release during the late spring and summer months. The operation strategy therefore augmented the river flow during low natural summer flows. The released stored water provided for irrigation needs during the growing season, and recreational opportunities within the reservoirs and rivers in the form of summertime boating and fishing.

The summertime releases from the reservoir provided much needed flows to assimilate the newly treated wastewater discharges. Without this flow, the tremendous efforts by cities and industries to install the required treatment facilities would not have been fully successful during the critical summer low flow period (U.S. Army Corps of Engineers, 1991).

CANNOT STAND STILL

It has been more than 20 years since the Willamette River was restored. In these two decades, Oregon has continued to grow. More people and industries have come, more houses have been built, more land has been disturbed and eroded, more impervious surfaces have been created (Water Resources Commission, 1992). Agricultural production continues to grow. Chemicals continue to be used. River quality monitoring has become more sophisticated. Analytical techniques allow for much more detailed inventories of chemicals present in water as well as fish tissues. As more advanced analysis is completed, more contaminants are found (Oregon Department of Environ-

FIGURE 5. "High and dry" in the Willamette Basin.

mental Quality, 1992). These contaminants may have been present decades ago, but now they are being detected, and questions are now being asked as to their effect on the health of the river.

Citizens, municipalities, and industries have again become concerned over water quality in the river. This concern for the river manifested itself in 1991, when the state legislature requested that the Department of Environmental Quality initiate a comprehensive water-quality study of the river (Oregon Department of Environmental Quality, 1993). This study was much too expensive for the state to implement on its own so it solicited the support of the municipalities and major industries along the river. A cooperative study was initiated in 1991. The study was to provide information to determine how well the river was doing. The study was divided into two phases. The Phase I reviewed and summarized existing data, developed and evaluated tools to measure water quality,

TABLE 2
Congressionally Authorized U.S. Army Corps of Engineers Dams in the Willamette River Basin

Hills Creek
Lookout Point (including the Dexter re-regulating dam)
Fall Creek
Cottage Grove
Dorena
Cougar
Blue River
Fern Ridge
Green Peter
Foster
Detroit (including the Big Cliff re-regulating dam)

FIGURE 6.   U.S. Army Corps of Engineers Detroit Dam.

and developed methods to predict water-quality conditions. This later task was the central part of the cooperative effort. All parties involved wanted to see whether the existing data when placed in models would show whether the river was being affected. Priorities were established on data collection in Phase I and this new information was integrated into models to see if present conditions could be described and future conditions predicted. Below is a summary of the results for the major study components.

TOXIC MODELING COMPONENT

The toxic model effort relied on existing data collected from several different studies with various levels of basin coverage. Consequently, the results were not very complete with numerous data gaps being identified. There was very little the researchers could conclusively state about toxicicants throughout the basin. High priority was given to the collection of toxic data in Phase II of the study.

DISSOLVED OXYGEN MODELING COMPONENT

This proved to be one of the most fruitful parts of the Phase I study. Dissolved oxygen data collected during a synoptic survey, was used to calibrate a DO-nutrient-phytoplankton growth model. After adjustments were made for sediment oxygen demand (SOD), the final model fit well the steady-state conditions. The model predicted that during low-flow periods:

- minimum DO conditions occur in the Newberg Pool, RM 27,
- the most significant DO uptake is from sediments and the degradation of organic compounds from point sources and tributaries, and
- the amount of river flow was a significant factor controlling DO concentrations and phytoplankton biomass.

Researchers want to specifically monitor SOD to confirm the rates used in the Phase I modeling.

BACTERIAL MODELING

This was one of the more disappointing study components. The data collected during Phase I were not sufficient to calibrate a model. Many questions remain unanswered as a result of the bacteria modeling effort. Many more bacteria were in the river than would have been predicted by the model. Considerable work remains to identify sources — both point and nonpoint — and rerun the model.

NONPOINT SOURCE LOADING

Considerable time and effort was spent evaluating and selecting nonpoint source models. Several different models were examined, each with its own advantages and disadvantages. The most important element of these models is the amount of data needed to accurately predict nonpoint source pollution loadings. The modeling of nonpoint sources was achieved by intensely developing nonpoint source information on one small watershed. This model was then extended to a sub-basin and then to the entire Willamette River Basin. The reduction in resolution for each next extrapolation, however, became a key sticking point for general acceptance of this modeling approach. It was apparent that more specific water-quality monitoring data was needed from many different sites throughout the basin to develop better modeling capabilities and predictions of nonpoint source impact.

ECOLOGICAL SYSTEM INVESTIGATION

Of the many Phase I study efforts the most telling may be the results from the ecological system investigation. Added to the study at the insistence of the environmental community representative, the ecological investigations were a series of different tests and mini-studies designed to gather information on the health of the aquatic systems. The original objection to these tests centered on the ability of the tests to actually determine the cause of different impacts if they are noted. But in the final analysis, the tests were very useful in indicating where potential ecological problems existed. Yes, researchers are still left with the task of finding the cause of some disturbing results; but, without the tests themselves, researchers would have had precious little to show for their efforts or little information on the overall health of the system. The river was divided into four regions for the ecological studies (Figure 7):

Region I   Eugene to Corvallis
Region II   Corvallis to head of Newberg Pool at Yamhill River confluence
Region III   Head of Newberg Pool to Willamette Falls
Region IV   Willamette Falls to mouth (Portland Harbor)

Benthic Community Assessment

The first ecologic system effort centered on the benthic communities. Traditional kicknet and sediment grab samples were collected. Both of these techniques used the U.S. EPA Rapid Bio-assessment Protocols (RBP) to reevaluate benthic community responses to habitat and water quality. The test seemed to indicate that the water-quality degradation and not physical-habitat degradation are impairing the communities.

Fish Community Health Assessment

Several different tests were conducted to indicate whether the fish communities were impaired. The first two looked at community diversity and fish health. Communities were sampled to determine type and quantity of fish species and these were recorded. Selected fish in each study region were also autopsied. The results showed some regions were affected by pollution but the results

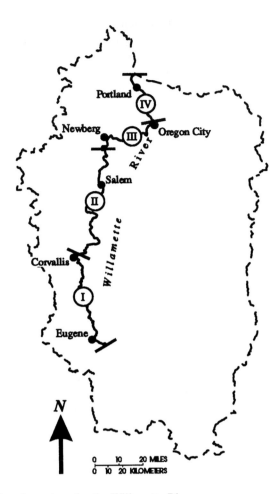

FIGURE 7. Ecological study regions for the Willamette River.

were inconclusive. The site considered to be least impacted by man's activities (background) showed impact as well as downriver sites. Questions about fish migration and specific point-source effects need to be answered. More sampling stations and additional species need to be studied in the future to provide better confidence in the results.

The third fish community test seemed to yield the most interesting and maybe the most challenging data. The test involved the examination of skeletal abnormalities. These tests showed skeletal deformities in 1 to 2% of those fish tested in Regions I, II, and IV. In Region III, however, deformities were found in more than 13% of the fish tested. These results are interesting for several reasons, one of which is that Region IV is downriver in the Portland Harbor and Region III is the first significant sink on the river. Why are test results in the harbor lower than the upper most site? Could the Newberg Pool, Region III, be the first point on the river where pollutants collect? These and other questions will need to be addressed in Phase II.

RIVER BASIN MANAGEMENT

All of the work conducted in the original restoration and the work currently being conducted during the most recent cooperative water-quality study, point to a simple fact: protection of the Willamette River and maintenance of its beneficial uses will need to be a comprehensive management approach. This includes not only those point sources discharging treated effluents directly to the river, but also the indirect discharges from nonpoint sources such as agriculture, forestry, and the urban areas.

Likewise, not only the dischargers need to be included, but also those taking water from the river for use as drinking water, industrial water, and irrigation water and those recreating in and on the water. All of these uses need to be considered together to identify needs and to evaluate what the river is capable of providing. Then each government agency and private organization must keep constant vigil on whether goals are being reached and expectations fulfilled.

SUMMARY

The cleanup illustrated how citizens, industries, municipalities, and government can work together to restore a tremendously polluted river. Salmon returned to a river that once suffocated them in a matter of minutes. Two decades later, attention has again focused on the Willamette River. In the past 20 years, the river basin's population has increased, its industries and urban areas have expanded, and its rural land has experienced continued use and development. A cooperative water quality study has been undertaken to determine the health of the river. Early results may indicate that we again need to examine impacts on the river and develop new management strategy if the river is to be truly restored and useful.

REFERENCES

Gleeson, G. W., The Return of a River — The Willamette River, Oregon, June 1972.

Oregon Department of Environmental Quality, Oregon Water Quality Status Assessment Report, April 1992.

Oregon Department of Environmental Quality, Willamette River Basin Water Quality Study — Summary Report, June 1993.

Oregon State Sanitary Authority (SSA), Report on Water Quality and Waste Treatment Needs for the Willamette River, May 1964.

Oregon State Sanitary Authority (SSA), Water Quality Standards, Willamette River, February 1967.

State Water Resources Board, Lower Willamette River Basin, June 1965.

State Water Resources Board, Oregon's Long Range Requirements for Water, June 1969.

U.S. Army Corps of Engineers, Willamette River Basin Reservoir System Operation, May 1989.

U.S. Army Corps of Engineers, Willamette River Basin Review Reconnaissance Study, June 1991.

Water Resources Commission, Willamette Basin Plan, January 1992.

5 Water Quality in the Vistula Basin: A River Under Stress

Jan R. Dojlido

The water quality of rivers in the Vistula River Basin has changed drastically since 1945 from clean water to heavily polluted water. Historical data show the continual deterioration of water quality in the basin up to approximately 1985. From that time, water consumption started to decrease, wastewater treatment began to improve, and the further decrease of water quality was stopped. Even with this water-quality trend reversal, the 1992 water-quality classification still shows that only 4% of the total river length belongs to the first class of water quality (good) and 50% of river length is still in class IV (very polluted). This condition existed in 1992 with 33% of wastewater produced in the Vistula basin still untreated and only 33% of the wastewater treated biologically.

INTRODUCTION

The Vistula River is the largest Polish river and the second largest river in the Baltic Sea drainage area. Its total length is 1047 km and total drainage area 194,700 km^2. The Vistula Basin covers 55% of Poland and is inhabited by about 21 million people.

The Vistula is an unregulated river for the major part of its length, with many meanders and islands. When most of the rivers in Europe were regulated in nineteenth century, Poland was in a political backwater and very few river improvements were ever made. The Vistula, however, is regulated to some small degree in the upper and lower parts. In the upper Vistula, there are two impoundments: Wisła Czarne (96.8 km) and Goczalkowice (42.8 km), and three water weirs: Laczany (38.5 km), Dabe (81.0 km), and Przewoz (92.2 km). The middle part of the Vistula is in a natural state. Very little regulation occurs in the lower part from control of the Włocławek reservoir located on the main course of the river at 674 km.

HISTORICAL MEASUREMENT OF VISTULA WATER QUALITY

The first information on measurement of Vistula water quality is from the nineteenth century. The famous scientist, Mendeleyew, in 1876, analyzed the Vistula water quality in connection with water treatment for municipal purposes. He determined mostly inorganic components of water (e.g., cations responsible for water hardness).

More complex monitoring of Vistula water was done in 1923 to 1924 by Kirkor. He measured 14 components every 2 weeks. He found: BOD$_5$: 2.2 mg/L; iron: 0.2 mg/L; and ammonia: 0.1 mg/L. His conclusions were that Vistula water in Warsaw was fairly clean, but that it was turbid and of yellowish color. The routine monitoring of Vistula water in Warsaw was started in 1926 by the laboratory of Water Works. From 1926 to 1939 the water was clean, with only the concentration of iron increasing.

The monitoring done from 1945 to 1948 showed that Vistula water was still clean. During this time, the mineralization of water was only just noticed. The Vistula River at Warsaw was of good enough quality for recreation, and many beaches were in use. During this time, the Vistula was used for swimming, water sports, and fishing by many inhabitants of Warsaw.

0-56670-138-4/97/$0.00+$.50
© 1997 by CRC Press, Inc.

TABLE 1
**Comparison of Selected Water-Quality Parameters: Data Collected in 1945 and
That Collected in 1950, 1960, and 1970, Vistula River at Warsaw**

Year	Odor	Color	COD-Mn	NH$_4$	Cl	Total residue	BOD$_5$
1950	0	70	2	65	80	80	—
1960	15	90	20	85	98	100	50
1970	100	95	40	100	100	100	95
1945 (conc.)	Natural	Natural	7 mg/L	0.01 mg/L	15 mg/L	250 mg/L	2.2 mg/L

Note: All comparison values are in percent for samples taken that exceeded the 1945 value. COD, chemical oxygen demand; BOD$_5$, biochemical oxygen demand, measured after 5 days.

The process of the industrialization of Poland started from 1945 to 1948. Many heavy-industry plants were constructed, mostly in the upper part of the Vistula basin. Coal mines were expanded, and petrochemical plants, artificial fertilizer plants, chemical plants, power plants with open cooling systems, and other plants and factories were built. The construction of treatment plants was always very far behind the construction of industrial plants and new towns, so the pollution of water started to increase rapidly.

Vistula Water Quality in Warsaw, 1945 and 1970

The year 1945 can be used as a base where the Vistula River water at Warsaw can be considered clean. By 1970, nearly 100% of the water-quality constituents that were being measured exceeded the concentrations measured in 1945 (Table 1).

The dissolved oxygen (DO) content was quite good, despite growing organic pollution. This situation could be explained by the natural character of Vistula, with many meanders, beaches, and islands that filtered and aerated the water. The lowest DO concentration found in winter, under the ice cover, was 2.4 mg/L (Figure 1).

The concentration of ammonia increased from 1945 to 1970 (Figure 2). The maximal value measured was 4.4 mg N/L. A seasonal variation was observed, with the highest values noticed in December and January, when water was covered by ice.

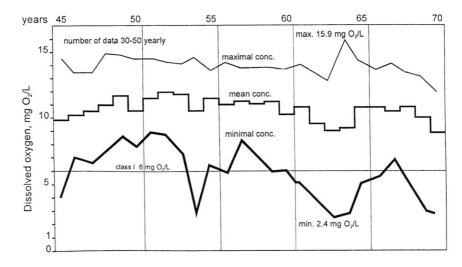

FIGURE 1.   Dissolved oxygen concentration in Vistula water in Warsaw from 1945 to 1970.

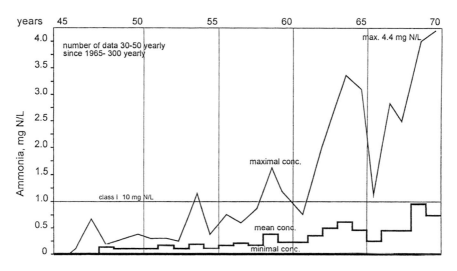

FIGURE 2.   Ammonia concentration in Vistula water in Warsaw from 1945 to 1970.

The most spectacular increase in a water-quality constituent was observed in the chloride concentration (Figure 3). The average concentration increased from 15 mg Cl/L in 1945 to 74 mg Cl/ L in 1969. The cause for the increase was the growth of the coal mine industry in upper part of the Vistula Basin in the Silesia region. With these data, a correlation between chloride concentration and water flow was observed. In "wet" years the concentration values of chloride were lower.

The chemical oxygen demand (COD) values (obtained using the permanganate method) were increasing, but irregularly. The highest value was during this period — 19 mg O_2/L — but the average oscillated around 10 mg/L.

CHANGES OF VISTULA WATER QUALITY FROM 1969 TO 1978

The water quality in 1968 and 1978 along the Vistula course were compared using actual data and calculated data that are within the 90% confidence limit. The 5-day biochemical oxygen demand

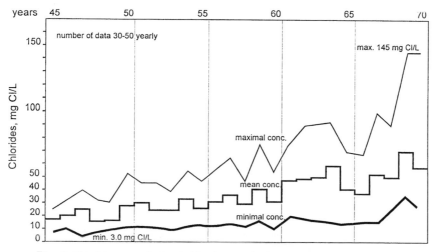

Fig. 3 Chloride concentration in Vistula water in Warsaw, 1945-1970

FIGURE 3.   Chloride concentration in Vistula water in Warsaw from 1945 to 1970.

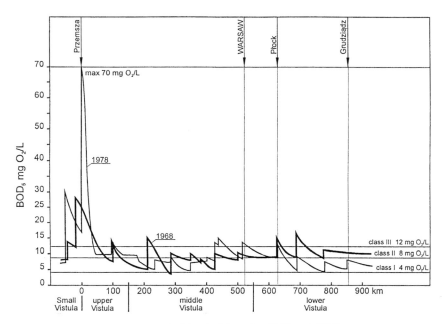

FIGURE 4.   Estimated BOD$_5$ values in the Vistula River for 1968 and 1978.

(BOD$_5$) generally increased along the Vistula between 1968 and 1978, but not in all courses (Figure 4). The BOD$_5$ increased in the middle part of the Vistula, while in the lower part of the river the BOD$_5$ decreased. From the BOD$_5$ data (Figure 4) it can be seen (within a 90% degree of confidence) that high pollution occurs at the upper part of the river, and there are also increases of pollution below the each big town or industrial plant along the Vistula River.

A large increase is noticed in chloride concentration between 1968 and 1978 (Figure 5). Concentrations are greater by several times in the upper part, double in the middle part, and 1.5 times in the lower part of the Vistula.

The increase was also noticed in total dissolved solids. Other water-quality parameters were in the similar range.

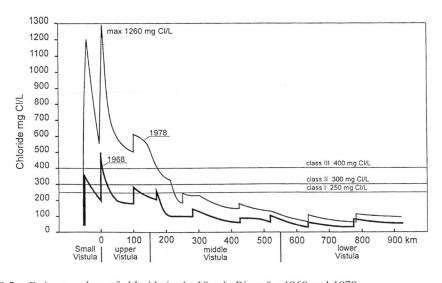

FIGURE 5.   Estimate values of chloride in the Vistula River for 1968 and 1978.

The general change of the Vistula pollution for 1968 and 1978 was similar. By 1978, the Vistula River could be divided into four sections and classifications:

1. Small Vistula, from source to about 40 km — clean water.
2. Small Vistula from about 40 km to 0 km and Vistula from 0 km to about 20 km — very highly polluted.
3. Vistula from about 20 km to about 200 km — rapid decrease of pollution.
4. Vistula from about 200 km to the Baltic Sea — a similar level of pollution to that of 1968, with decreases below each additional pollution source.

VISTULA WATER QUALITY FROM 1989 TO 1992

This evaluation is based on intensive water quality monitoring. Measurements were made at five stations:

- Tyniec, 63.7 km (upper part of Vistula, upstream from Cracow)
- Warsaw, 510.0 km (middle Vistula, upstream from sewage effluent of Warsaw)
- Wyszków, 33.0 km (Bug River, upstream from Zegrzynskie Lake)
- Pultusk, 63.0 km (Narew River, upstream from Zegrzynskie Lake)
- Kiezmark, 926.0 km (lower Vistula, near Baltic Sea)

The water quality was measured two times a week in 1989, 1990, and 1991 and once a week in 1992.

During 4 years of measurement, approximately 350 data sets were collected and more than 40 parameters were determined. The following is a summary of several of the parameters that were analyzed:

1. pH of water was stable, from 6.6 to 9.7.
2. Dissolved oxygen was the lowest in the Vistula at Tyniec; the minimum value was 0 mg/L. In the Vistula in Warsaw, DO concentrations varied from 5.4 mg/L up to 17.4 mg/L (oversaturation). The highest concentrations were found in Bug River at Wyszków, with a value as high as 20.3 mg/L. In general, there is a good oxygenation of Vistula River water; however, some problems occurred in the upper part of the river during the winter.
3. BOD_5 was the highest in the Vistula at Tyniec with the maximal value of 31.8 mg/L, and the annual average of about 8 mg/L. Along the Vistula, BOD_5 values generally decreased in a downstream direction.
4. Chloride concentrations are very high reaching 2140 mg/L in Tyniec. The lowest value was 46 mg/L, which was sampled in Warsaw.
5. Dissolved solids concentrations were very high in the upper part of the Vistula (maximal value 4160 mg/L), then much lower downstream and quite low in the Bug and Narew Rivers (Figure 6).
6. The sulfate concentrations were very high in the Vistula at Tyniec (maximum value 314 $mgSO_4/L$), and decreased rapidly in the distance to Warsaw. The Bug and Narew Rivers had lower values than Vistula.
7. Sodium concentrations were very high in upper Vistula. The annual average was 700 mg/L in the Vistula at Tyniec, but only 80 mg/L in the lower Vistula at Kiezmark.
8. Hardness of Vistula water varied from 150 to 1000 mg $CaCO_3/L$. The highest values were at the upper part of the river.
9. Calcium and magnesium concentrations were distributed similarly as hardness; calcium concentration varied from 33 to 260 mg/L and magnesium from 6.8 to 116 mg/L. The concentrations decreased along the course of the river.

Dynamics

6 Rain and Rocks: The Recipe for River Water Chemistry

Gerald Best

INTRODUCTION

Throughout the world, samples of river water are being analyzed for a variety of chemical determinants in order to assess the quality of the water.

The results of these measurements are usually compared against earlier values or against a baseline standard to decide whether there has been a shift in the quality. If that shift reflects deterioration, then steps are taken to find the source of the pollution to effect an improvement. An essential part of this judgment is an understanding of the 'background' state of the particular sample of water. This requires an understanding of the various factors that affect the chemistry of the water before it arrives at the point where it is sampled. This evaluation is important because a water authority usually establishes a target quality which the water should achieve.

This target must take into account the natural state of the water, otherwise an unrealistic target could be set that would result in dischargers being asked to install unnecessary treatment processes for wastewater. "Natural" water contains trace quantities of a wide range of substances and the concentration of these in uncontaminated river water varies according to the nature of rainfall, the geology of the catchment, and the vegetation.

RAINFALL

Even before water arrives on the earth's surface (whether as rainfall, hail or snow), it dissolves substances present in the atmosphere. These originate from sea spray, natural emissions such as gases from volcanoes and swamps, and from emissions put into the atmosphere by human activities.

The influence of salt from sea spray on the chemistry of rainfall is demonstrated by data in Table 1. These figures were reported by Harriman and Wells (1985) who collected rainfall on a daily basis at a site in Pitlochry, Scotland. The site is approximately 120 km from the West Coast from where the gale originated; it is also well removed from industrial sources of pollution.

The results clearly show that sea salt can be transported well inland from the coast and is then solubilized by the rainwater which brings it to earth. This process is shown by the data in Table 2 for mean values of constituents in rainfall from various publications. These show that the values of chloride in precipitation for a maritime country like Scotland are considerably greater than for precipitation collected in inland areas such as in Denver, CO (central U.S.), or in the Czech Republic (central Europe).

Another feature of the data in Table 1 is the difference in the concentration of sulfate ion between precipitation originating from the Southeast compared with that from the West. Sea spray contains sulfate ion but the air mass entering Scotland from the Southeast comes from the land mass of Europe and from England. The sulfate is formed in the precipitation from sulfur dioxide which is emitted into the atmosphere primarily from power plants and from industry. Figure 1

TABLE 1
Change in Chemical Composition
of Rainfall Wind Direction
(Pitlochry, Scotland)

Determinants	Origin of wind (concentrations in $\mu g\ L^{-1}$)	
	Southeast	Westerly gale
Na^+	2100	22,300
Cl^-	3200	37,000
SO^{2-}	11,500	4800
NO^{3-}	700	100
Amount (mm)	5	15

shows the influence of wind direction on the acidity of rainfall (Burns et al., 1984). The reports by Overrein et al. (1980), Edwards et al. (1990), Last and Whatling (1991), and Morrison (1994) deal with this topic. From these investigations, a clear gradation emerges of a declining concentration of sulfate ion in precipitation when going from a southeast to a northwest direction in the U.K.

As well as the major constituents, precipitation also contains trace quantities of metals and organic substances, mostly from authropogenic sources. Even precipitation falling in remote areas, far removed from the sources of metals can contain contaminants discharged into the atmosphere from human activity.

Figure 2 gives data obtained by Murozumi et al. (1969) of concentrations of lead in a core of Greenland snow. The layers of snow were dated by radioisotope techniques and show the rapid increase in concentration from 0.08 to 0.20 $\mu g\ L^{-1}$ in an approximately 20-year period. These results should be compared with recently published data by Gorlach et al. (1991) who examined the concentration of lead and other metals in Greenland snow which had fallen over the last 20 years. As shown in Figure 3, their analyses indicate that there had been a decline in lead levels from 0.20 to about 0.025 $\mu g\ L^{-1}$. With such a dramatic reduction, one would expect that there would also be evidence of a reduction in emissions of lead to the atmosphere. The graph in Figure 2 shows that the greatest increase in lead occurred in the 1950s, coinciding with the increase in road traffic using lead as an antiknock additive in the fuel. Recently there has been a distinct shift by motorists to using lead-free petrol, which has reduced lead emissions. It has been calculated that the lead emitted into the atmosphere from vehicles in Europe in 1975 amounted to 124,000 t but this was halved to 68,000 t by 1982. Depending on the amount of lead-free petrol used, the predictions for the year 2000 range from 32,400 to 10,900 t (Pacyna et al., 1991).

TABLE 2
Major Chemical Constituents of Precipitation

Determinants	Scotland	Washington State	Denver (snow)	Czech Republic
Na^+	4370	55–480	90–800	50–320
Ca^{2+}	820	25–210	100–1340	500–2500
Mg^{2+}	480	8–48	<10–130	40–240
SO^{2-}	8100	<700–2500	200–6200	300–11,000
Cl^-	5900	—	50–850	200–900

Note: Mean or range concentrations in $\mu g\ L^{-1}$.

Rair

el
a
fi
h
r
d
c
t
t

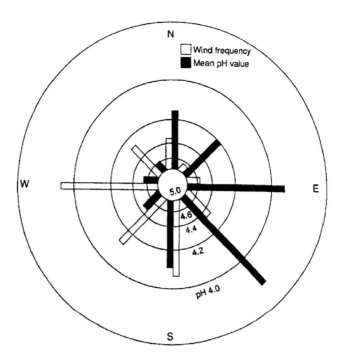

FIGURE 1.   Influence of wind direction on the acidity of rainfall.

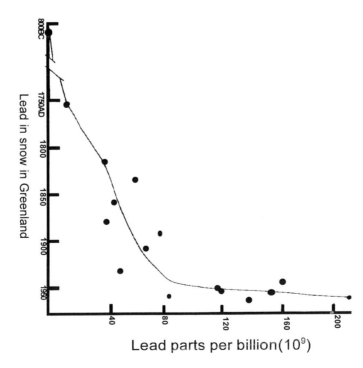

FIGURE 2.   Concentrations of lead in Greenland snow (800 BC–1965 AD).

GEOLOGY

The concentrations of inorganic and organic substances in precipitation are substantially altered as the precipitation passes from the receiving ground to the drainage stream. Only negligible amounts of rainfall will actually arrive at the stream or lake without first passing over or through the ground in the catchment area.

The rainfall is in contact with the ground for varying periods of time depending on such factors as the intensity and duration of the rainfall, the porosity of the ground, and its moisture content. The duration of contact obviously affects the chemical composition of the drainage water. Flood water, after an intense storm or rainfall falling onto saturated ground, has maybe only a few hours of contact with the ground, whereas a base flow into a river is of water which has been stored in the soil and rocks for weeks, months, or even longer since it first fell to earth. These differences are illustrated in Figure 4a and b. In Figure 4a, the rainfall permeates through the ground, undergoes various chemical transformations because of ion exchange, adsorption, biochemical interaction, weathering and dissolution, and then emerges from the river bank. In Figure 4b, however, the ground is saturated with water from earlier rainfall and the pathway of the rainfall to the stream is mainly over the surface of the ground. In this latter case, the drainage water will be of similar composition to the rainfall.

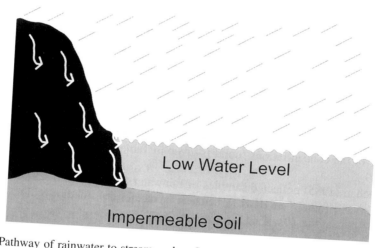

FIGURE 4a. Pathway of rainwater to stream — low flow.

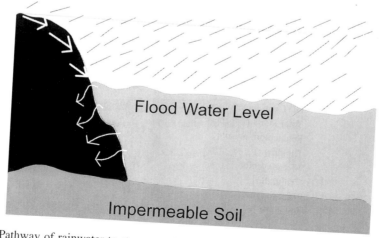

FIGURE 4b. Pathway of rainwater to stream — high flow.

TABLE 5
Chemical Composition of Water
Draining from Different Types of Rock
(mg L^{-1} except pH)

Determinant	Rock type		
	Granite	Sandstone	Limestone
pH	6.6	6.8	7.9
Ca^{2+}	0.8	1.8	51.2
Mg^{2+}	0.4	0.8	7.7
Na$^+$	2.0	1.2	0.8
K$^+$	0.3	0.8	0.5
SO$_4^{2-}$	1.5	4.6	4.1
HCO^{3-}	7.8	7.6	195.0
SiO$_2$	9.0	9.0	6.0

As has already been mentioned, an important source of some substances in stream water is their presence in rainfall. This particularly applies to material of marine origin. Rainfall is the predominant source of Na$^+$ and Cl$^-$ in stream water and is a major contributor of Mg^{2+} and K$^+$, particularly in rivers with catchments close to the sea. However, Walling and Webb (1986) calculated that the average dissolved solids contents of the rivers throughout the world is 120 mg L^{-1} and this is comprised largely of the ions HCO^{3-}, SO$_4^{2-}$, Ca^{2+}, and SiO$_2$ — all of which originate principally from the weathering of the rocks. The nature of the geology of a catchment will have a marked effect on the chemical characteristics of the drainage water because of the differing weathering rates of rock types. Old igneous rocks such as granite and gneiss are resistant to breakdown and dissolution; limestone, on the other hand, is much more readily broken down and solubilized. This contrast is shown by data in Table 5 which gives the average chemical composition of unpolluted river waters draining from different types of geology (Webb and Walling, 1992).

From these data, one would expect that there will be marked differences in the chemical composition of river waters which are uncontaminated by artificial sources if they drain from catchments with different geologies. In the U.K. there is a pronounced difference in the geology across the country, from the old igneous rocks in the Scottish Highlands to the limestone areas in Southeast England (Figure 5). One way of summing the total ionic concentration of a river water is to measure the electrical conductivity. As an approximate guide, the total dissolved solids in a river water sample in mg L$^-$ is between 0.55 and 0.7 of the conductivity when expressed as µS cm^{-1}.

The data in Figure 6 show the variation in conductivity for some of the major rivers in U.K. in 1992 (Institute of Hydrology) which range from 41 µS cm^{-1} for the River Beauly in the Scottish Highlands to 955 µS cm^{-1} for the River Nene in Eastern England.

In U.K., perhaps the greatest change in geology occurs in Scotland with the ancient granitic rocks in the northwest and the sedimentary and limestone areas of the southeast. In Scotland, there are also two distinct geological faults that traverse the whole country. The Highland boundary fault dissects the country in the middle in a northeast to southwest direction while the southern uplands fault is more east-northeast to a west-southwest direction (Figure 7).

The data in Figure 7 also show the concentration of dissolved salts as measured by the conductivity while the alkalinity values indicate the calcium concentration. The measurements show the anomaly of the River Tweed in Southeast Scotland when compared with other rivers in Scotland. This is because the water chemistry of this river in its lower reaches is dictated mainly by the limestone outcrops in the catchment.

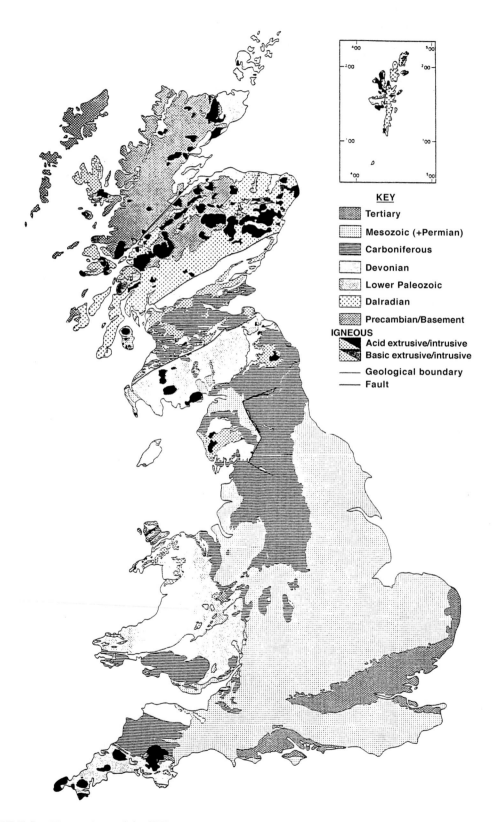

FIGURE 5.   The geology of the U.K.

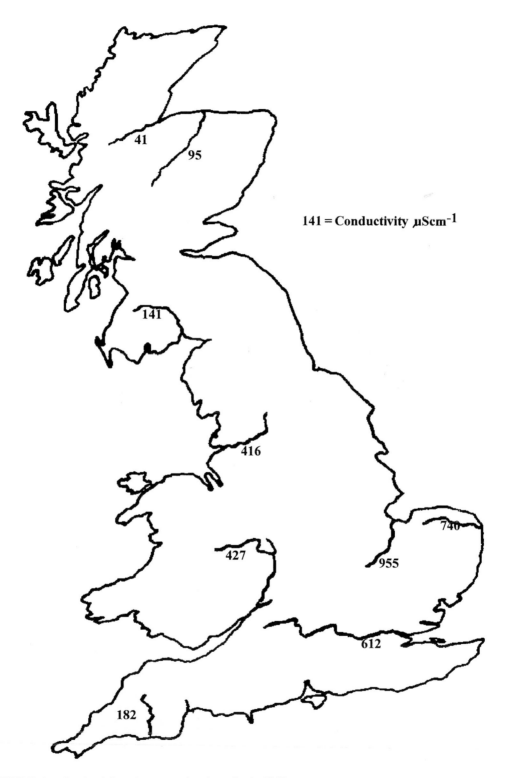

FIGURE 6. Conductivity of some major rivers in the U.K.

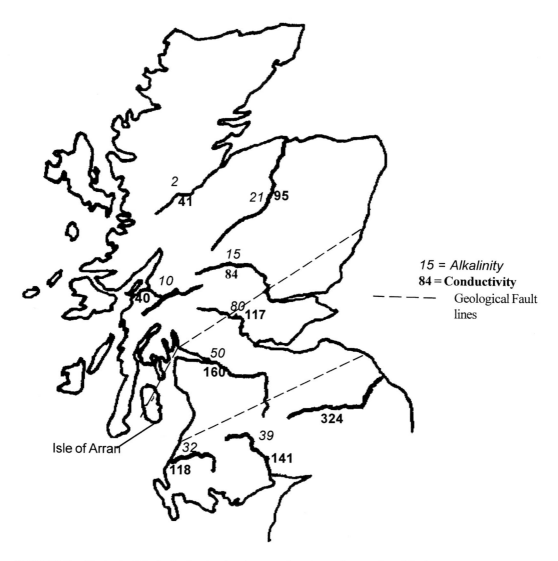

FIGURE 7.   Geological faults in Scotland and water chemistry of some Scottish rivers.

When the fault lines in Figure 7 are examined in detail, it will be noticed that the Highland boundary fault passes midway through the Isle of Arran in the Firth of Clyde and this results in there being two distinct geological areas on the island. In fact, Arran is sometimes described as "Scotland in miniature" because, within its 32 km length, there are examples of a typical Highland area in the north with its main peak of Goatfell (874 m), while the southern part of the island is similar to much of Scotland's Lowlands. The chemistry of the pristine rivers on the island reflect the geology with conductivity values of about 50 μS cm^{-1} in the north compared with values of more than twice this in the south (Figure 8). The northern granitic rocks are deficient in base cations as shown by the alkalinity of less than 5 mg L^{-1} compared to concentrations of about 20 to 30 mg L^{-1} in the south.

AQUATIC BIOTA

The different chemical environments of unpolluted rivers caused by the nature of the geology is reflected in the various organisms that inhabit the waters. This is particularly evident in the invertebrate species that are found in the stream bed. Some of these require minimal levels of

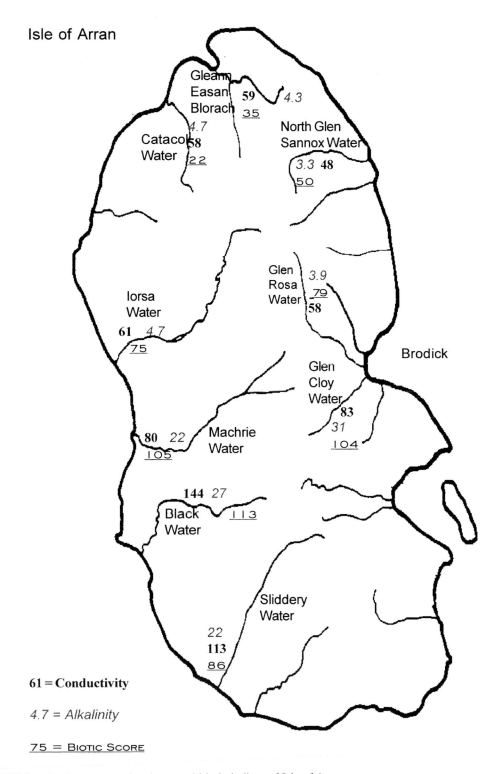

Isle of Arran

Gleann
Easan **59** *4.3*
Blorach 35

4.7
Cataco **58**
Water 22

North Glen
Sannox Water

3.3 **48**
50

Glen
Rosa *3.9*
Water 79
58

Iorsa
Water

61 *4.7*
75

Brodick

Glen
Cloy
Water
83
31
104

80 *22*
105

Machrie
Water

144 *27*
Black 113
Water

Sliddery
Water

22
113
86

61 = Conductivity

4.7 = Alkalinity

75 = BIOTIC SCORE

FIGURE 8. Geology, water chemistry, and biotic indices of Isle of Arran.

TABLE 6
Acid Indicator Freshwater Invertebrate
Species and Their Distribution in
Relation to Mean pH

Mean pH	Group	Species absent
<7	Crustacea	*Gammarus pulex*
<6	Mayflies	*Baetis muticus*
		Caenis rivulorum
	Stoneflies	*Perla bipunctata*
		Dinocras cephalotes
	Beetles	*Esolus parallelepipedus*
	Caddisflies	*Glossosoma* spp.
		Philopotamus montanus
		Hydropsyche instabilis
		Sericostoma personatum
<5.5	Mayflies	*Baetis rhodani*
		Rhithrogena sp.
		Ecdyonurus spp.
		Heptagenia lateralis
	Stoneflies	*Perlodes microcephala*
		Chloroperla tripunctata
	Caddisflies	*Hydropsyche pellucidula*

calcium to be present in the water to construct a shell or an exoskeleton, for example snails, limpets and shrimps, while others require an environment to live in of fine muds which are deposited on the stream bed in readily eroded catchments, e.g., midge larvae, oligochaete worms.

In the U.K., the diversity of the invertebrate population in running waters is expressed as a biotic index. The index system that has been standardized throughout the country by the freshwater biologists of the U.K. is the Biological Monitoring Working Party (BMWP) score. The higher the score, the greater the diversity and number of invertebrate species present in the water and thus the cleaner the water. There is no fixed upper limit the score and unpolluted rivers may have widely differing BMWP scores for reasons unrelated to pollution. For example, stream draining areas in a base rich catchment, such as the chalk rivers of Southern England, may have BMWP scores of over 200. By contrast the nutrient-poor highland rivers in Scotland may have an optimum score of only 70 to 80. In the west of Scotland, the maximum score achievable is about 180.

The BMWP scores for the unpolluted rivers in Arran are shown in Figure 8 and range from 22 in the north of the island to 113 in the south. This difference is mainly dictated by the geology of the catchment and the chemistry of the water.

The scores in the northern stream, however, are also suppressed because of the effect of acid rain. The catchment areas of the northern rivers are characterized by resistant geology and thin, nutrient-poor soils. Any acidity in the rainfall passes with little change to the streams. As stated earlier, the west of Scotland occasionally receives precipitation that is carried to the area on southeasterly winds (10% of precipitation events originate from this wind quadrant) and the rain is markedly more acidic than that arriving from the southwest. Daily values of pH have varied from 2.6 to 6.6, almost three orders of magnitude of acidity (Fowler and Irwin, 1989). The streams in the south of the island originate from catchments richer in base cation and the incoming acidic rainfall is neutralized more effectively before it arrives in the streams. The acidity in the stream adversely affects the invertebrate biota but some organisms are more sensitive to the acidity than others. Doughty (1990) summarized the relationship between mean pH values and invertebrate species survival (Table 6). As well as invertebrate organisms, the fish populations in the northern rivers are also reduced. Some have no salmonid species because the acidity in the rainfall releases aluminium from the granite producing toxic levels in the stream water. Mean concentrations of aluminum in the northern streams are about 100 μg L^{-1} compared with 45 μg L^{-1}, for example, in the Machrie Water.

CONCLUSIONS

The chemistry of water in streams that have received no direct discharges of polluting substances differ markedly according to several factors. The major influences on water chemistry are the closeness to the sea and the nature of the geology but trace organic and inorganic substances can be brought to a pristine catchment after being emitted into the atmosphere from sources far removed from the river.

REFERENCES

Burns, J. C., Coy, J. S., Tervert, D. J., Harriman, R., Morrison, B. R. S., and Quine, C. P., The Loch Dee project: a study of the ecological effects of acid precipitation and forestry management on an upland catchment in South West Scotland. 1. Preliminary investigations. *Fish. Manage.*, 15, 145, 1984.

Doughty, C. R., Acidity in Scottish rivers, A report by the Scottish River Purification Boards Association, Department of Environment, London, 1990.

Edwards, R. W., Gee, A. S., and Stoner, J. II., Acid Waters in Wales, Kluwer Academic Publishers, Dordrecht, Netherlands, 1990.

Fowler, D. and Irwin, J. G., The pollution climate of Scotland, Acidification in Scotland. Symposium proceedings, Scottish Development Department, Edinburgh 1989.

Gregor, D. J., Ph.D. thesis, Faculty of Science, University of Geneva, Switzerland, 1989.

Gorlach, U., Candelone, J. P., and Boutron, C., Changes in heavy metals concentration in Greenland snow during the past twenty years. Heavy metals in the environment, International Conference, Edinburgh, CEP Consultants, Edinburgh, 1991.

Harriman, R. and Wells, D. E., Causes and effects of surface water acidification in Scotland, *Water Pollut. Control*, 84, 215, 1985.

Institute of Hydrology, Hydrological data U.K. 1992 Yearbook, Institute of Hydrology, Wallingford, Oxon, 1993.

Last, F. T. and Whatling, R., eds., *Acidic Deposition, Its Nature and Impacts.* Proceedings of the Royal Society of Edinburgh, Section B, 97, 1991.

Miller, E. K. and Friedland, A. J., Recent reductions of the atmospheric lead flux to high elevation forests in North Eastern U.S.A. Heavy metals in the environment, International Conference, Edinburgh, CEP Consultants, Edinburgh, 1991.

Morrison, B. R. S., Acidification, in *The Fresh Waters of Scotland,* Maitland, P. S., Boon, P. J., and McLusky, D. S., eds., Wiley & Sons, Chichester, Chap. 24, 1994.

Murozumi, M., Chow, T. J., and Patterson, C. C., Chemical concentration of pollutant Pb aerosols, terrestrial dusts and sea salts in Greenland and Antarctic snow strata. *Geochim Cosmochim Acta,* 33, 1247, 1969.

Overrein, L. N., Seip, H. M., and Tollan, A., Acid precipitation — Effects on forests and fish. Final report of the SNSF Project 1972–1980, 1980.

Pacyna, J. M., Munch, J., Alcamo, J., and Anderberg, S., Emission trends for heavy metals in Europe. Heavy metals in the environment, International conference, Edinburgh, CEP Consultants, Edinburgh, 1991.

Radojevic, M., Tyler, B. J., Pendergherst, N. J., and Hall, S. A., Trace metals in cloudwater and their influence on SO_2 oxidation. Heavy metals in the environment, International conference, Edinburgh, CEP Consultants, Edinburgh, 1991.

Strachan, W. M. J., Organic substances in the rainfall of Lake Superior, 1983. *Environ. Toxicol. Chem.,* 4, 677, 1985.

Walling, D. E. and Webb, B. W., Solutes in river systems, in *Solute Processes,* Trudgill, S. T., ed., Wiley & Sons, Chichester, 1986.

Webb, B. W. and Walling, D. E., Water quality. II. Chemical characteristics, in *The Rivers Handbook*, Vol. 1. Calow, P. and Petts, G. E., eds., Blackwell Scientific Publications, London, 1992.

Welch, H. E., Muir, D. C. G., Billeck, B. N., Lyle Lockart, W., Brunskill, G. J., Kling, H. J., Olson, M. P., and Lemoine, R. M., Brown Snow: A long-range transport event in the Canadian Arctic, *Environ. Sci. Technol.*, 25, 280, 1991.

7 A Geomorphic Basis for Interpreting the Hydrologic Behavior of Large River Basins

Gordon E. Grant

Many of the processes controlling runoff at both small and large scales are linked to the underlying geomorphology of a region. The westward-flowing tributaries of the Willamette River in western Oregon flow perpendicular to regional geologic trends, affording the opportunity to examine effects of geomorphology on streamflow. The interaction of geologic substrate, topography, and climate determines the overall surface water discharge regime, including the shape and timing of the annual hydrograph. Drainage density, reflecting the hydraulic transmissivity of the underlying rocks, influences the efficiency of the channel network to transmit water during individual storm events. An understanding of the physical and biological responses of watersheds to human modifications, including reservoir and forest management, requires appreciation of the broader geomorphic framework in which such changes occur.

INTRODUCTION

For much of its history, the hydrologic sciences have concentrated on relatively modest space and time scales. During this century, much hydrologic research has focused on plot, field, or small watershed studies, and addressed problems of water input, throughput, storage, and consequences for physical and biological processes that could be measured over time scales of minutes to seasons to years. Despite the fact that human activities, including reservoir management, irrigation, shifting land use, and urbanization were modifying flow regimes of rivers and landscapes at unprecedented rates, scant research examined these impacts and their consequences over spatial scales of large watersheds and regions, or timescales of decades to centuries. The result has been that our ability to transform river systems has far outdistanced our ability to understand the implications of those transformations.

The past two decades, however, have witnessed a dramatic increase in the scales and objectives of hydrologic science. Driven by concerns of impending global climate change, loss of biodiversity, fragmentation of river systems (i.e., Dynesius and Nilsson, 1994), and deteriorating water quality (i.e., Smith et al., 1987), hydrologists are beginning to examine the behavior of river systems at regional, continental, and even global spatial scales, and to consider the effects of environmental changes well into the next century. The difficulty of interpreting cause-and-effect relations increases markedly as the temporal and spatial scales of inquiry expand, however, because the hiearchy of controlling processes shifts with scale, and opportunity for experimentation and direct observation commonly diminish with increasing scale. Processes that strongly influence streamflow generation at the scale of small- to moderate-sized catchments, e.g., subsurface flow and channel routing, all but disappear at the scale of larger catchments, where they are replaced by other processes, such as regionally and topographically controlled precipitation patterns. Understanding controlling processes at appropriate scales has become a fundamental challenge to hydrologists (National Research Council, 1991).

Many of the processes controlling runoff at both small and large scales are linked to the underlying geomorphology of a region. This chapter examines how geomorphology controls runoff generation processes in watersheds with drainage areas of 10^2 to 10^4 km^2. Using examples from the Willamette River Basin in western Oregon, I consider how the flow regime of channels is determined by the interaction of climate and intrinsic geomorphic controls. Finally, I discuss the utility of this geomorphic perspective for evaluating human modifications of the fluvial system.

STUDY AREA

The area of the Willamette River Basin is more than 11,000 km^2 and includes several large tributary river systems draining the central Oregon Cascades: the Clackamas, N. Santiam, S. Santiam, McKenzie, and Middle Fork Willamette Rivers (Figure 1). Within the Willamette River Basin, sharp contrasts in climate, geology, and topography are expressed along an east-west transect (Figure 1). These zonations are also reflected in the soils and vegetation type and productivity. Major tributaries of the Willamette, including the McKenzie, Santiam, and Clackamas Rivers, generally flow westward; the upper reaches of the McKenzie and Santiam trend north-south, however, as they follow the western margin of the fault-bounded High Cascade province (Sherrod and Smith, 1989). Because of their orientation orthogonal to major topographic and geologic trends, the rivers cross three biogeoclimatic zones: (1) the High Cascades, with elevations >1200 m where most precipitation falls as snow, is underlain by glacial deposits and <2 million year (MY) old, porous, volcanic rocks; (2) the Western Cascades, with elevations of 400 to 1200 m where precipitation falls as rain and snow, is underlain by 3.5 to 25 MY old, deeply weathered but relatively impervious, volcanic rocks; and (3) the Cascade foothills and Willamette Valley, with elevations of less than 400 m where most precipitation falls as rain, is underlain by alluvium and > 25 MY old sedimentary and volcanic rocks (Figure 1). Portions of the High and Western Cascades basins have been modified since 1950 by timber harvest, roads, and dams; the Willamette Valley has been extensively modified since the middle of the last century by agriculture and urbanization. Timber harvest and road construction have affected little of the high-elevation zone, which includes extensive federal wilderness, but up to 25% of mid-elevation public forestlands and 100% of some low-elevation private lands have been harvested. Large dams have been constructed on Willamette River tributaries mostly below 200 to 400 m. Each of the five large river systems has historical gaging records dating back to the early part of this century from up to 20 nested subbasins, each ranging from 1 to 5000 km^2.

GEOMORPHIC CONTROLS ON HYDROLOGY AT THE PROVINCIAL SCALE

Geomorphology, including both geomorphic processes and landforms, exerts both direct and indirect controls on the pattern, timing, and volume of runoff generated within a basin (Figure 2). At the scale of physiographic provinces, the natural (i.e., uninfluenced by human activities) flow regime is determined by two factors: (1) the broad-scale interaction between geology and climate establishes the overall pattern of runoff at the annual scale (i.e., amount, seasonality) and (2) drainage network structure and longitudinal organization of the river determines the timing and rate of runoff for individual storm events.

EFFECTS OF CLIMATE, TOPOGRAPHY, AND GEOLOGY

The regional climate determines the overall volume, seasonality, form (i.e., rain vs. snow), and areal distribution of precipitation. Geology strongly influences the topography and landforms of a region and determines the moisture holding and transmissivity properties of the soil and regolith; these properties also interact with climate to determine the distribution of vegetation. The interaction of climate and geology, as mediated by topography, soils, and vegetation, determines the distribution and intensity of precipitation by elevation, the precipitation state (i.e., rain, snow), the potential for

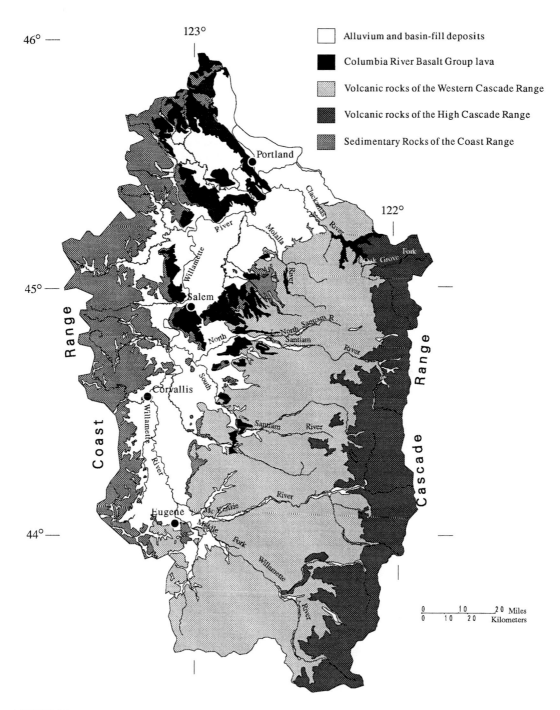

FIGURE 1.   Location map showing geology, topography, and trends of major rivers, Willamette River system, Oregon.

storage on the landscape or in the soil mantle, the loss of moisture through evapotranspiration, and the rate that precipitation is transformed into runoff.

The broad geographic setting, as defined by climate, geology, soils, topography, and vegetation, therefore strongly influences riverine systems over ecologically relevant time scales (Schumm and Lichty, 1965). Because all of these factors operate simultaneously in most rivers, we have a poor understanding of how they interact to affect hydrologic behavior and stream and riparian zone

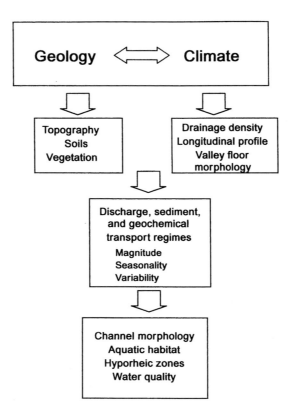

FIGURE 2. Conceptual model of effects of geomorphology on hydrology and other watershed processes.

conditions. At the provincial scale, however, their separate influences on the resultant hydrograph can be disentangled.

This can be observed in a comparison of mean monthly discharges for five unregulated streams draining watersheds with different geologic and topographic conditions in the upper Willamette Basin, using streamflow data from Moffat et al. (1990), and geological mapping and characterization by Ingebritsen et al. (1991) (Table 1; Figure 3). The five watersheds all have approximately the same drainage area, ranging from 250 to 500 km². Two distinct patterns of annual runoff are observed: the Little North Santiam, Molalla, and South Santiam Rivers all have sharp increases in runoff in the late fall with peak runoff during the winter months, a long period of declining runoff

TABLE 1
Topographic and Geologic Characteristics of Five Watersheds
Shown in Figure 1

Watershed	USGS station No.	Drainage area (km²)	Elevation (m)	Geology
Little North Santiam	14182500	290	805	17–25 ma andesite
Molalla River	14198500	250	887	7–17 ma andesite
South Santiam River	14185000	450	875	17–25 ma andesite
Oak Grove Fork	14209000	330	1140	<7 ma andesite
Clackamas River	14209500	500	1088	7–17 ma andesite

Data from Ingebritsen, S. E., Sherrod, D. R., and Mariner, R. H., *J. Geophys. Res.*, 97, 4599, 1991.

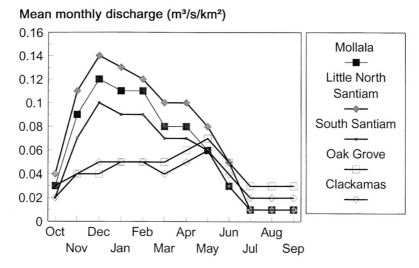

Mean monthly discharge (m³/s/km²)

FIGURE 3.   Mean monthly unit discharges for Western and High Cascade basins listed in Table 1. (Data from Moffat et al., 1990.)

during the spring, and constant low flows during the summer. The upper Clackamas and Oak Grove Fork of the Clackamas, on the other hand, have much less variable annual hydrographs with much less pronounced rises during the fall and winter months, late spring peak flows, and more sustained, higher low flows during the summer months. During the winter peak flow months, mean monthly flows for the Clackamas streams are half of the mean monthly flows for the other streams on a unit area basis; conversely, average summer low flows in the Clackamas streams are more than twice as high (Figure 3).

These differences can be explained by the interaction among climatic, topographic, and geologic factors. The Little North Santiam, Molalla, and South Santiam drainages are at lower elevations (Table 1), and tend to accumulate less snow during the winter than the two Clackamas streams. At elevations lower than 1200 m, winter storms are often a mix of both rain and snow, and peak flows occur from rapid melting of snowpacks during warm rainstorms from November to March (Harr, 1981). In contrast, most winter precipitation is snow at elevations higher than 1200 m, and peak flows occur during the spring snowmelt. With mean elevations ranging from 805 to 887 m, the three lower elevation watersheds have a greater proportion of their area subject to rain-on-snow melting and produce higher winter peaks than the two higher elevations, Clackamas watersheds that peak in the spring (Figure 3). The geology also contributes to this pattern in that the two higher elevation watersheds are underlain by younger, more porous volcanic rocks (i.e., thick piles of aa lava flows), which act as geologic reservoirs, storing groundwater during the melt season and releasing it slowly during the low flow summer months. The importance of this effect is readily observed in a comparision of flow duration curves for the five watersheds for mean daily discharges during August (Figure 4). The two Clackamas basins have significantly higher mean daily discharges than the other three watersheds.

EFFECTS OF DRAINAGE NETWORK STRUCTURE

Hydrologic regimes of large river basins also reflect the underlying architecture of the drainage basins themselves. The areal distribution of streams determines the rate at which water accumulates with distance downstream and the overall efficiency of the landscape for transforming precipitation and groundwater into streamflow. One measure of the drainage network structure is the drainage density, or total length of streams per unit area (km/km²). The drainage density has been shown in other studies to reflect the hydraulic transmissivity of the underlying rocks (Carlston, 1963). Where

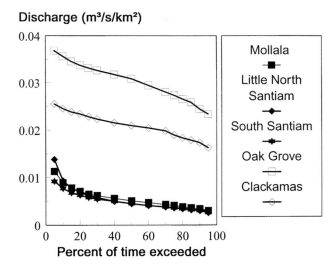

Discharge (m³/s/km²)

Percent of time exceeded

FIGURE 4. Flow frequency curves for mean daily flows during August for Western and High Cascade Basins listed in Table 1. (Data from Moffat et al., 1990).

rocks are highly permeable, transmissivity is high and a greater proportion of total precipitation is transmitted as groundwater. Conversely, where rocks are impermeable and transmissivity low, a greater proportion of incoming precipitation cannot infiltrate and must take a surface-water path downslope. Through geologic time, this greater proportion of surface-water flow develops discrete channel networks and higher drainage densities. Areas with highly permeable rocks should therefore support lower drainage densities than areas with less permeable rocks.

This is borne out in a comparison of drainage densities between the High and Western Cascades (Figure 5). The young, porous volcanic rocks of the High Cascade Province have very high hydraulic permeabilities and support a poorly integrated trellis to dendritic network of surface streams. Measured drainage densities from a random sampling of 1:24,000 U.S. Geological Survey ranged from 1.14 to 2.85 km/km², averaging 1.73 (FEMAT, 1993). In contrast, streams draining the older, less permeable rocks of the Western Cascade have well-developed dendritic drainage patterns and drainage densities ranging from 1.49 to 7.15 km/km², averaging 4.10.

Because the stream network is less dense and well integrated in the High Cascades, peak discharges tend to be less per unit area than in the Western Cascades for the same frequency of flow. A comparison of the rate of increasing discharge with drainage area for the 5% exceedance probability flow (Q_5) for 9 High and 44 Western Cascade Basin shows that unit discharges of this frequency are approximately 43% higher for the same drainage area in the Western Cascades (Figure 6). Assuming a zero-intercept, the corresponding empirical equations for the Q_5 flows (measured in m³/s) as a function of drainage area (DA) in km² are

$$\text{Western Cascades:} \quad Q_5 = 0.13 \text{ DA} \quad (R^2 = 0.94, n = 44) \tag{1}$$

$$\text{High Cascades:} \quad Q_5 = 0.09 \text{ DA} \quad (R^2 = 0.98, n = 9) \tag{2}$$

Since the large trunk streams, such as the McKenzie or Santiam, drain the High Cascade province before entering the Western Cascade Province, the rate at which they accumulate discharge with distance might be expected to increase with distance downstream. Hence a curve of drainage area vs. discharge should be concave upward. This can be seen in the drainage area — discharge relation for the unregulated 5% exceedance probability flow for the McKenzie River (Figure 7). The McKenzie turns west out of the High Cascade Province at a drainage area of approximately 1000 km² and enters the Western Cascades (Figure 1). The upward concavity of this relation

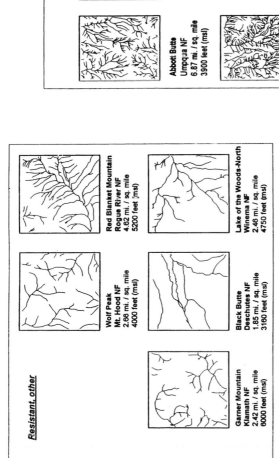

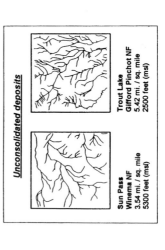

FIGURE 5. Examples of contrasting drainage densities and patterns in the High and Western Cascades. (After Forest Ecosystem Management Assessment Team, U.S. Dept. of Interior, 1993.)

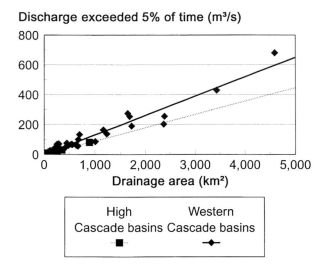

FIGURE 6. Relation between Q_5 (discharge exceeded 5% of time) based on mean daily flows and drainage area for selected Western and High Cascade Basins.

demonstrates the distinct nonlinearity between discharge and drainage area as province boundaries are crossed. The opposite trend is apparent in the low flow data (Figure 7). Although the data are limited, the downward inflection at a drainage area of 1000 km² shows that contribution to low flow is higher for the High Cascades relative to the Western Cascade portions of the landscape.

EFFECTS OF HUMAN ACTIVITIES ON HYDROLOGY

Human activities affect streamflow in this region in several ways. Direct human manipulation of flow regimes through dam regulation is a primary mechanism for altering the magnitude, frequency, and timing of runoff. These flow modifications occur at discrete locations (dam sites) and are well characterized by comparison of flow frequency relations before and after dam construction. A more subtle mechanism is land-use activities distributed throughout a watershed that alter flow regimes through various processes. These activities are not uniformly distributed through the landscape, and data on the extent of streamflow alteration due to land use is usually limited or absent. Forest

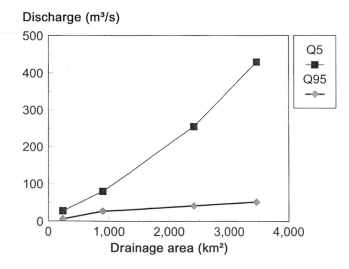

FIGURE 7. Relation between Q_5 and Q_{95} and drainage area for the McKenzie River watershed.

cutting and roads are the dominant land-use activity in the region considered here. This next section considers how the effects of both dam regulation and forest harvest activities may vary in relation to large-scale geomorphic controls.

RESERVOIR MANAGEMENT

Major flood control reservoirs were constructed on many of the tributaries of the Willamette River system from 1940 to 1970. Most of these reservoirs were located in the Western Cascades, although several small projects on the McKenzie (i.e., Carmen Smith, Trailbridge Dams) are located in the High Cascade Province. These reservoirs are primarily used for flood control during the winter and low flow augmentation of the Willamette during the summer, with hydroelectric power generated throughout the year at some sites. The common regulating schedule for the flood control reservoirs is to begin drawdown during the summer and early fall and continue until November when lowest pool volume is reached. The reservoirs then store flood flows during the winter months and are refilled during the spring to achieve full pool volume by early summer.

The effect of these flow alterations is to reduce the highest winter peaks and increase summer base flows. A comparison of flow frequency curves for the McKenzie River at Vida, which is located below three large dams, reveals that post-dam low flows increased by 47%, as indexed by the 95% exceedance probability discharge (Figure 8). High flows, as indexed by the 5% exceedance probability discharge, also increased slightly (8%) following dam construction. The major effect on peak flows, however, was to reduce the size of the annual instantaneous peak flows by 44%, from an average of 895 m³/s between 1925 and 1962 to 504 m³/s between 1968 and 1992 (Figure 9) (Minear, 1994). Overall, post-dam flows are less variable than pre-dam flows (Figure 8). Analysis of climate records during this period indicates that climate variability was not a factor in these changes (Minear, 1994).

From a regional perspective, the effect of dam regulation is to impose a more High Cascade type of flow regime on regulated Western Cascade streams. This is true both in terms of the hydrology and temperature regimes. Release of cold water from thermally stratified reservoirs mimics the contribution of cold water from groundwater sources in the High Cascades. The geomorphic and ecological consequences of propagating a colder and more uniform flow regime further downstream are not well understood. Potential consequences might include an extension of channel geometries and sediment sizes characteristic of High Cascade streams further downstream as channels adjust their dimensions. Reduced high flows and consequent reductions in transport of coarse sediment and large woody debris below dams might be expected to result in more stable

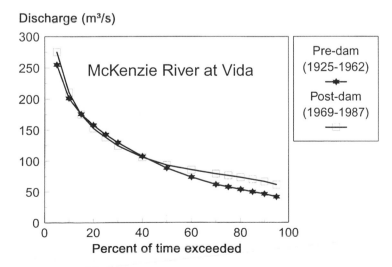

FIGURE 8.   Pre- and post-dam flow frequencies based on mean, daily discharges, McKenzie River at Vida, OR (USGS gauge number 14162500). (Data from Moffat et al., 1990).

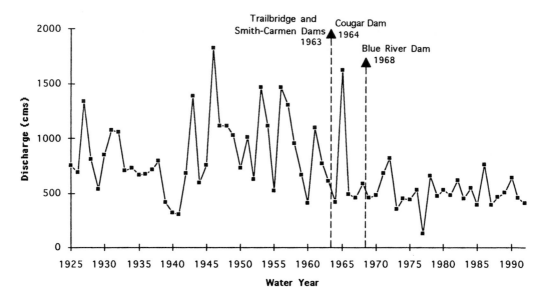

FIGURE 9. Maximum annual instantaneous peak discharges, McKenzie River at Vida, OR, 1925–1992 showing timing of upstream dam construction. (From Minear, P. J., M.S. thesis, Dept. of Fisheries and Wildlife, Oregon State University, 1994, 173.)

riparian surfaces and vegetation, and a similar downstream extension of aquatic diversity representative of the High Cascades. The latter effect may be most pronounced for aquatic invertebrates, which are highly sensitive to thermal regulation. Potential downstream changes in aquatic biota are complicated by the physical barriers to migration posed by the dams and reservoirs themselves in the fluvial system.

Effects of Forest Harvest Activities

The effects of forestry activities on streamflow have been hotly debated in this region, due in part to the equivocal results of more than 30 years of research into this question. Recent studies from more than 40 years of streamflow records in small and large basins with different levels of harvest activities have demonstrated that clearcutting and road construction can increase peak flows from 20 to 50% for small- to moderate-sized storms (less than a 2-year return period) (Jones and Grant, 1996). While the small watersheds examined (<1 km^2) were exclusively located in the Western Cascades, the larger basin pairs (up to 630 km^2) straddled both the Western and High Cascade Provinces. Much of the High Cascade Province is managed as wilderness where no logging is allowed, however, most of the cutting is concentrated in Western Cascade portions of the larger watersheds, and the observed changes in peak flows can only be applied with confidence there.

One reason for the sometimes contradictory results of research into effects of forest management on streamflow (e.g., Ziemer, 1981; Wright et al., 1990) is differences in the larger geomorphic context in which these studies occur. The response to forest cutting and road construction is likely to be different in watersheds whose hydrologic regimes are dominated by rain vs. rain-on-snow processes (Wright et al., 1990; McDonald et al., 1991). Differences in hydraulic transmissivity of the substrate and drainage density will also play a role. For example, a major mechanism proposed by Jones and Grant (1996) to explain the increases in peak flows observed in the Western Cascades was the extension and integration of the drainage network due to impervious forest roads and associated ditches. The effect of the road network potentially increased the drainage density by as much as 40%, thereby increasing the efficiency by which water moved downslope (Wemple, 1994; Wemple et al., in press). These density increases were measured for Western Cascade basins where the unaltered drainage density for the winter high-flow stream network was 3.0 km/km^2 (Wemple, 1994). If the same absolute increases in density after roading occurred in the High Cascades, where natural densities are half as high (1.5 km/km^2), the corresponding drainage density after roading would be approximately 2.59 km/km^2 or an increase of 73%. On the other hand, the gentler slopes and greater infiltration rates in the High Cascades might be expected to decrease the efficiency of roads in capturing subsurface flow, thereby offsetting the effect of roads.

Although speculative, these comments underscore the importance of understanding the regional coupling between geomorphology and hydrology in interpreting effects of human modifications to the landscape. At a minimum they suggest the need for extreme caution in applying results of studies conducted within one physiographic province to other areas.

CONCLUSIONS

These examples demonstrate that the large-scale geologic and topographic setting strongly controls important properties of hydrographs. Depending on the intrinsic hydrologic properties of geologic provinces, drainage density, and orientation of streams relative to province boundaries, the rate of change of discharge with drainage area may be either linear or nonlinear. Furthermore, this rate may vary with discharge and season. The consequences of these results for interpreting stream geomorphology and ecology, particularly the effects of human activities on streams are not well understood. Recognizing the importance of geomorphic controls on streamflow is the first step toward placing human alterations to hydrology in their proper context.

REFERENCES

Carlston, C. W., Drainage density and streamflow, U.S. Geological Survey Prof. Pap. 422-C, 1963, 8.
Dynesius, M. and Nilsson, C., Fragmentation and flow regulation of river systems in the northern third of the world, *Science*, 266, 753, 1994.
Forest Ecosystem Management Assessment Team (FEMAT), Forest ecosystem management: an ecological, economic, and social assessment, Portland, OR, U. S. Dept. of Agriculture, U. S. Dept. of Interior [and others], [Irregular pagination], 1993.
Harr, R. D., Some characteristics and consequences of snowmelt during rainfall in western Oregon, *J. Hydrol.*, 53, 277, 1981.
Ingebritsen, S. E., Sherrod, D. R., and Mariner, R. H., Rates and patterns of groundwater flow in the Cascade Range volcanic arc, and the effect on subsurface temperatures, *J. Geophys. Res.*, 97, 4599, 1991.
Jones, J. A. and Grant, G. E., Long-term stormflow responses to clearcutting and roads in small and large basins, western Cascades, Oregon, *Water Resour. Res.*, 32, 959, 1996.
McDonald, L. H., Smart, A. W., and Wissmar, R. C., Monitoring guidelines to evaluate effects of forestry activities on streams in the Pacific Northwest and Alaska, U. S. Environmental Protection Agency Report EPA/910/9-91-001, Seattle, WA, 1991, 166.
Minear, P. J., Historical change in channel form and riparian vegetation of the McKenzie River, Oregon, M.S. thesis, Dept. of Fisheries and Wildlife, Oregon State University, Corvallis, OR, 1994, 173.
Moffat, R. L., Wellman, R. E., and Gordon, J. M., Statistical summaries of streamflow data in Oregon. Vol. 1. Monthly and annual streamflow, and flow duration values. U. S. Geological Survey Open-File Report 90-118, Portland, OR, 1990, 413.
National Research Council (U.S.), Committee on Opportunities in the Hydrologic Sciences, *Opportunities in the Hydrologic Sciences*, National Academy Press, Washington, DC, 1991, 348.
Schumm, S. A. and Lichty, R. W., Time, space and causality in geomorphology, *Am. J. Sci.*, 263, 110, 1965.
Sherrod, D. R. and Smith, J. G., Preliminary map of upper Eocene to Holocene volcanic and related rocks of the Cascade Range, Oregon, U.S. Geological Survey Open-File Report 89-14, 1989.
Smith, R. A., Alexander, R. B., and Wolman, M. G., Water-quality trends in the nation's rivers, *Science*, 235, 1607, 1987.
Wemple, B. C., Hydrologic integration of forest roads with stream networks in two basins, Western Cascades, Oregon, M.S. thesis, Dept. of Forest Science, Oregon State University, Corvallis, OR, 1994, 88.

Wemple, B. C., Jones, J. A., and Grant, G. E., Hydrologic integration of forest roads with stream networks in two forested basins in the western Cascades of Oregon. *Water Resourc. Bull.*, in press.

Wright, K., Sendek, K. H., Rice, R. H., and Thomas, R. B., Logging effects on streamflow: storm runoff at Caspar Creek in northwestern California, *Water Resourc. Res.*, 26, 1657, 1990.

Ziemer, R. R., Storm flow response to road building and partial cutting in small streams of northern California, *Water Resourc. Res.*, 17, 907, 1981.

8 Precipitation-Runoff and Streamflow-Routing Modeling as a Foundation for Water-Quality Simulation in the Willamette River Basin, Oregon

Antonius Laenen and John C. Risley

Precipitation-runoff and streamflow-routing models were constructed for the Willamette River Basin as part of a water-quality study. The 11,500 mi^2 of the Willamette River Basin were partitioned into 21 major basins and 253 subbasins. For each subbasin, spatial coverages of land use, soils, geology, and topography were used to define unique hydrologic response units (HRUs). Subbasin flows were simulated with a Precipitation-Runoff Modeling System (PRMS) using inputs of daily precipitation and air temperature. PRMS model output was used as input to the streamflow-routing model, Diffusion Analogy Flow model (DAFLOW). Time-of-travel measurements and channel cross-section data were used to describe the geometry of about 760 miles of main stem and tributary channels in streamflow routing. Modeling consists of 21 streamflow-routing networks that can be run either separately or together to determine flow at any location on the stream system. In each network, observed streamflow data are used as input where streamgauges exist and the precipitation-runoff model is used to simulate ungauged inputs. Accuracy is a function of how much flow is simulated by PRMS (errors associated with the PRMS model are largely attributed to uncertainties in precipitation and air temperature). The model can provide estimates of streamflow at almost 500 locations on the mainstem and major tributaries. Relative contributions of surface runoff, subsurface flow, and groundwater flow can be assessed at nearly 1000 locations, corresponding to the number of individual HRUs identified for precipitation-runoff modeling. Model output has been used to determine discharge at ungauged water-quality sampling locations, to determine water-quality sampling in a Lagrangian sampling scheme, to examine flow separation, and to simulate the unsteady-state movement of dye in the Pudding River network as an initial effort to study and model water quality in the Willamette River Basin.

INTRODUCTION

In 1991, the U.S. Geological Survey (USGS) began a cooperative program with the Oregon Department of Environmental Quality (ODEQ) to evaluate water-quality conditions in the main stem and major tributaries of the Willamette River. The hydrologic models developed during this study will provide a foundation for future long-term management decisions in the basin. In addition to being part of a larger ODEQ study, the USGS program was designed to complement the Willamette Basin National Water Quality Assessment (Wentz and McKenzie, 1991).

This chapter describes how the constructed basin hydrologic models can be used in water-quality analysis; using descriptions of the watershed and channel-hydraulic components of the models, specific streamflow-routing network applications, and a hypothetical application for future use in water quality modeling. A watershed model based on physical precipitation-runoff processes can be used to answer questions regarding the effects of human activities on basin hydrology, and a streamflow-routing model can be used to determine local velocity information and to provide unsteady-state flow data for soluble transport water quality models.

PURPOSE AND SCOPE OF THE MODELING EFFORT

Basin modeling is part of a larger effort to build a comprehensive picture of water quality in the Willamette River Basin. The model study described in this report was designed to be open-ended, providing only the hydrology necessary for a water quality model. No work was done to collect data or model the 26.5 mi tidal-affected reach of the Willamette River from Willamette Falls at Oregon City to the mouth.

STUDY AREA DESCRIPTION

The Willamette River Basin is a watershed of approximately 11,500 mi^2 (see Figure 3 in the Introduction). The basin contains the State's three largest cities, Portland, Eugene, and Salem, and approximately 1.96 million people, representing 69% of the state's population (1990 census). The basin supports an economy based on agriculture, manufacturing, timber, and recreation and contains extensive fish and wildlife habitats. Elevations range from higher than 10,000 ft in the Cascade Range to less than 10 ft near the Columbia River. The Willamette Valley — generally considered to be the part of the basin with elevations below 500 ft — is about 30 miles wide and 117 miles long; it represents about 30% of the basin area.

The main stem Willamette River from river mile (RM) 187 near Eugene to RM 26 at Oregon City can be divided into three distinct reaches: (1) an upper reach characterized by a meandering and braided channel with many islands and sloughs; (2) a middle reach characterized by a meandering channel deeply incised into the valley; and (3) the Newberg Pool, which is deep and slow-moving, and can be characterized hydraulically as a reservoir (Willamette Basin Task Force, 1969).

There are 11 major reservoirs in the Willamette River Basin, with a combined usable capacity of nearly 1.9 million acre-ft. The reservoirs are designed for multipurpose use, but their primary function during the summer is to provide a minimum flow of 6000 ft^3/s at Salem to sustain state water quality standards. Reservoir regulation has a profound impact on low flow in the Wilamette River and has curtailed the length of the low-flow period, which typically occurs from mid-July through mid-October, prior to regulation, and from mid-July through mid-August, following regulation.

STRUCTURE OF THE WILLAMETTE BASIN MODELS

Models are constructed as 21 individual streamflow-routing networks that can be run either separately or together to determine unsteady-state flow in a daily time step. Precipitation-runoff modeling was used to provide simulated inflows from ungauged basins in conjunction with gauged inflows to drive network simulations using a streamflow-routing model. The PRMS and the DAFLOW are the hydrologic computer model programs, and they are linked together using operating system control commands that can be made transparent. PRMS and DAFLOW use the same water data-management (WDM) input and output file system.

The entire Willamette River system can be simulated in one model run if necessary, but this task would require several weeks to set up and is not likely to be often requested. Instead, simulations are normally done on a network basis. Flow is not routed through reservoirs. PRMS is used to simulate flows downstream to reservoirs, and the observed streamflow record is used to start the model below reservoirs. Most network simulations use an inflow hydrograph as a starting boundary condition and the precipitation-runoff simulations for ungaged tributary inflow. This configuration allows for the greatest accuracy in the simulation of instream flows. For some situations, it might

TABLE 1
Statistical Results of Combined PRMS and DAFLOW Model Calibration and Verification for 10 Representative Calibration Basins and 10 Streamflow-Routing Network Applications for 1972 to 1978 Water Years

Basin/Network	Total drainage area (mi^2)	Drainage area simulated (%)	Coefficient of determination[c] R^2	Absolute error[d] (%)
Basins[e]				
Johnson Creek	28.2	all	0.79	43.6
Gales Creek	33.2	all	0.80	28.8
Silver Creek	47.9	all	0.81	25.8
Butte Creek	58.7	all	0.92	22.3
Molalla River	97.0	all	0.81	30.8
Rickreal Creek	27.4	all	0.82	28.9
Thomas Creek	109	all	0.73	40.2
Marys River	159	all	0.89	23.6
Mohawk River	177	all	0.87	25.9
Lookout Creek	24.1	all	0.66	39.4
Networks				
Johnson Creek[a]	51.3	49	0.98	9.4
Clackamas River	936	28	0.98	7.1
Tualatin River	706	82	0.95	17.6
Molalla River	323	70	0.94	18.5
Santiam River	1790	27	0.98	6.9
McKenzie River[b]	1340	17	0.94	8.3
Willamette River				
Jasper to Harrisburg	3420	21	0.96	7.7
Harrisburg to Albany	4840	3	0.98	5.3
Albany to Salem	7280	9	0.99	3.7
Salem to Wilsonville[b]	8400	4	0.97	3.3

[a] Period of record from May 1, 1989 to August 31, 1992.
[b] Period of record is 1972 water year.
[c] Coefficient of determination, $R^2 = 1 - \Sigma\ e^2/\Sigma\ eM^2$
     $e = S - O$, where S is simulated runoff; and O is observed runoff.
     $eM = O - O$, where O is mean observed runoff for full period of simulation.
[d] Absolute error, $= 100 \times \Sigma\ |S - O|/\Sigma\ O$.
[e] Refer to Figure 3 for basin locations.

be desirable to have all flows simulated by precipitation. However, instream flow accuracy is decreased by this technique.

Table 1 shows the absolute error of daily mean discharge at simulation points for those basins used in calibration and for those stream networks used in validation. Accuracy is a function of how much of the stream network is simulated using PRMS, which, in turn, is directly related to the uncertainty of the precipitation and air temperature inputs. Figure 1 shows the results of routing the input flow on the Molalla River 26.2 mi downstream to Canby, and simulating runoff from rainfall at ungauged intermediate locations, where runoff is compared with the observed flow at the Molalla River at Canby stream-gauging station (14,200,000). PRMS was used to simulate flows in 226 mi^2 of intervening drainage area (70% of the basin), which resulted in an absolute error of 18.5% in daily mean discharge. Low-flow simulations were least accurate, because no attempt was made to account for summertime irrigation. Water-use information indicates that about 30 ft^3/s may be consumed by agriculture in this part of the basin in the summertime. The results of routing the input flow for the Willamette River at Albany 35.2 mi downstream to Salem indicate the extent to

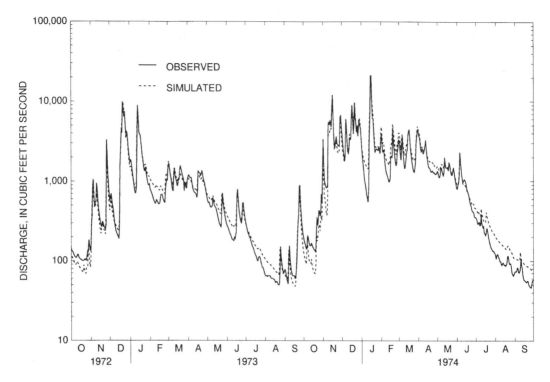

FIGURE 1.   Observed and simulated discharge for the Molalla River at Canby, 1973 to 1974.

which accuracy is improved by the reduction of PRMS simulation in the basin. PRMS was used to simulate flows in only 650 mi^2 of the intervening drainage area (9% of the basin); this simulation resulted in an absolute error of only 3.7% in daily mean discharge.

For Willamette Basin modeling, a total of 253 sub-basins can be simulated separately. In addition, each sub-basin contains several individual hydrologic response units (HRUs), for which separate output can also be obtained. For each subbasin, spatial coverages of precipitation, land use, soils, geology, and topography were used in a geographic information system (GIS) to define the homogeneous HRUs. These spatial data were converted to model input values representing interception, evapotranspiration, infiltration, and subsurface and groundwater flow rates, by an Arc Macro-Language (AML) program. The AML program can be used in the future to redefine model parameters with newer and more well-defined spatial coverages. It is envisioned that a forthcoming step in evaluating water-quality data differences will be the redescription of HRUs, such as the division of agricultural land into finer units that delineate crop types.

PRECIPITATION-RUNOFF MODELING

PRMS is a physical-process, deterministic, distributed-parameter modeling system designed to analyze the effects of precipitation, climate, and land-use on streamflow and general basin hydrology (Leavesley et al., 1983). Heterogeneity is accounted for by partitioning the basin into specific areas on the basis of elevation, slope, aspect, land use, soil type, geology, and precipitation distribution. Each specific area, assumed to be hydrologically homogeneous, is designated an HRU. A water balance is computed during each time step for each HRU and for the entire basin (Figure 2).

Delineation of Basin Physical Characteristics

Data layers that were used to create HRUs include basin and sub-basin boundaries, annual precipitation, land use, slope, aspect, geology, and soils. Table 2 defines the specific basin categories used

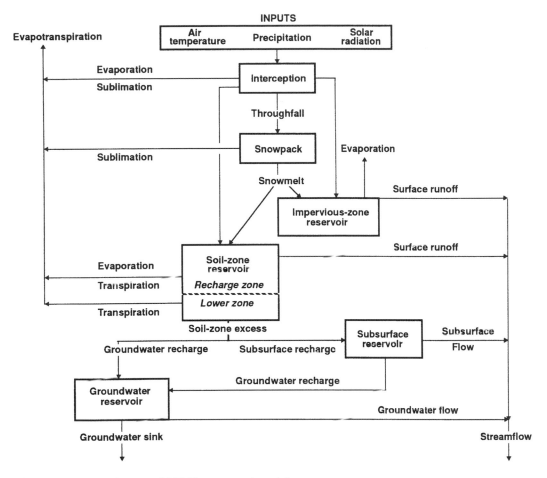

FIGURE 2.   Flow diagram of PRMS conceptual model.

in the study and their corresponding codes in the GIS. The basin boundary coverage of the Willamette River Basin is partitioned into 21 major basins and 253 subbasins.

The precipitation data layer contains contour lines of annual precipitation derived from grid cell output of the Precipitation elevation Regressions on Independent Slopes Model (PRISM) (Daly and Neilson, 1994). Annual average precipitation values were determined from 5-inch contour intervals for individual HRUs.

The land-use coverage data used in the study were acquired from the USGS National Mapping Division (NMD) (Fegeas et al., 1983) as part of a national land-use database coverage. Land classes included are forest, agriculture, urban, wetlands, wetlands lakes, wetlands rivers, and rangeland.

A slope coverage for the entire river basin was created from elevation data contained in 1:250,000 scale Digital Elevation Model (DEM) coverages (produced by the NMD). The slope coverage contains three classes: 0–5%, 5–30%, and greater than 30%. For slopes of more than 30%, polygons were also created to represent the four aspects of north, east, south, and west.

The surficial geology coverage was digitized from a 1:500,000-scale USGS aquifer-units map (McFarland, 1983). The coverage included five classes of aquifer units found in the Willamette River Basin; each unit has varying permeability rates. Tertiary-Quaternary sedimentary deposits, located in the flatter regions of the basin near the river, have the highest permeability rates with well yields as high as 2000 gallons/min. Tertiary rocks of the Coast Range have the lowest permeability rates with well yields of less than 10 gallons/min.

The soils coverage was digitized from a 1:500,000-scale U.S. Soil Conservation Service (SCS) general map of Oregon soil series (U.S. Soil Conservation Service, 1986). The coverage uses the

TABLE 2
Basin Spatial-Coverage Categories and Corresponding Code Numbers Used in Geographic Information System

Category	Code	Description
Land use	1###	Forest
	2###	Agriculture
	3###	Urban
	4###	Wetlands
	41##	Wetlands, lakes
	42##	Wetlands, rivers
	5###	Rangeland
Slope and aspect	#0##	0 to 5%, aspect assigned to basin
	#3##	5 to 30%, aspect assigned to basin
	#6##	> 30%, 0 degrees aspect
	#7##	> 30%, 90 degrees aspect
	#8##	> 30%, 180 degrees aspect
	#9##	> 30%, 270 degrees aspect
Geology	##0#	Tertiary-quaternary sedimentary deposits (I)
	##1#	Tertiary rocks of the Coast Range (V)
	##3#	Tertiary-quaternary volcanic rocks of the High Cascade Range (II)
	##6#	Columbia River Basalt Group (IV)
	##9#	Tertiary volcanic rocks of the Western Cascade Range (III)
Soils	###1	Group A[a] (0.45–0.30)
	###2	Group B[a] (0.30–0.15)
	###3	Groups B and C[a] (0.30–0.05)
	###4	Group C[a] (0.15–0.05)
	###5	Groups C and D[a] (0.15 to less than 0.05)
	###6	Group D[a] (less than 0.05)

Note: #, code number from one of the other categories.

[a] Infiltration rate in inches per hour.

SCS four-group classification for infiltration properties, ranging from approximately 0.45 in./h (Group A) to less than approximately 0.05 in./h (Group D).

The land use, slope, soils, and geology coverages were merged to create a basinwide HRU coverage for each major basin and subbasin in the Willamette River Basin. A four-digit code was assigned to all polygons, which identified the combination of landuse, slope and aspect, geology, and soils classes within the polygon (Table 2). More than 1000 individual HRUs were created for the Willamette River Basin.

PRMS Parameters

It is necessary to assign appropriate values to a wide range of parameters within PRMS reflecting the various physical processes that are modeled. Many of the initial parameter values used in this study were determined from regionalized parameters developed in an earlier USGS study that described effects of timber harvesting in the Oregon Coast Range (Risley, 1993). PRMS parameters are both distributed and nondistributed. Distributed parameters contain specific values for each HRU, subsurface reservoir, or groundwater reservoir; they allow a better representation of varying basin surface conditions than the nondistributed, or lumped, parameters, which are applied over the entire basin.

Most of the values assigned to HRU-related parameters were directly determined from the GIS coverages; however, some additional analysis was required to determine appropriate precipitation and temperature adjustments required for each individual HRU. In all, 42 parameters are used to reflect the physical characteristics of an HRU in the PRMS model parameter input file. Figure 2 shows a schematic of how water balance is computed for each HRU.

Two hundred sixty basinwide and forty-three distributed PRMS parameter values were used in the daily mode of the model; however, four basinwide parameter values could not be predetermined from laboratory and field data, and six distributed parameter values could not be predetermined from GIS coverages. The parameters that could not be predetermined needed to be optimized in the calibration process. Of the optimized parameters, all basinwide parameters, and one distributed parameter were given a regionally constant value. Parameter values for the remaining five distributed parameters describing subsurface and groundwater flow, were classified into basin type and location categories as part of a regional analysis and subsequent report (Laenen and Risley, unpub. data).

The mean annual precipitation value calculated for each HRU was used to determine appropriate precipitation adjustment weights for each HRU. Precipitation adjustments are needed because elevation and orographic differences exist between the input precipitation gauge(s) used and the HRU, which is usually at a different physical location. In addition, a more general basinwide adjustment was also applied to the precipitation input, to accommodate the overall water balance. Precipitation weights were almost always adjusted upward — even when the raingauge was located within an HRU, because raingauges do not catch all precipitation.

Minimum and maximum air temperatures were used in PRMS to compute evapotranspiration rates at each HRU. To account for the air-temperature difference between the elevation of the temperature station and the mean elevation of the HRU, minimum and maximum air temperature lapse rates (change in degrees per thousand feet change in elevation) must be included as a PRMS parameter input. Using the 30 year (1960 to 1990) mean monthly minimum and maximum air temperature data, 18 lapse rates were computed for various locations throughout the Willamette River Basin.

By calibrating and validating PRMS with observed discharge time-series data, it was possible to transfer calibrated PRMS parameter values to nearby ungauged basins. Ten basins located throughout the Willamette River Basin were used for calibration and validation of the PRMS model (Figure 3). The basins were selected to provide an adequate representation of Willamette River Basin headwater streams that have no significant regulation. The most critical model parameters were related to evapotranspiration, subsurface flow, groundwater flow, snow and snowmelt processes:

The Hamon method was used to determine evapotranspiration by multiplying minimum and maximum air temperature with one of 12 monthly coefficients. The coefficients are typically adjusted during the calibration process to help replicate an appropriate water balance for the basin.

Subsurface flow is considered to be the relatively rapid movement of water from the unsaturated zone to a stream channel. Subsurface flow is computed using a conceptual reservoir routing system. Two parameters control the rate of subsurface flow as a nonlinear function of storage volume in the subsurface reservoir. Although values for parameters could not be determined through field measurement, initial values were determined by using graphical flow separation techniques on observed data. These values were later readjusted during the calibration process.

The groundwater system in PRMS is conceptualized as a linear reservoir and baseflow leaving the groundwater reservoir is controlled by only a single parameter. As with subsurface parameters, this parameter could not be determined through field measurement, and its initial value was determined using graphical flow-separation techniques on observed data. This derived value was also readjusted during the calibration process.

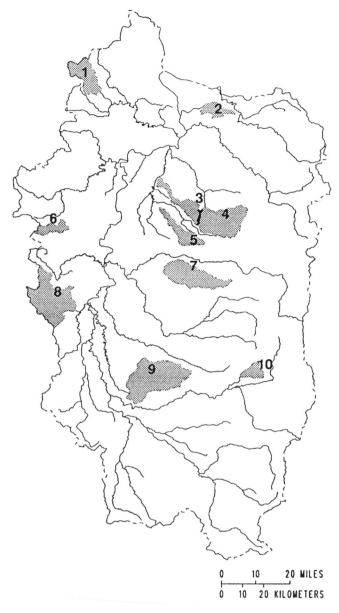

1 Gales Creek (14204500)
2 Johnson Creek (14211500)
3 Butte Creek (14201500)
4 Molalla River (14198500)
5 Silver Creek (14200300)
6 Rickreall Creek (14190700)
7 Thomas Creek (14188800)
8 Marys River (14171000)
9 Mohawk River (14165000)
10 Lookout Creek (14161500)

FIGURE 3.   Location of PRMS calibration basins.

The snow component of PRMS can simulate the initiation, accumulation, and depletion of a snowpack on each HRU. Critical snow parameters used in PRMS include: a transmission coefficient for the vegetation canopy over the snowpack; a maximum air temperature that causes all precipitation to become rain; a convection-condensation energy coefficient; the temperature below which precipitation is snow, and above which it is rain; and emissivity of air on days without precipitation. Only one calibration basin had enough snow cover for these parameters to be calibrated.

STREAMFLOW-ROUTING MODELING

Precipitation-runoff modeling was used to provide inflows from ungaged basins for DAFLOW, which is a one-dimensional, unsteady-state model that requires minimal cross-section data (Jobson,

1989). The model is designed to be used in conjunction with the solute-transport Branched Lagrangian Transport Model (BLTM). The DAFLOW model is used to solve the energy equation and a simplified version of the momentum equation, and it can be easily calibrated with time-of-travel data. "One-dimensional" refers to flow that is modeled in one plane. In the case of DAFLOW, flow direction is also limited to downstream flow only, because acceleration terms are neglected. One-dimensional models based on the Lagrangian reference frame have been found to be very accurate and stable. The simplicity of solving the diffusion analogy in a Lagrangian-reference frame greatly reduces computation time. "Unsteady-state" is sometimes referred to as "dynamic." Flow computations are made at prescribed time intervals, and flood waves (which include even small perturbations) are routed downstream using inputs of channel geometry and conveyance (ability of the stream channel to convey flow) to control attenuation — time varying input is allowed, time varying output is produced.

The DAFLOW model requires channel input values for effective cross-sectional area, average channel width, and average wave diffusion at selected grid intervals (Jobson, 1989). Channel roughness is not a required model input. Instead, low-flow values of area (discharge divided by water velocity) were primarily defined from time-of-travel studies, and high-flow values were determined from flood-study cross sections. Time-of-travel data, which are the result of a dye-tracer studies, yield velocity information for low-flow situations and provide the only means by which pool and riffle channels can be readily characterized. Widths were determined from discharge measurements, 1:2400-scale topographic maps, and flood-study cross sections. Diffusion coefficients were computed from channel width and discharge data.

Special Studies to Document Low Flow

It is difficult to calibrate low-flow model parameters and to relate streamflow to physical properties of the basin, because important low-streamflow data often are missing at critical locations. Time-of-travel studies and low-flow measurements (to define groundwater gains and losses) were made during the study to identify model low-flow parameters and to gain an understanding of the low-flow system.

Dye-tracer studies to determine stream time-of-travel and dispersion were conducted in the Willamette River and nine tributaries from April 1992 to July 1993 during low to medium stream-flows (Lee, unpub. data). Time-of-travel data from the dye-tracer studies were used to define the low- and medium-flow range of cross-sectional area/discharge relations required in the DAFLOW model. Results of these studies define dispersion rates in solute-transport models.

A special study was also conducted on the Willamette and Santiam Rivers to determine gains and losses from groundwater. Stream discharge measurements were made at a summertime low flow from August 17–28, 1992, and at a springtime (medium) flow from June 21–30, 1993. A water-use inventory was made at the time of the streamflow measurements to account for with-drawals from the main stem.

Gain-loss measurements identify (1) the seasonality of groundwater inflow to the main stem, and (2) the magnitude and general location of hyporheic flow that can occur. Measurements made in summer indicate little or no groundwater contribution, and measurements made in spring indicate an approximate 2000 ft³/s groundwater contribution between RM 195.0 and RM 55.0 on the Willamette River. The upper Willamette River between RM 195.0 and 140.0 is a system of braided streams with many islands, sloughs, and gravel bars. In this reach of river, gain-loss measurements showed as much as 1000 ft³/s; 15% of the total river flow can be considered to be in the hyporheic flow zone. The word "hyporheic" means "under river," and the hyporheic zone is defined as the subsurface area where stream water and groundwater mix. From a water-quality standpoint, important chemical and biological processes occur in the hyporheic zone. Large increases of 9 and 7% in Willamette River flow were also noted adjacent to the alluvial fans of the Santiam and Molalla Rivers, respectively.

MODEL APPLICATIONS

As it exists, the model can be used to: (1) determine discharge at ungauged water-quality sampling locations, (2) help design a water-quality sampling network and a sampling schedule, (3) distinguish flow sources from spatial perspectives, and (4) simulate the unsteady-state movement of conservative chemicals in the Pudding River as an initial effort to do water-quality modeling.

DETERMINING DISCHARGE AT UNGAGED SITES

There are 54 climate stations (temperature and precipitation) and 30 streamgauges that can be used to simulate hydrographs at nearly 500 model grid points. Mostly 1972 to 1978 data now reside in the system data base, because this period has the most operating streamgages. More than half of the streamgauges in operation from 1972 to 1978 in the Willamette River Basin are no longer in operation. Climate and streamflow data for the appropriate stations for at least three months prior to the period of required simulation (to account for antecedent conditions) needs to be input into the model data base, before simulation can be accomplished for any other time frame at any of these locations.

Model use in determining real-time discharge for water-quality sampling can therefore be limited, because data generally are not readily available for most flow simulation queries. Preliminary rainfall, temperature, and streamflow data can take as long as 3 to 6 months to obtain, and final data are not usually available for a year.

DESIGNING A WATER-QUALITY SAMPLING SCHEME

Sampling water quality constituents in unsteady-state flow often requires the use of a Lagrangian scheme. A Lagrangian scheme requires sampling a parcel of water as it moves downstream through the channel system. Hydrologic modeling can help define sampling times for a given flow condition, thereby providing more information about processes and eliminating the uncertainty of relating of one sample to another.

DISTINGUISHING FLOW SOURCES FROM A SPATIAL PERSPECTIVE

Analysis of source and fate of water-quality constituents is dependent on a good understanding of flow paths from a spatial perspective. The hydrologic model allows analysis of separate flows from different areas, different stream reaches, and different vertical layers.

Area distribution of flows in the hydrologic model is done by the descretization of the basin into major subdivisions, smaller divisions (subbasins), and ultimately into HRUs. For the Willamette River Basin models, flow can be obtained at nearly 1000 locations corresponding to the number of HRUs that have been delineated.

Linear distribution of flow in the model is accomplished by descretization of the streams into branches that create stream linkages, and grid points that identify unique stream geometry and flow characteristics. The mapped and schematized Molalla River stream network is shown in Figure 4. The Molalla River is schematized as a five branch network with 13 tributary inflows. Discharge hydrographs from observed data at the USGS stream-gauging station on the Molalla River near Wilhoit (14198500) at RM 32.2, and at the stream-gauging station on Silver Creek at Silverton (14200300 discontinued) at RM 52.8 on the Pudding River, are used as upstream boundary inputs. Subbasin hydrographs for the tributaries shown on Figure 4 were first simulated by PRMS modeling and input to the Willamette WDM file. These hydrographs were then input to the appropriate grid locations for DAFLOW modeling using transparent operating-system control commands.

Vertical distribution of flow in the model is done by simulating separate flows as surface runoff, subsurface flow, and groundwater flow (Figure 2). As an example, Figure 5 shows the relative magnitude of those flows simulated for Butte Creek in the Molalla River Basin. The flow separation shown in this figure is typical of that occurring throughout the basin. Precipitation in the basin

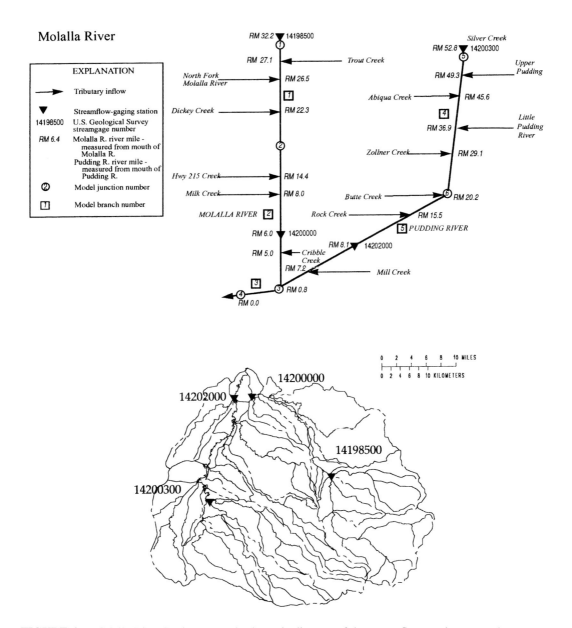

FIGURE 4. Molalla River Basin map and schematic diagram of the streamflow-routing network.

rarely exceeds 3 in./day, even at the higher elevations. This relatively low precipitation intensity, coupled with predominantly forested areas with loosely compacted soils and forest litter, results in a flow component that is predominantly from the subsurface zone. Overland runoff is a small component of the total flow and only occurs during higher intensity events.

SIMULATING THE UNSTEADY-STATE MOVEMENT OF DYE

An important goal for precipitation-runoff simulation and channel flow-routing is to accomplish water-quality modeling in a basin context. As an example application of the model, flow data supplied by the Willamette River Basin models for the Pudding River have been used to drive the Branched Lagrangian Transport Model (BLTM) by Jobson and Schoellhamer (1987) and have simulated the dispersion of a dye injection. The model is then ready to simulate transport of

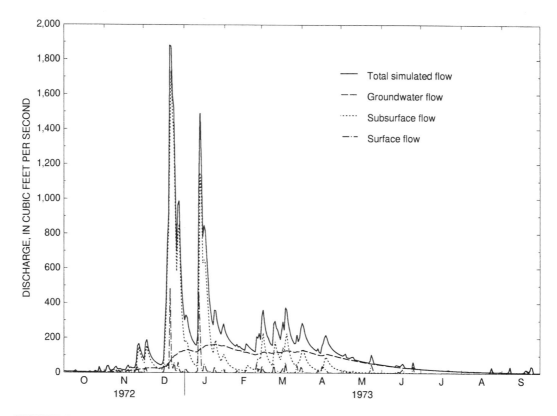

FIGURE 5. Typical simulated flow separation output from PRMS — Butte Creek.

conservative chemicals. Other constituents, such as nutrients or dissolved oxygen, whose fate and transport are linked to their cycling by biota and chemical reactions, will need further calibration with field data to define sources, sinks, and transformations identified in model simulation algorithms.

PUDDING RIVER BASIN

The Pudding River is a highly sinuous stream incised in silt, and it drains primarily agricultural land (Willamette Basin Task Force, 1969). There is considerable interest in major nonpoint sources of contamination in this basin, which has been targeted for intensive study by both ODEQ and the USGS. Point sources of contamination are also important in the basin, although their effects on stream water quality are not well known.

BRANCHED LAGRANGIAN TRANSPORT MODEL

A one-dimensional water-quality model based on the Lagrangian reference frame was developed for use in simulating the transport of conservative constituents and applying reactions among constituents for branched river systems, tidal canal systems, and deltaic channels (Jobson and Schoellhamer, 1987). The model, BLTM, solves the convective–dispersion equation by using a Lagrangian reference frame in which the computational nodes move with the flow. Unsteady flow hydraulics must be supplied to the model and constituent concentrations are assumed to have no effect on the hydraulics. A flow model such as DAFLOW can be used to supply the hydraulic information to the model. Reaction kinetics for nonconservative constituents, however, need to be supplied by the user.

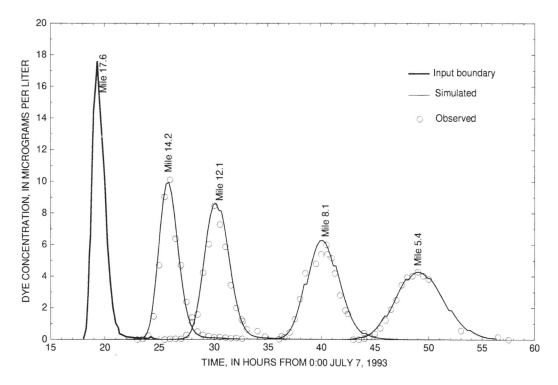

FIGURE 6. BLTM simulation of dye transport from Pudding River mile 17.6 to 5.4 at a low river flow.

Calibration of the Pudding River Flow-Routing Model

Data were collected for time-of-travel studies that define the water discharge and the travel and dispersion of dye in the Pudding River on July 7 to 9, 1993. These data were used to calibrate a solute-transport model for the transport of conservative constituents. Figure 6 shows the observed dye concentrations compared to model simulated concentrations. The timing of the concentration curve and the volume under the curve have both been calibrated to the observed data. In calibration, it was noted that the volume under the curve was very sensitive to tributary inflow values, and the use of measured flows is much preferred over simulated PRMS discharges.

FUTURE WATER-QUALITY MODELING

The next logical sequence in water quality modeling is to incorporate a solute-transport model into the existing dynamic streamflow model of the Willamette River Basin for the various stream networks as interest dictates. This can be done by configuring and calibrating BLTM to operate with DAFLOW using dye-study data.

Further calibration can include providing reaction dynamics for the simulation of dissolved oxygen, bacteria, and water temperature, but this would require additional data collection. The model could then be used to determine the relations of dissolved oxygen and water temperature to such variables as nutrient loading, sediment oxygen demand, reaeration rates, flow conditions, climatic conditions, algal growth rates, and riparian vegetation cover during summer months and to assess the potential effectiveness of various management options.

Simulation of fate and transport of toxic substances and trace elements would be another logical use of the model, but these processes are still not well understood. Further research and data collection will eventually provide definition of the processes involved with the transport and reaction dynamics of these substances. Some of these substances will be transported in the suspended phase; therefore, a sediment-transport model will have to be linked to the flow model (Laenen, 1995). When all the relevant processes are known and adequate data for calibration have been collected, the unsteady-state flow simulation capability of this model will be available to be used to predict the results of management scenarios.

Eventually, overland and subsurface transport processes carrying substances into the stream will have to be defined so that source modeling can be done. These model elements remain in the research stage, but can be incorporated in future modeling when they become usable.

REFERENCES

Daly, C. and Neilson, R. B., A statistical-topographic model for estimating climatological precipitation over mountainous terrain, *J. Appl. Meteorol.*, 33 (2), 140, 1994.

Fegeas, R. G., Claire, R. W., Guptill, S. C., Anderson, K. E., and Hallam, C. A., Land use and land cover digital data, U.S. Geological Survey Circular, 895-E, 21, 1983.

Jobson, H. E. and Schoellhamer, D. H., Users manual for a branched Lagrangian transport model, U.S. Geological Survey Water-Resources Investigations Report 87-4163, 1987, 73.

Jobson, H. E., Users manual for an open-channel streamflow model based on the diffusion analogy, U.S. Geological Survey Water-Resources Investigations Report 89-4133, 1989, 73.

Laenen, A., Willamette River Basin water-quality study — analysis of sediment transport in the main stem and major tributaries, in *The Cutting Edge of Water Research*, Vol. 1, Oregon Water Resources Research Institute, 1995, 4.

Leavesley, G. H., Lichty, R. W., Troutman, B. M., and Saindon, L. S., Precipitation-Runoff Modeling System: User's Manual, U.S. Geological Survey Water-Resources Investigation Report 83-4238, 1983, 207.

McFarland, W. D., A description of aquifer units in Western Oregon, U.S. Geological Survey Open-File Report 82-165, 1983, 35.

Risley, J. C., Use of precipitation-runoff model for simulating effects of forest management on streamflow in 11 small drainage basins, Oregon Coast Range, U.S. Geological Survey Water-Resources Investigations Report 93-4181, 1993, 20.

USSCS, General soil map, state of Oregon, U.S. Soil Conservation Service, Portland, OR, 1986.

Wentz, D. A. and McKenzie, S. W., National Water-Quality Assessment Program — The Willamette Basin, Oregon, U.S. Geological Survey Water Fact Sheet, Open-File Report 91-167, 1991, 2.

Willamette Basin Task Force, Hydrology, Willamette Basin Comprehensive Study: Pacific Northwest River Basins Commission Report, 1969, 163.

9 Groundwater and Surface-Water Relations in the Willamette Valley, Oregon

Marshall W. Gannett and Dennis G. Woodward

The Willamette Valley occupies a broad structural and erosional lowland between the Coast and Cascade Ranges in Western Oregon. This lowland has accumulated a variety of basin-fill deposits including several hundred feet of Columbia River Basalt Group lava and up to 1600 ft of sediment. Five hydrogeologic units have been defined in the region: (1) the basement confining unit, (2) the Columbia River Basalt aquifer, (3) the Willamette confining unit, (4) the Willamette aquifer, and (5) the Willamette Silt unit. The basement confining unit consists of sedimentary and volcanic rocks of the Coast and Cascade Ranges that underlie the entire region and have generally low permeability. The remaining four units occur within the basin-fill deposits.

Groundwater recharge in the valley originates primarily as precipitation on the valley floor and on basalt uplands in and adjacent to the valley. Numerical simulation suggests that most groundwater flow occurs within the Willamette aquifer in the upper part of the basin-fill deposits. Some water, however, flows through other units along deeper flow paths. Most groundwater eventually discharges to the Willamette River and its tributaries, many of which are intermittent. Some groundwater in the Portland area discharges directly to the Columbia River.

Particle-tracking simulations indicate that groundwater flow paths range in length from hundreds of feet to tens of miles and traverse a variety of rock types. The chemical evolution of groundwater is likely to be different along these widely different paths. Groundwater discharging to streams comes from a multitude of converging flow lines, each with potentially different water chemistry. Because groundwater is likely to acquire different chemical characteristics along different flow paths, groundwater discharging to streams is likely to exhibit spatial variations in quality.

Precipitation, and hence recharge, in the Willamette Valley is seasonal, and the water-table elevation fluctuates accordingly. The fluctuating water table causes seasonal variations in the baseflow to streams. If the relative proportions of groundwater discharging from shallow and deep flow paths varies with the total baseflow, the chemical characteristics of discharging groundwater may exhibit seasonal variation as well.

INTRODUCTION

The Willamette River drainage basin lies between the crests of the Cascade and Coast Ranges in western Oregon, encompassing approximately 11,500 mi^2 (Figure 1). The central part of the basin consists of a broad, flat-bottomed, sediment-filled lowland known as the Willamette Valley. This chapter describes the geologic framework of regional groundwater flow in the Willamette Valley, a conceptual model of the regional groundwater flow system, the results of cross-sectional numerical simulation of groundwater flow, and discusses implications for surface-water quality. This work was part of a comprehensive hydrogeologic investigation of the Puget-Willamette Lowland in

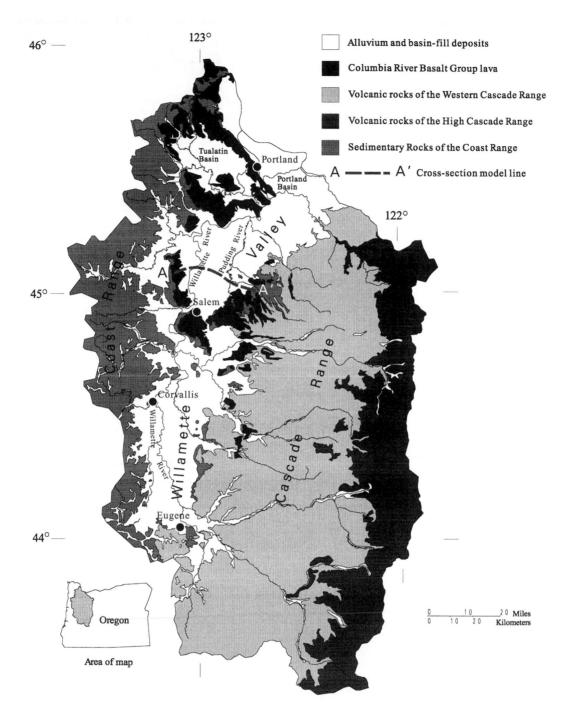

FIGURE 1. Generalized geology and geographic features in the Willamette Basin.

Oregon, Washington, and British Columbia, Canada, conducted as part of the U.S. Geological Survey Regional Aquifer-System Analysis program.

GEOLOGIC FRAMEWORK

The Willamette Valley is a sediment-filled structural and erosional lowland between the Coast and Cascade Ranges (Figure 1). The Coast Range, to the west of the Willamette Valley, consists of

Tertiary marine sandstone, siltstone, shale, and associated marine volcanic and intrusive rocks. The Cascade Range, to the east of the valley, is an active volcanic arc consisting of lava flows, ash-flow tuffs, and pyroclastic and epiclastic debris. Continental and marine strata interfinger beneath and adjacent to the Willamette Valley.

Very early in the development of the Willamette Valley, during middle Miocene time, several hundred feet of Columbia River Basalt Group lava flowed into the area, covering what is now the northern two thirds of the valley. Folding and faulting during and after incursion of the basalt formed a series of uplands dividing the valley into a series of subbasins. These subbasins, separated in most places by uplands capped by the basalt lava, have locally accumulated more than 1600 ft of fluvial sediment derived from the adjacent Cascade and Coast Ranges or transported into the region by the Columbia River.

During Pleistocene time, large-volume glacial-outburst floods, which originated in western Montana, periodically flowed down the Columbia River drainage and inundated the Willamette Valley. These floods deposited as much as 250 ft of silt, sand, and gravel in the Portland Basin, and as much as 130 ft of silt, known as the Willamette Silt, elsewhere in the Willamette Valley.

The geologic units in the Willamette Valley have been grouped into five regional hydrogeologic units (Gannett and Caldwell, in press). These units are (1) the basement confining unit, (2) the Columbia River Basalt aquifer, (3) the Willamette confining unit, (4) the Willamette aquifer, and (5) the Willamette Silt unit (Figure 2).

The basement confining unit consists of the marine sedimentary rocks and associated marine volcanic and intrusive rocks of the Coast Range and the volcanic rocks of the western Cascade Range. These rocks have generally low permeability and are considered to act primarily as a lower confining unit with respect to regional groundwater flow in the basin-fill deposits. A sharp contrast between the chemical nature of the water in these rocks and the overlying rocks suggests that the volume of water moving between these units is relatively small.

The Columbia River Basalt aquifer consists of accordantly layered basalt flows which generally are characterized by low vertical permeability and high lateral permeability. The Columbia River Basalt aquifer is locally capable of producing large amounts of water to wells.

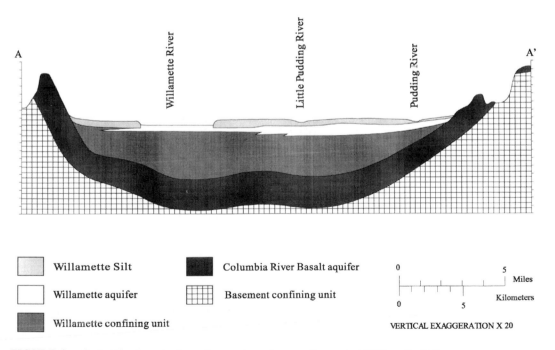

FIGURE 2. Generalized geologic section east-west across the central Willamette Valley.

The Willamette confining unit and the Willamette aquifer occur within the basin-fill sedimentary deposits of the Willamette Valley. The Willamette aquifer, the principal water-bearing unit in the region, is composed predominantly of sand and gravel with lesser amounts of silt and clay, whereas the Willamette confining unit is dominated by silt and clay with substantially less sand and gravel. The Willamette aquifer includes regions of predominantly coarse-grained material from 200 to 400 ft thick occurring where major drainages debouch into the Willamette Valley from the Cascade Range. In most places, these thick deposits are hydraulically connected by thinner, but widespread, gravel deposits that occur at or near the top of the preflood, basin-fill deposits. The coarse-grained deposits that comprise the Willamette aquifer are interpreted as the proximal facies of alluvial fans that existed for much of the depositional history of the valley, and which prograded across much of the valley floor during Pleistocene time. Elsewhere, the basin-fill deposits are dominated by fine-grained materials assigned chiefly to the Willamette confining unit. The fine-grained deposits formed primarily as distal fan facies and deposits of low-gradient streams on the valley floor.

The Willamette Silt unit consists of silt and fine sand deposited in the central and southern Willamette Valley by late Pleistocene glacial-outburst floods. In the Portland Basin, the flood deposits are coarser grained and are considered part of the Willamette aquifer. In the Tualatin Basin, the fine-grained flood deposits lie directly over the lithologically similar Willamette confining unit and are considered part of that unit. The Willamette silt unit has been entirely eroded from the modern floodplain of the Willamette River and its major tributaries.

GROUNDWATER FLOW SYSTEM

Regional groundwater flow in the Willamette Valley is thought to be largely restricted to the basin-fill deposits, including the Columbia River Basalt. Recharge to the regional flow system is primarily from precipitation in the valley. Lesser sources of recharge include subsurface inflow in alluvial deposits of tributary river valleys and local losses from streams. Inflow from the Coast and Cascade Range rocks that form the basement confining unit are considered to be insignificant when compared to the groundwater budget as a whole. The occurrence of isolated saline springs and areas of saline groundwater, however, indicate that inflow from the basement units is locally important.

Groundwater recharge has been estimated for the entire Willamette Valley by Woodward and others (in press) based on work in specific parts of the valley by Hart and Newcomb (1965), Price (1967), Foxworthy (1970), Hampton (1972), Frank (1973, 1974, 1976), Gonthier (1983), and Snyder et al. (1994). Recharge estimates range from 16 to 20 inches/year in most of the valley. Precipitation is the primary source of groundwater recharge in the region. Precipitation is seasonal and recharge shows a corresponding variation manifested as a seasonal fluctuation of the water-table elevation.

Once water infiltrates to the saturated zone, it moves toward discharge areas — primarily through coarse-grained materials in the upper few hundred feet of the basin-fill deposits. Most groundwater discharges to the Willamette River and its tributaries, including small intermittent streams. In the Portland subbasin, however, there is substantial discharge to the Columbia River. Because groundwater levels are locally only a few feet below the ground surface during parts of the year, direct evapotranspiration is also a significant avenue of discharge. Pumping from wells is locally an important discharge mechanism.

NUMERICAL SIMULATION

Cross-sectional numerical flow models were applied in selected parts of the Willamette Valley to test and refine the conceptual model of the groundwater flow system, to test estimates of hydraulic properties, and to provide information on the groundwater flow budget (Woodward et al., in press). This paper discusses one of the model applications.

The location of the modeled cross section is shown in Figure 1, and the modeled cross section is shown in Figure 2. The cross section was located perpendicular to potentiometric surface contours,

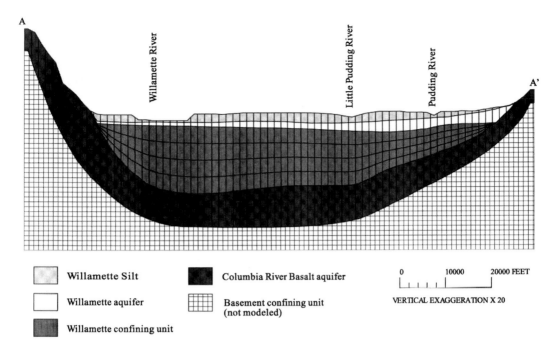

Willamette Silt

Willamette aquifer

Willamette confining unit

Columbia River Basalt aquifer

Basement confining unit
(not modeled)

0   10000   20000 FEET

VERTICAL EXAGGERATION X 20

FIGURE 3.   Cross-sectional numerical flow-model grid.

thereby approximating a flow line. The model section was also located so the subsurface geology and surface drainage network were more or less typical of the region. The flow system was simulated in steady state using the U.S. Geological Survey modular three-dimensional finite-difference groundwater flow model (McDonald and Harbaugh, 1988). Flow paths were calculated and plotted using the programs MODPATH and MODPATH-PLOT (Pollock, 1989). Water budgets for individual aquifer units were determined using the program ZONEBUDGET (Harbaugh, 1990).

The model grid is shown in Figure 3. Each hydrogeologic unit was represented by a single model layer except the Willamette confining unit, which was modeled using four layers because of its thickness. Layer thicknesses varied from cell to cell depending on the thickness of the hydrologic unit at the cell location. Model cells were 1500 ft long in the direction of model section and one foot wide perpendicular to the section.

The base and ends of the model section represent the contact between the Columbia River Basalt aquifer and the underlying basement confining unit and are therefore modeled as no-flow boundaries. Because the model section is assumed to represent a plane parallel to flow directions, there is no flow into or out of the plane of the model section; hence, the boundaries parallel to the model section were also treated as no-flow boundaries. The upper boundary of the model section was the water table and was treated as a free surface. The uppermost active layer in the model section was the Willamette Silt (layer 1), the Willamette aquifer (layer 2), or the Columbia River Basalt aquifer (layer 7) depending on the location within the model.

All perennial and ephemeral streams present on USGS 1:24,000 topographic maps were included in the model. Streams were simulated as head-dependent flux boundaries. Intermittent streams were modeled in a manner that did not allow water to enter the aquifer from the streams when head in the aquifer dropped below the streambed elevation.

Because the water table in the modeled area is within a few feet of the ground surface during certain times of the year, direct evapotranspiration from the water table was accounted for in the simulation. Evapotranspiration parameters were based on the maximum rooting depths and water requirements for ground covers assumed typical for the modeled area — a mixture of grass (grown for seed), grain, or pasture.

The uppermost active cells received a recharge flux that was determined using the regression relation between precipitation and elevation determined by Snyder et al. (1994) for the Portland Basin. Where the Columbia River Basalt aquifer was the uppermost unit, a recharge rate of 14 inches/year was used. Although this is lower than predicted by the regression relation, it is considered reasonable because the low vertical hydraulic conductivity of the basalt relative to other units may act to inhibit recharge. Recharge values used in the model ranged from 14.0 to 21.2 inches/year.

Estimates of horizontal and vertical hydraulic conductivity for the hydrogeologic units included in the simulation were taken from previous work done in the region by Price (1967), Hansen et al. (1994), and Morgan and McFarland (1994); from published values for similar materials (Bureau of Reclamation, 1985; Driscoll, 1986); and from analysis of well-test data from well logs. Initial estimates of streambed conductances were based on the aquifer properties of the cell containing the stream.

The model was not rigorously calibrated. Initial aquifer parameters were adjusted within acceptable limits until simulated heads and discharge fluxes agreed favorably with field observations. Simulated heads were evaluated using water-level elevations and seasonal water-level fluctuations measured in wells. Discharge fluxes were evaluated using stream discharge estimates from Price (1967) and gain/loss measurements by Laenen and Risley (Chapter 8). Agreement between simulated and observed heads and between simulated and measured or independently estimated stream discharge fluxes indicates that the conceptual model of the regional groundwater flow system is generally valid.

Water budgets from the numerical simulation indicate that approximately 16% of the groundwater in the simulation area discharges directly to evapotranspiration, 40% discharges to the Willamette River, 24% discharges to the Pudding River, 14% discharges to the Little Pudding River, and the remaining 6% discharges to ephemeral streams. Of the total steady-state groundwater flow budget, 81% is restricted to the Willamette silt and Willamette aquifer. Flow through the Willamette confining unit and Columbia River Basalt aquifer accounts for only 19% of the total flow.

PARTICLE TRACKING

The particle-tracking and plotting programs MODPATH and MODPATH-PLOT (Pollock, 1989) were used to help visualize the movement of water through the simulated flow system. The particle-tracking program calculates and records the paths of hypothetical particles of water as they move through the simulated system. MODPATH-PLOT plots the paths of these particles. Flow lines through the model section were plotted by tracing the paths of the water particles from the point where they enter the groundwater system at the water table to the point where they discharge from the system.

The paths of single particles that started at the water table in the center of each cell and were tracked to their discharge points are shown in Figure 4. Simulated flow paths show water moving downward through the Willamette Silt unit into the underlying Willamette aquifer. Most water entering the Willamette aquifer moves laterally to discharge to the Willamette River or major tributaries. Water infiltrating near the drainage divides in the valley, where horizontal head gradients are low, may percolate through the Willamette aquifer into the underlying Willamette confining unit, and possibly into the underlying Columbia River Basalt aquifer. Water recharged to the Columbia River Basalt aquifer in upland areas moves through the basalt toward the basin, eventually discharging to the Willamette aquifer either directly or through the Willamette confining unit, and ultimately to streams.

WATER QUALITY IMPLICATIONS

The simulated flow paths shown in Figure 4 provide insight into the history and chemical evolution of groundwater entering streams that may, in turn, have implications for surface water quality. Streams are the locus of many converging groundwater flow paths. These paths have a wide variety

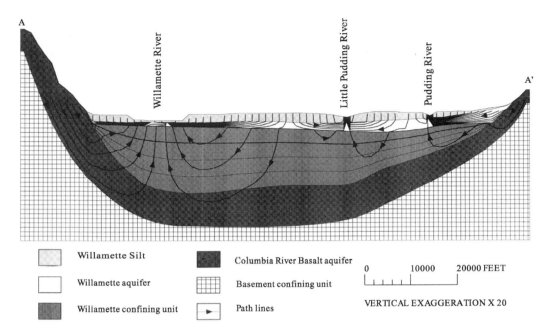

Willamette Silt	Columbia River Basalt aquifer
Willamette aquifer	Basement confining unit
Willamette confining unit	Path lines

0   10000   20000 FEET

VERTICAL EXAGGERATION X 20

FIGURE 4. Simulated path lines of particles started at the water table surface in the center of each cell.

of lengths and geometries. As a result, groundwater following these paths is exposed to a variety of rock types over a wide range of travel times, and therefore acquires different chemical characteristics. Water following the shortest flow lines in Figure 4 is exposed to only the Willamette silt and Willamette aquifer sands and gravels, and has simulated travel times of a few to several years. Flow lines of intermediate length have simulated travel times of several tens of years and some traverse the silts and clays of the Willamette confining unit. Water following the longest flow lines, which originate near drainage divides, may be exposed to all the hydrogeologic units, and has simulated travel times of several hundred to 2000 years. It is reasonable to expect that water traveling along these longer flow lines will have chemical characteristics substantially different from that following short flow paths because of the greater time available for dissolution of minerals and the wider variety of rock types to which it is exposed.

Rounds et al. (1994) installed several piezometers through the streambed of the Tualatin River, a tributary to the Willamette. Water levels in the piezometers were higher than the stream level, suggesting that groundwater is discharging to the stream. Groundwater discharge to streams was confirmed by seepage measurements. The water sampled from the piezometers exhibited a wide range of chemical characteristics. These variations in water chemistry likely reflect the variety of groundwater flow paths converging in the vicinity of the Tualatin River.

In addition to spatial variations in groundwater quality described above, temporal variations are likely as well. The water table in the Willamette Valley rises and falls seasonally in response to recharge from precipitation (Figure 5). Annual water-level fluctuations range from a few feet to a few tens of feet. As the water table rises and falls, the hydraulic gradients toward the streams, and hence the rate of groundwater discharge to the streams, vary accordingly. Gain/loss studies conducted by Laenen and Risley (Chapter 8) demonstrate the seasonal variation in groundwater discharge to the Willamette River. Measurements along the Willamette River main stem in the summer of 1992 indicate little or no groundwater contribution to streamflow. Measurements made during the spring of 1993 indicate groundwater was discharging to the river at a rate of approximately 2000 ft³/s between river miles 195 and 55. This amounts to approximately 18% of the total streamflow during that time.

Seasonal variations in groundwater discharge due to the fluctuating water table may affect shallow flow paths more than deep flow paths. This would cause seasonal variations in the propor-

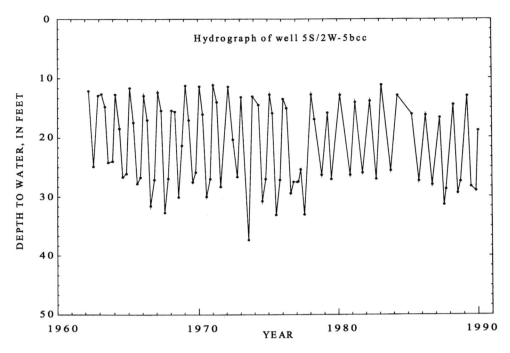

FIGURE 5. Hydrograph showing seasonal water-level fluctuations in part of the Willamette Valley.

tions of the total discharge from deep and shallow flow paths. Such variations in these proportions would result in corresponding variations in the chemical characteristics of groundwater discharging to streams.

SUMMARY

Five regional hydrogeologic units have been defined and mapped in the Willamette Valley, OR: (1) the basement confining unit, (2) the Columbia River Basalt aquifer, (3) the Willamette confining unit, (4) the Willamette aquifer, and (5) the Willamette Silt unit. Regional groundwater flow appears to be largely restricted to basin-fill deposits, including the Columbia River Basalt Group lava, above the basement confining unit. The main source of groundwater recharge to the regional flow system is precipitation onto the valley floor and the Columbia River Basalt uplands in and around the valley. Most groundwater flow occurs in the Willamette aquifer and the overlying Willamette Silt. Water moves laterally through the Willamette aquifer and discharges to surface streams. Precipitation in the Willamette Valley is seasonal and results in corresponding water-table fluctuations. The fluctuating water table causes seasonal variations in groundwater discharge to streams as well. Such seasonal baseflow variations have been measured by Laenen and Risley (see Chapter 8).

Cross-sectional numerical simulation of the regional groundwater flow system supports the conceptual understanding of the regional flow system. Numerical simulation also provides insight into the flow paths and potential variations in chemical evolution of groundwater. Groundwater discharging to streams comes from a multitude of converging flow paths that traverse a variety of rock types over widely differing travel times. It is reasonable to assume that water traveling along these different flow paths will acquire different chemical characteristics. Studies along the Tualatin River, a tributary to the Willamette River, by Rounds et al. (1994) support this concept. The seasonal variation in precipitation and groundwater discharge is likely to affect shallow flow paths more than deep flow paths. Because of this, the relative proportions of water following different flow paths may vary seasonally, resulting in seasonal variations in the chemical characteristics of groundwater discharging to streams.

A thorough understanding of the groundwater flow system and its relation to the surface water system is of great value in evaluating surface water quality. Numerical simulation of the groundwater system can help develop this understanding.

REFERENCES

Bureau of Reclamation, Ground Water Manual, U.S. Government Printing Office, Denver, 1985, 480.

Driscoll, F. G., *Groundwater and Wells,* Johnson Division, St. Paul, MN, 1986, 1089.

Foxworthy, B. L., Hydrologic conditions and artificial recharge through a well in the Salem Heights area of Salem Oregon, U.S. Geological Survey Water-Supply Paper 1594-F, 1970, 56.

Frank, F. J., Groundwater in the Eugene-Springfield area, southern Willamette Valley, Oregon, U.S. Geological Survey Water Supply Paper 2018, 1973, 65.

Frank, F. J., Groundwater in the Corvallis-Albany area, central Willamette Valley, Oregon, U.S. Geological Survey Water-Supply Paper 2032, 1974, 48.

Frank, F. J., Groundwater in the Harrisburg-Halsey area, southern Willamette Valley, Oregon, U.S. Geological Survey Water-Supply Paper 2040, 1976, 45.

Gannett, M. W. and Caldwell, R. R., Geologic framework of the Willamette Lowland, aquifer system, Oregon and Washington, U.S. Geological Survey Professional Paper 1424-A, in press.

Gonthier, J. B., Ground-water resources of the Dallas-Monmouth area, Polk, Benton, and Marion Counties, Oregon, Oregon Water Resources Department Ground Water Report 28, 1983, 50.

Hampton, E. R., Geology and ground water of the Molalla-Salem Slope area, northern Willamette Valley, Oregon, U.S. Geological Survey Water-Supply Paper 1997, 1972, 83.

Hansen, A. J., Vaccaro, J. J., and Bauer, H. H., Ground-water flow simulation of the Columbia Plateau Regional Aquifer System, Washington, Oregon, and Idaho, U.S. Geological Survey Water-Resources Investigations Report 91-4187, 1994, 101.

Harbaugh, A. W., A computer program for calculating subregional water budgets using results from the U.S. Geological Survey modular three-dimensional finite-difference ground-water flow model, U.S. Geological Survey Open-File Report 90-392, 1990, 46.

Hart, D. H. and Newcomb, R. C., Geology and ground water of the Tualatin Valley, Oregon, U.S. Geological Survey Water-Supply Paper 1697, 1965, 172.

Laenen, A. and Risley, J. C., Precipitation-runoff and streamflow-routing modeling as a foundation for water-quality simulation in the Willamette Basin, Oregon, Proceedings of the Poland-USA International River Quality Symposium, Portland, OR, March 21–25, 1994.

McDonald, M. G. and Harbaugh, A. W., A modular three-dimensional finite-difference ground-water flow model, U.S. Geological Survey Techniques of Water-Resource Investigations, Vol. 6, Chap. A1, 1988.

Morgan, D. S. and McFarland, W. D., Simulation analysis of the ground-water flow system in the Portland basin, Oregon and Washington, U.S. Geological Open-File Report 94-505, 1994, 85.

Pollock. D. W., Documentation of computer programs to compute and display pathlines using results from the U.S. Geological Survey modular three-dimensional finite-difference ground-water flow model, U.S. Geological Survey Open-File Report 89-381, 1989, 188.

Price, D., Geology and water resources in the French Prairie area, northern Willamette Valley, Oregon, U.S. Geological Survey Water-Supply Paper 1833, 1967, 98.

Rounds, S. A., Lynch, D. D., and Caldwell, J. M., Phosphorous in Tualatin River basin ground water: International River Quality Symposia, Programs and Abstracts, Portland State University, Portland, OR, 1994, 62.

Snyder, D. T., Morgan, D. S., and McGrath, T. S., Estimation of ground-water recharge from precipitation, runoff into drywells, and on-site waste-disposal systems in the Portland basin, Oregon and Washington, U.S. Geological Survey Water-Resources Investigations Report 92-4010, 1994, 34.

Woodward, D. G., Gannett, M. W., and Vaccaro, J. J., Hydrogeologic framework of the Willamette Lowland aquifer system, Oregon and Washington: U.S. Geological Survey Professional Paper 1424-B, in press.

10 Regional Distribution of Nitrite plus Nitrate in Shallow Groundwater from Alluvial Deposits of the Willamette Basin, Oregon

Stephen R. Hinkle

Water samples were collected from 70 randomly distributed domestic wells in the Willamette Basin, OR, during summer 1993 as part of the U.S. Geological Survey's National Water-Quality Assessment Program. All wells were less than 25 m deep and were completed in alluvial deposits. Analysis of the data provides a preliminary assessment of factors affecting the regional distribution of nitrite plus nitrate in shallow groundwater in the Willamette Basin. Concentrations of nitrite plus nitrate ranged from less than the analytical detection limit (0.05 mg/L as N) to 26 mg/L as N. Groundwater underlying irrigated agricultural land generally contained higher nitrite-plus-nitrate concentrations than did groundwater underlying nonirrigated agricultural land. These differences may reflect hydrogeologic factors that control solute advection (movement), in addition to factors associated with land use. Distributions of nitrite-plus-nitrate and dissolved-oxygen concentrations suggest that denitrification may be an important pathway for nitrate loss in shallow groundwater in the Willamette Basin.

INTRODUCTION

The National Water-Quality Assessment (NAWQA) Program of the U.S. Geological Survey (USGS) has goals of describing the status of and trends in water quality of large, representative portions of the nation's water resources and providing an understanding of the various factors affecting the quality of these resources (Hirsch et al., 1988). The Willamette Basin (Figure 1), a 31,000 km^2 area in northwestern Oregon drained by the Willamette and Sandy Rivers, is one of 60 NAWQA study units. Ongoing data collection and analysis in the Willamette Basin study unit is beginning to allow NAWQA program goals to be addressed. This chapter presents results and preliminary interpretations of a regional survey of dissolved nitrite plus nitrate (henceforth called nitrate) in shallow groundwater in the Willamette Basin.

STUDY DESIGN AND METHODS

Several local, state, and federal agencies conduct groundwater-quality monitoring in the Willamette Basin, and extensive, targeted groundwater sampling efforts have been conducted by the Oregon Department of Environmental Quality (e.g., Fortuna et al., 1988). These data and interpretations have provided valuable information about groundwater quality in many areas in the Willamette Basin, but lacking in these efforts has been an attempt to collect and interpret water-quality data

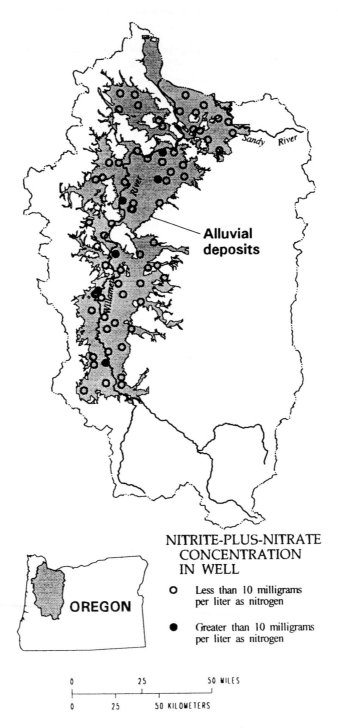

FIGURE 1. Location of Willamette Basin study unit, extent of alluvial deposits, and distribution of wells sampled.

representative of the overall resource. In order to begin to address this need, the current study set the objective of sampling an approximately random distribution of shallow wells in alluvial deposits.

Seventy shallow (less than 25 m deep) domestic wells were sampled during summer 1993 for a suite of organic and inorganic analytes. The median well depth was 18 m, and the median depth of penetration of wells into the water table was 12 m. Alluvial deposits were chosen because water-supply data from Broad and Nebert (1990) and Gonthier (1985) indicated that more than 80% of groundwater used in the Willamette Basin is pumped from alluvium. Wells were chosen using a grid-based, random-selection process. The following criteria were established for initial well selection: (1) presence of a submersible or water-lubricated, vertical turbine pump, (2) existence of a well-driller's report, (3) presence of iron casing and plumbing, and (4) ability to collect raw water samples (before exposure to pressure tanks, filters, chlorinators, or water softeners).

Standard NAWQA sample collection and processing protocols were followed (Koterba et al., 1995). Samples were analyzed for nitrite plus nitrate by using the cadmium reduction method at the USGS National Water Quality Laboratory in Arvada, CO.

A standard NAWQA procedure was used to evaluate land use associated with each well: land use within a 0.4-km radius of each well was characterized by field observations at the time of sampling. If more than 50% of the land use within this circle was of one land-use category, that category was considered to represent the land use associated with that well. If no one category of land use predominated, the land use associated with that well was considered mixed. Land-use categories in the Willamette Basin include urban, nonirrigated agriculture, irrigated agriculture, and forest. The land use within 0.4 km of a well may not represent the land use associated with the recharge area of that well. Furthermore, land use in any given area usually changes over time. Thus, interpretations of relations between nitrate concentrations observed in well water and land use assigned to those wells should be viewed with this limitation in mind.

Geologic materials present at land surface were assumed to affect solute advection (movement) to wells. A modification of the generalized geologic map by Gannett and Caldwell (in press) was used to characterize surficial geologic materials. Geologic units were grouped into two informal categories: Holocene alluvium and older (Pleistocene, Pliocene, and Miocene) alluvium.

Nonparametric statistical tools facilitated data analysis and presentation. The rank-sum (also called Wilcoxon rank-sum, or Mann-Whitney) test was used to compare data groups. The rank-sum test is a t-test procedure applied to rank-transformed data (Helsel and Hirsch, 1992). The Spearman test was used to test for correlation between variables. The Spearman test (correlation coefficient, rho) is a regression analysis applied to rank-transformed data (Helsel and Hirsch, 1992). In this chapter, a significance level (p-value, or α-value) below 0.05 was considered statistically significant, and the word "significant" is used exclusively in a statistical sense. Categorical frequency distributions of nitrate data are shown as boxplots.

The results presented in this chapter are subject to at least two limitations with respect to data and interpretations. First, the selection of shallow, existing wells with well-driller's reports on record results in a preliminary assessment of nitrate in only shallow groundwater in alluvial deposits. These data can neither be used to evaluate nitrate in deeper groundwater, nor can they be used to evaluate nitrate in water at the water table. Public water-supply wells generally pump groundwater from deeper portions of the alluvial deposits; thus, the data presented in this report may not be representative of nitrate concentrations in groundwater associated with many public water-supply wells. Alternatively, water at the water table is likely to be most impacted by surficial activities. Groundwater at the water table generally is not utilized by the population of newer domestic wells for which well-driller's reports exist, but is utilized by many older hand-dug wells. Thus, the data presented in this chapter may not be representative of nitrate concentrations in groundwater pumped from hand-dug domestic wells.

A second limitation is that 70 sites constitute a sparse areal coverage that can, at best, only begin to characterize regional water quality in the targeted resource. Analysis of these data, therefore, represents a broad-brush attempt to characterize the status of water quality in the shallow resource and a first cut at understanding factors affecting the quality of this resource.

In spite of the above limitations, the study design has produced an internally comparable set of data from an approximately random distribution of wells within the targeted resource. Such a data set is more likely to represent the targeted resource than is a collection of nonrandomly distributed data collected for different purposes and by different methods.

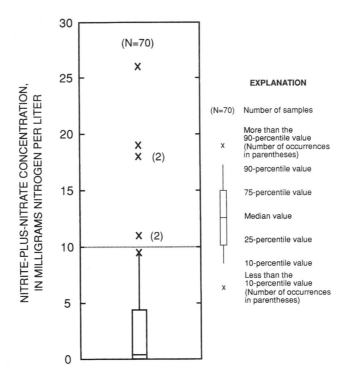

FIGURE 2.   Distribution of nitrite-plus-nitrate concentrations for 70 wells sampled for the 1993 Willamette NAWQA groundwater survey. (Dotted line indicates the U.S. Environmental Protection Agency Maximum Contaminant Level for nitrite plus nitrate in drinking water.)

RESULTS AND DISCUSSION

A frequency distribution of nitrate concentrations from the 70 wells sampled is shown in Figure 2. Nitrate concentrations ranged from below the analytical detection limit (0.05 mg/L as N) to 26 mg/L as N. (For plotting purposes in this and subsequent figures, nitrate concentrations below the detection limit were arbitrarily plotted at one-half the detection limit.) Nitrate at 6 of the 70 sites exceeded the U.S. Environmental Protection Agency Maximum Contaminant Level of 10 mg/L as N.

An assessment of factors controlling the distribution of nitrate in groundwater should take into consideration source, transport, and fate of nitrate. These factors are considered below.

SOURCE

Nitrate may enter groundwater from several point and nonpoint sources, including natural soil nitrogen, atmospheric deposition, septic systems, land disposal of municipal wastes, fertilizers, manure, and fixation of atmospheric nitrogen. Agricultural activities typically represent the most important nonpoint source of nitrate to water in aquifers with elevated concentrations of nitrate (Madison and Brunett, 1985). Agricultural activities include application of fertilizer and manure, as well as farming practices such as irrigation. Shallow groundwater underlying areas dominated by irrigated agriculture generally receives greater inputs of nitrogen than underlying areas dominated by nonirrigated agriculture. This condition reflects the generally greater nitrogen application rates on irrigated farmland relative to nonirrigated farmland, and the flushing effect of irrigation water (Hallberg and Keeney, 1993). Thus, a comparison of nitrate concentrations in shallow groundwater underlying areas dominated by irrigated and nonirrigated agricultural land use can provide insight into the apparent relation between nitrogen source (as represented by land use) and groundwater nitrate concentrations in the Willamette Basin. This apparent relation is shown in

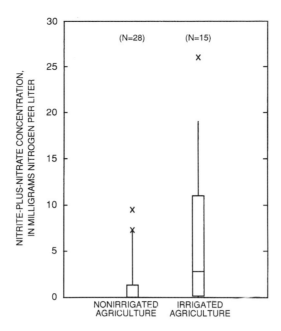

FIGURE 3.   Distributions of nitrite-plus-nitrate concentrations for wells in land-use categories "non-irrigated agriculture" and "irrigated agriculture."

Figure 3. Only those sites classified as dominantly irrigated or nonirrigated agriculture are compared. Nitrate concentrations in wells associated with irrigated agriculture are significantly greater than those associated with nonirrigated agriculture.

Transport

Nitrate concentrations in groundwater may be related to hydrogeologic factors that affect transport of nitrate in groundwater. Inverse relations between nitrate concentration and well depth, depth to top of open interval, penetration of water table by top of open interval, or cumulative thickness of clay above open interval were statistically significant but weak (absolute value of Spearman rho < 0.35).

Nitrate concentrations in groundwater associated with Holocene alluvium were compared with those associated with older alluvium. In the Willamette Basin, permeability of Holocene alluvium is generally greater than that of the older alluvium (Piper, 1942), so faster advection of nitrate might be expected in the Holocene alluvium. However, nitrate concentrations in water underlying Holocene alluvium were not statistically greater than those underlying older alluvium (Figure 4).

Fate

Biological denitrification is the most important nitrate sink in most regional groundwater systems (Hallberg and Keeney, 1993). Denitrification generally occurs in anoxic and low dissolved-oxygen environments. A statistically significant, but weak (Spearman rho = 0.62), relation between dissolved-oxygen and nitrate concentrations (Figure 5) is consistent with denitrification pathways for nitrate fate in this study.

The possibility of nitrate loss through denitrification compels reevaluation of interpretations presented above. Boxplots of nitrate concentrations in oxic water (dissolved-oxygen concentrations ≥1.0 mg/L) underlying areas dominated by irrigated and nonirrigated agricultural land use are shown in Figure 6, along with the land-use boxplots from Figure 3 for comparison. Removing the low dissolved oxygen water (<1.0 mg/L) shifts the boxplots higher on the concentration scale, but

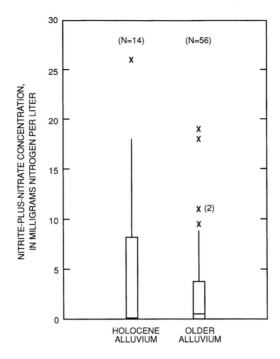

FIGURE 4. Distributions of nitrite-plus-nitrate concentrations for wells located under Holocene alluvium and older alluvium.

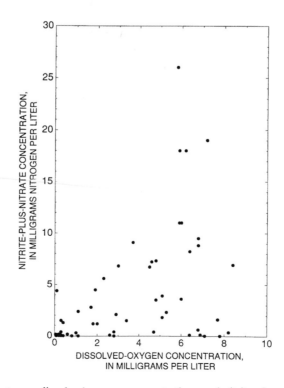

FIGURE 5. Relation between dissolved-oxygen concentrations and nitrite-plus-nitrate concentrations.

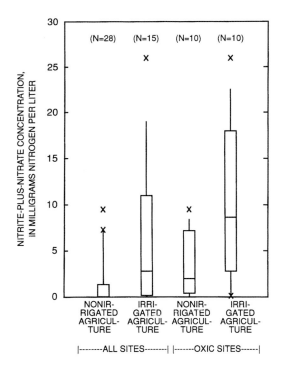

FIGURE 6. Distributions of nitrite-plus-nitrate concentrations for wells in land-use categories "nonirrigated agriculture" and "irrigated agriculture," for all sites (left) and sites (right) with oxic water (dissolved-oxygen concentrations ≥1.0 mg/L).

nitrate concentrations associated with irrigated agriculture remain significantly greater than those associated with nonirrigated agriculture. Thus, although the possibility exists that denitrification has occurred in some of the groundwater sampled, data reselected to account for potential denitrification still demonstrate a similar apparent relation with land use.

Similarly, nitrate data for oxic water underlying Holocene alluvium and older alluvium are shown in Figure 7, along with the boxplots from Figure 4 for comparison. Again, the removal of the low dissolved-oxygen samples shifts the nitrate values higher on the concentration scale. The data associated with oxic water underlying Holocene alluvium have a median value greater than that of the data associated with oxic water underlying older alluvium. This difference is not significant at the 0.05 significance level (p-value for the one-sided test = 0.08). The lack of a more significant p-value may reflect, in part, the small size of the data set.

RELATIONS AMONG SOURCE, TRANSPORT, AND FATE

The possibility of a relation between nitrate concentrations and surficial geologic materials warrants further analysis because land use and surficial geologic materials in the Willamette Basin exhibit a degree of interdependence. For example, none of the sites in the category "oxic water/nonirrigated agriculture" are associated with Holocene alluvium, whereas 6 of the 10 sites in the category "oxic water/irrigated agriculture" are associated with Holocene alluvium. The frequent association of irrigated agriculture with Holocene alluvium in the Willamette Basin is not coincidental. Areas underlain by Holocene alluvium generally have more favorable drainage, irrigation wells of higher capacity, and closer proximity to major sources of surface water than areas underlain by older alluvium. If nitrate is advected more quickly in Holocene alluvium than in older alluvium, nitrate observed in water underlying irrigated agricultural land may be more recent than nitrate observed in water underlying nonirrigated agricultural land.

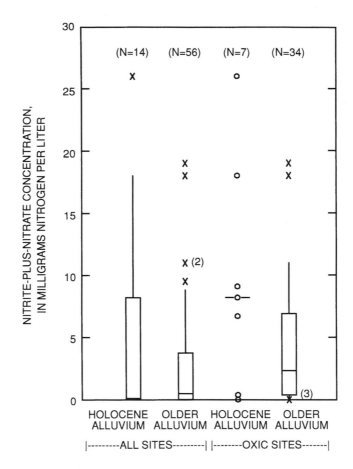

FIGURE 7. Distributions of nitrite-plus-nitrate concentrations for wells located under Holocene alluvium and older alluvium, for all sites (left) and sites (right) with oxic water (dissolved-oxygen concentrations ≥1.0 mg/L). For distribution with N < 10, individual data were plotted as open circles, and the median was identified by a horizontal bar.

   Evaluation of data from Alexander and Smith (1990) indicates that nitrogen fertilizer application rates in the Willamette Basin steadily increased from 1,000,000 kg of nitrogen per year (kg N/yr) in 1945 to 40,000,000 kg N/yr in 1985. Thus, nitrate concentrations in groundwater are a function not only of spatial variations in nitrate source (assumed to be represented by spatial variations in land use) and of nitrate fate (e.g., denitrification), but also of temporal variations in nitrate source as they relate to transport or time elapsed since recharge. The implication is that, assuming unchanging land use over time, nitrate concentrations currently observed in association with non-irrigated agricultural land may reflect a greater time lag than those currently observed in association with irrigated agricultural land and, therefore, differences in nitrogen application rates. In other words, samples collected in 1993 may, as a result of transport kinetics, contain nitrate associated with sources that vary both spatially (e.g., as a function of land use category) and temporally (e.g., as a function of fertilizer application rate over time).

   Future work on this project will include analysis of tritium to help characterize groundwater residence times. That information may facilitate an understanding of the relative importance of source and transport on nitrate concentrations observed in this study.

SUMMARY AND CONCLUSIONS

Seventy randomly distributed shallow wells in alluvial deposits were sampled for nitrate as part of an appraisal of regional groundwater quality in the Willamette Basin. Groundwater underlying irrigated agricultural land generally contained higher nitrate concentrations than did groundwater underlying nonirrigated agricultural land. These differences may reflect sources of nitrate related to local land use, or may reflect transport kinetics related to hydrogeologic factors. Distributions of nitrate and dissolved-oxygen concentrations suggest that denitrification may be an important pathway for nitrate loss in shallow groundwater in the Willamette Basin.

The analysis of differences in nitrate concentrations as a function of land use presented in this chapter may provide useful information regarding the distribution of nitrate in water currently being pumped by shallow wells. However, historical changes in fertilizer application rates, coupled with effects of groundwater transport kinetics, may cause the magnitude and distribution of nitrate concentrations to change in future years. Effective nitrogen management will require an understanding of factors controlling source, transport, and fate of nitrogen in groundwater, in addition to current groundwater nitrate concentrations.

REFERENCES

Alexander, R. B. and Smith, R. A., County-Level Estimates of Nitrogen and Phosphorus Fertilizer Use in the United States, 1945 to 1985, U.S. Geological Survey Open-File Report 90-130, 1990, 12 pp.

Broad, T. M. and Nebert, D. D., Oregon water supply and use, in National Water Summary 1987 — Hydrologic Events and Water Supply and Use, U.S. Geological Survey Water-Supply Paper 2350, 1990, 425–432.

Fortuna, S. M., McGlothlin, M. L., Parr, J. L., and Pettit, G. A., Assessment of Oregon's Ground-Water for Agricultural Chemicals, Department of Environmental Quality, Portland, OR, 1988, 104 pp.

Gannett, M. W. and Caldwell, R. R., Geologic Framework of the Willamette Lowland Aquifer System, Oregon and Washington, U.S. Geological Survey Professional Paper 1424-B, in press.

Gonthier, J. B., Oregon ground-water resources, in National Water Summary 1984 — Hydrologic Events, Selected Water-Quality Trends, and Ground-Water Resources, U.S. Geological Survey Water-Supply Paper 2275, 1985, 355–360.

Hallberg, G. R. and Keeney, D. R., Nitrate, in *Regional Ground-Water Quality*, Alley, W. M., Ed., Van Nostrand Reinhold, New York, 1993, 297–332.

Helsel, D. R. and Hirsch, R. M., *Statistical Methods in Water Resources*, Elsevier, Amsterdam, 1992, 522 pp.

Hirsch, R. M., Alley, W. M., and Wilber, W. G., Concepts for a National Water-Quality Assessment Program, U.S. Geological Survey Circular 1021, 1988, 42 pp.

Koterba, M. T., Wilde, F. D., and Lapham, W. W., Ground-Water Data-Collection Protocols and Procedures for the National Water-Quality Assessment Program: Collection and Documentation of Water-Quality Samples and Related Data, U.S. Geological Survey Open-File Report 95-399, 1995, 113 pp.

Madison, R. J. and Brunett, J. O., Overview of the occurrence of nitrate in ground water of the United States, in National Water Summary 1984 — Hydrologic Events, Selected Water-Quality Trends, and Ground-Water Resources, U.S. Geological Survey Water-Supply Paper 2275, 1985, 93–105.

Piper, A. M., Ground-Water Resources of the Willamette Valley, Oregon, U.S. Geological Survey Water-Supply Paper 890, 1942, 194 pp.

Wastewater Treatment

The four major WWTPs in the basin vary considerably both in size and their effect on water quality in the river. The two smaller plants are at Forest Grove (RM 55.2) and Hillsboro (RM 43.8); the larger plants are at Rock Creek (RM 38.1) and Durham (RM 9.3). Removal of ammonia from effluent by nitrification is maintained throughout the warm summer months at the Rock Creek and Durham plants. As water temperatures cool at the end of the summer, however, the nitrification process becomes progressively less efficient and eventually ceases completely, resulting in large increases in ammonia concentrations in the effluent from these two plants. At the same time, at the beginning of November, the two smaller WWTPs begin discharging their effluent, also containing large quantities of ammonia, directly into the river. Consequently, loads of ammonia in the river increase markedly in November and remain high throughout the winter season.

CONDITIONS DURING NOVEMBER 1992

The 14-day period from November 6 to 20, 1992 was characterized by Tualatin River streamflow averaging less than 300 ft³/s at West Linn (Figure 3). Estimated time of travel from the top of the study reach (RM 58.8) to the river mouth was about 16 days, at a mean flow of 219 ft³/s. Water temperature throughout the main stem ranged from approximately 8 to 13°C. Temperatures at all sites decreased during this period; nonetheless they remained above 10°C in the lower river most of the time. A reasonable estimate of the cut-off temperature for significant nitrification activity (Thomann and Mueller, 1987) is 10°C.

Ammonia Loads

The total mean load of ammonia discharged to the Tualatin River in November 1992, more than 2000 lb/d, represented an increase of nearly 17-fold over the mean ammonia load in October. This load was dominated by WWTP sources, with tributary sources accounting for only 2% of the total

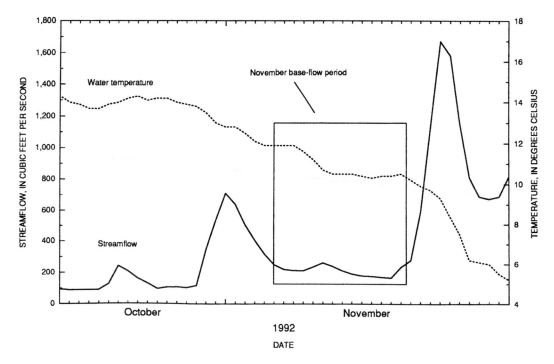

FIGURE 3. Daily mean streamflow and water temperature in the Tualatin River at West Linn, OR, October 15 through November 29, 1992.

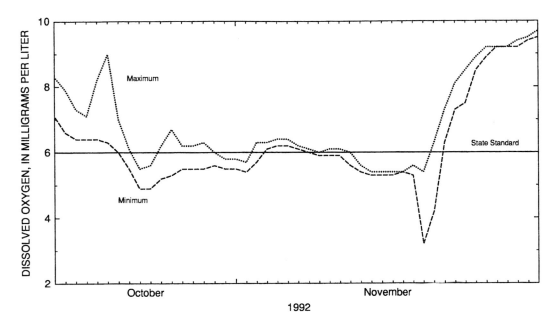

FIGURE 4. Daily maximum and minimum DO concentrations in the Tualatin River, OR at Lake Oswego Dam (October and November 1992).

load. More than one half of the total load (1300 lb/d), was discharged from the Rock Creek WWTP at RM 38.1. Ammonia concentrations in the river increased significantly from October to November, especially downstream of the Rock Creek WWTP, and exceeded 1 mg/L at some sampling sites.

Dissolved Oxygen

During October 1992, DO concentrations reflected the diel variation characteristic of significant levels of algal productivity (Figure 4). Concentrations sometimes varied by more than 2 mg/L throughout the day and ranged from 5 to 10 mg/L over the course of several days. After a storm at the end of October, which increased streamflow to around 800 ft³/s at the mouth (Figure 3) and presumably flushed the system of planktonic algal communities, diel variation in DO was reduced. These observations indicate that the effect of photosynthesis on DO concentrations in the river was not very significant during November 1992.

During the first 2 weeks of November, DO concentrations at RM 3.4 were approximately 6 mg/L, the minimum state standard concentration; during the third week of November, concentrations fell below 5.5 mg/L. The temporary decrease in DO on November 19 was the result of a fire that occurred at a compost facility on November 13. Runoff of the water used to extinguish the fire entered the river at RM 15.2 and exerted a significant instream oxygen demand as it moved downstream. The dramatic drop in DO illustrates the low reaeration potential of the lower river.

MODEL CONFIGURATION AND CALIBRATION

Flow and water quality in the lower part of the Tualatin River were simulated during November 1992 using CE-QUAL-W2, a two-dimensional model originally developed by the U.S. Army Corps of Engineers (Cole and Buchak, 1995). The hydrodynamic component of the model calculates water velocity and surface elevation; the water-quality component can simulate heat flow and as many as 22 different constituents, of which only five were modeled during this study. These included a tracer (chloride), ammonia, nitrite plus nitrate, CBOD (carbonaceous biochemical oxygen demand), and DO. Model simulations began on October 15, 1992, about 2 weeks prior to the base-

TABLE 1
Rate Constants Used in the Model
CE-QUAL-W2, Calibrated to
Conditions During November 1992

Process	Rate constant
Nitrification — RM 38.5–29.7	0.095
Nitrification — RM 29.7–3.4	0.024
CBOD	0.020
SOD	1.80

Note: Nitrification and carbonaceous biochemical oxygen demand (CBOD) decay rates as day^{-1} (base 10) at 20°C; sediment oxygen demand (SOD) rate in grams oxygen per square meter day^{-1} at 20°C.

flow period of interest (November 6 to 20, 1992). Results of the calibration process for DO only are presented here; a complete discussion of model configuration and calibration of other parameters is provided elsewhere (Kelly, 1996).

ASSUMPTIONS

The simulated reach extended from RM 38.4 to the LOC diversion dam at RM 3.4. Because of the negligible photosynthetic activity during this period, the algal dynamics were omitted. The concentration of DO during this period was considered to depend only on oxygen input at the model boundaries and the processes of reaeration, nitrification, sediment oxygen demand (SOD), and CBOD. Reaeration was simulated using coefficients calculated from the mean velocity and mean depth (Bennett and Rathbun, 1972). Two rates of nitrification were applied in different reaches of the river; the distinction between them was based on channel geometry, or the availability of habitat for bacterial growth (Table 1). A higher rate was applied to the shallow reach from the upstream boundary to RM 29.7, where the water depth began to consistently exceed 6 ft; the second rate was applied from that point to the downstream boundary. The SOD rate was assumed constant throughout and was based on *in situ* measurements of SOD during October 1992. The decay rate for CBOD was determined from observed laboratory decay rates. All rates were corrected for temperature (McCutcheon, 1989).

Calibration of DO was evaluated by comparing simulated DO concentrations at the downstream boundary with daily mean measurements from the mini-monitor that continuously recorded DO just upstream of the LOC diversion dam at RM 3.4. Good agreement was obtained between the model simulation and the observed data (Figure 5). These results indicate that the calibrated CE-QUAL-W2 model simulates the important instream processes with an acceptable degree of variability. This calibrated model is useful for predicting water-quality conditions, subject to the limitations of its assumptions. The most important of these include (1) the two nitrification rates, resulting from the relatively stable ammonia loading to the upper river, and (2) the absence of algal activity despite the low streamflow and relatively warm temperature conditions, resulting from the flushing effect of the storm at the end of October.

Relative Impact of Deoxygenation Processes

The relative significance of the various deoxygenation processes in the Tualatin River during November 1992 was evaluated by using the model in several steps. First, time-of-travel information for a parcel of water entering the model reach at the upstream boundary on November 10 was

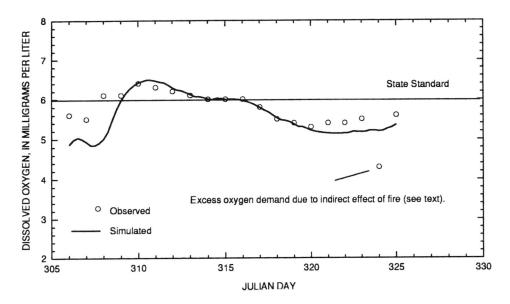

FIGURE 5. Observed and simulated dissolved oxygen at RM 3.4 in the Tualatin River, OR, during November 1992.

determined; the parcel required approximately 9 to 10 days to move through the model reach, arriving at the downstream boundary just before the end of the base-flow period on November 19. The model was used to track this water parcel and document the rates of change in DO for each major process as it moved downstream. These rates were then converted to cumulative losses or gains for selected subreaches, allowing the loss or gain of DO resulting from each process to be isolated during the time period the parcel was moving through each reach. Because reaeration and SOD are surface-dependent phenomena, and therefore highly dependent on channel characteristics, these terms of the DO budget were calculated by difference based on model runs that did not simulate these processes. When calculated in this manner, the DO budget was subject to 1 to 2% error due to the nonlinear nature of the reaeration process. A summary of this analysis for each of these processes is given in Figure 6 and Table 2.

Sinks for Oxygen

The results shown in Figure 6 and Table 2 demonstrate that SOD was the dominant factor governing deoxygenation throughout the model reach, accounting for 53% of the total DO loss. Even though a spatially constant rate was used, the effect of SOD was not the same for all the subreaches because it depends on the depth of the overlying water and the travel time. In the shallow upper river reach, SOD was exerted on the traveling water parcel at a rate of approximately 0.6 mg/L per day, but the net effect was reduced by the relatively short residence time. In the lower river, the rate of oxygen depletion due to SOD was reduced to about 0.2 to 0.4 mg/L per day as a result of the increased water depth, whereas the increase in travel time caused the overall DO loss to increase. In assessing DO management options in the Tualatin River, this high level of SOD is significant because it represents an "intrinsic" oxygen demand that is not directly amenable to treatment.

Nearly 30% of the total loss of DO during this period (November 10 to 19, 1992) was due to nitrification, which was especially pronounced in the upper river. The relatively high rate of deoxygenation due to nitrification, about 0.5 mg/L per day, was possible because of the abundance of habitat for nitrifying bacteria coupled with elevated concentrations of ammonia; oxygen loss from nitrification accounted for nearly one half the oxygen lost within this reach. In the lower river, because of the presumed decrease in suitable microbial habitat and the associated reduction in nitrification rate, this process became less important — although it still accounted for approximately

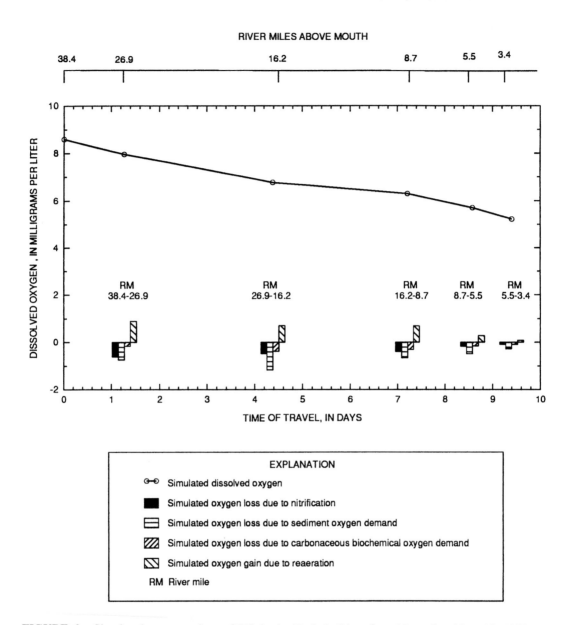

FIGURE 6. Simulated concentrations of DO in the Tualatin River from November 10 to 19, 1992, as a function of time of travel, and losses and gains in oxygen concentrations in selected river reaches caused by various deoxygenation processes.

20 to 25% of DO loss. In contrast, the overall effect of CBOD on DO in the river during this period was relatively small, accounting for less than 20% of the total DO loss.

Sources of Oxygen

The total input of oxygen from reaeration was relatively low, and was insufficient to offset the loss of DO as the water parcel moved through each reach. Even in the upper river, the relatively higher reaeration caused by the shallow water and swift current velocity did not compensate for the oxygen demand from nitrification and SOD. As the water moved downstream, the reaeration potential was reduced by the reservoir-like characteristics of the backwater reach. These data reflect the susceptibility of the lower Tualatin River to oxygen-demanding processes during periods of reduced streamflow, particularly when algae are not producing oxygen.

TABLE 2
Summary of Simulated Gains and Losses in DO, Tualatin River, OR (RM 38.4 to 3.4, November 10 to 19, 1992)

Factor causing change in DO	Change in DO (mg/L)	Percentage of total loss or gain in DO
Nitrification	–1.8	28
SOD	–3.4	53
CBOD	–1.2	19
Total loss	–6.4	100
Reaeration	2.7	90
Tributary input	0.30	10
Total gain	3.0	100
Net change	–3.4	

Note: CBOD, carbonaceous biochemical oxygen demand; SOD, sediment oxygen demand.

HYPOTHETICAL SIMULATIONS

Following calibration of the November base-flow period, the model was used in a series of hypothetical simulations to evaluate the effect of various WWTP ammonia loading scenarios over a range of base-flow and water-temperature conditions. The WWTPs included in this analysis were the two large plants at Durham and Rock Creek and the smaller plant in Hillsboro. The modeling effort focused on evaluating the various management strategies for achieving required DO concentrations during the transition period from late fall to early winter.

ASSUMPTIONS

Steady-state hydrologic and loading conditions were assumed; therefore, the results of these simulations should only be used in a semiquantitative manner to compare various loading strategies. The streamflow and temperature conditions for the hypothetical simulations were selected to bracket conditions under which DO problems would likely occur during November. Tested conditions included 150, 300, and 500 ft^3/s at RM 33.3 and constant water temperatures of 6, 12, and 18°C. Various loading scenarios for ammonia from the WWTPs were selected to reflect different management strategies. The intent was to assess the effect of *maximum* ammonia loads, characteristic of WWTP effluent that is not receiving nitrification treatment, interacting with a range of in-river conditions.

BASELINE OXYGEN DEMAND

Initial or baseline simulations were performed for the full range of flow and temperature conditions assuming no loading of CBOD or ammonia from any of the three WWTPs. This analysis provided a context for later loading scenarios by describing the natural or background oxygen demand in the river that is not subject to direct management control. Boundary conditions were defined by mean data observed during November 1992 (Kelly, 1996).

Results from the baseline simulations reflect the high level of intrinsic oxygen demand expected under base-flow conditions when no significant algal activity is present (Figure 7). Simulated DO concentrations at RM 3.4 were always less than the minimum state standard of 6 mg/L under flow

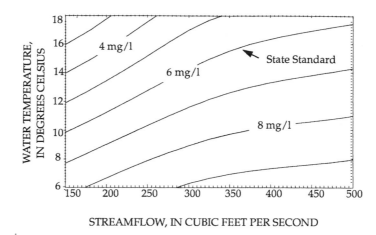

STREAMFLOW, IN CUBIC FEET PER SECOND

FIGURE 7. Hypothetical baseline dissolved oxygen concentrations in the Tualatin River at the Lake Oswego Dam, as a function of temperature and streamflow at RM 33.3.

conditions approximating 150 ft³/s when water temperature was greater than 10°C; conversely, simulated DO concentrations at flows of 500 ft³/s were almost always greater than 6 mg/L when water temperature was less than 18°C. In general, as water temperature increases, baseline DO concentrations remain above the minimum state standard only as long as flow is also increased. The implication for management is that the Tualatin River is inherently very vulnerable to oxygen demand under base-flow conditions, especially when water temperatures are warm. This baseline vulnerability is due primarily to the effect of SOD, which decreases as streamflow increases.

INTERACTION OF AMMONIA LOADS WITH STREAMFLOW AND WATER TEMPERATURE

Streamflow and water temperature have a marked impact on the effects of ammonia loading (Figure 8). In this plot, depicting the effect of ammonia loads from the Rock Creek WWTP, the DO concentration at RM 3.4 is presented because the greatest oxygen depletion is expected to occur at that location. Figure 8 allows the integration of all three factors (streamflow, temperature, and loading), thereby providing a useful tool for the management of water quality under November base-flow conditions. The graph is composed of three sets of curves, corresponding to the three temperatures used — 6, 12, and 18°C. For each temperature, the DO concentrations from each loading scenario are plotted as a function of streamflow. A spline technique (P. Turner, Oregon Graduate Institute, written communication, 1993) was used to estimate DO concentrations between the simulated streamflow values; linear interpolation is possible between the various temperature curves. Within each set of curves, there are four individual curves that represent the range of loading conditions.

Streamflow and water temperature have a characteristic effect on the shape of these curves. The slopes of the curves decrease and the vertical spread within each set of curves, representing the effect of increased loads, becomes less pronounced at higher flows. A decrease in water temperature also flattens the curve sets and increases the overall DO concentrations. These effects result from the inhibitory effect of lower temperature on biological deoxygenation processes and the increased saturation capacity of colder water for dissolved oxygen. A good demonstration of the interaction of streamflow and temperature in determining the extent of oxygen depletion in the lower river is provided by three selected combinations of streamflow, temperature, and ammonia load, as illustrated by the points A–H.

1. Points A, B, and C represent changes in ammonia loads (2000 lb/d, 1000 lb/d, and 0 lb/d, respectively) under unchanging conditions of streamflow (250 ft³/s) and temperature (12°C). Points A and B both lie below the minimum state standard DO concentration of

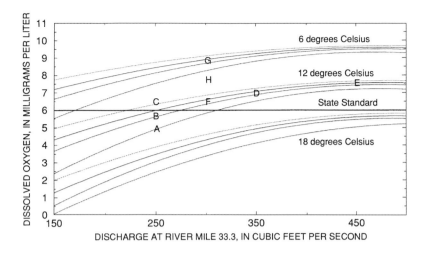

FIGURE 8. Influence of ammonia loads from Rock Creek WWTP on DO concentrations in the Tualatin River, OR at RM 3.4 under a range of streamflow and temperature conditions.

6 mg/L; the baseline concentration (C), at approximately 6.3 mg/L, lies scarcely above the state standard. These points clearly demonstrate that, under these flow and temperature conditions, the assimilative capacity of the river for ammonia loads to the upper river is very low. In this case, only by reducing ammonia loading essentially to the baseline level can violations of the state standard be avoided.

2. Points B, D, and E illustrate the effect of increases in streamflow, at a constant temperature of 12°C, when the ammonia load equals 1000 lb/d. The greatest benefit in DO concentrations occurs between points B and D, with an increase in streamflow from 250 to 350 ft³/s providing a concomitant increase in DO of nearly 1.5 mg/L. As streamflow is increased further, however, the relative benefit decreases gradually; an increase from 350 to 450 ft³/s results in a DO gain of approximately 0.6 mg/L, from about 6.9 to 7.5 mg/L. These results suggest that flow augmentation is most effective as a means to deal with increased ammonia loads for flows less than 300 to 350 ft³/s.

3. Points F, G, and H show the effect of temperature on a particular flow and loading condition and illustrate the method of interpolation for temperature values not included in the simulations. Points F and G represent reference loads (1000 lb/d) at 300 ft³/s, at 12 and 6°C, respectively. Point H is calculated as the mid-point on a straight line connecting F and G, and represents the same flow and loading condition at 9°C. In this case, a change in temperature of 3°C alters the expected DO concentration by about 1.2 mg/L. This result reflects the significant effect that temperature has on the capacity of the river to assimilate increased loads, especially at base flow.

SUMMARY

The focus of this study was to describe the capacity of the Tualatin River to assimilate oxygen-demanding material during early winter conditions when streamflow is often less than the normal

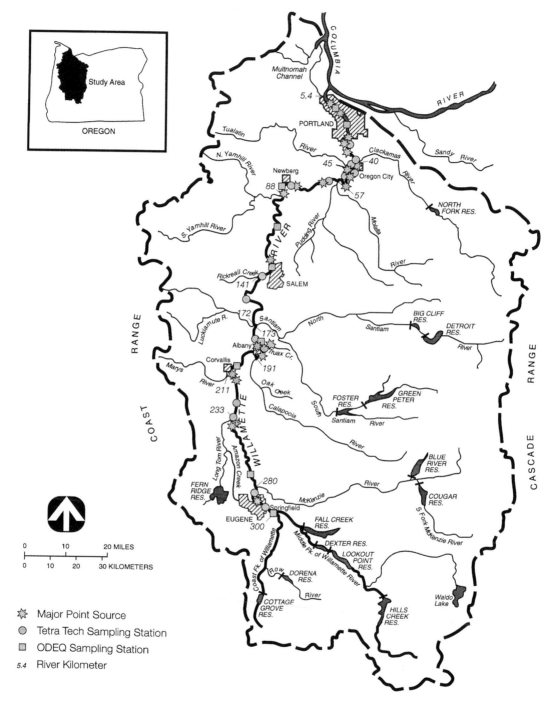

FIGURE 1. The Willamette River Basin study area and the locations of major point source discharges along the main stem of the river and the locations of the August 1992 Synoptic Survey stations.

mately 160 km (100 mi) to the west. However, due to the great distance to the ocean, flow reversals in this 42 km (26.5 mi) reach cause intrusion of only fresh Columbia River water into the Willamette. Flow in the reach below the falls is further complicated by the presence of two channels: the main channel which enters the Columbia River at RK 162 (RM 101.5) and the Multnomah Channel which passes between Willamette RK 5 (RM 3.4) and Columbia RK 140 (RM 88).

Based on hydraulic and physical characteristics, the main stem of the river may be divided into three distinct reaches (Rickert et al., 1976). Briefly, the upstream reach is 217 km (135 mi) long and is characterized by fast-moving currents flowing over a shallow, meandering riverbed composed of cobbles and gravel. The middle reach is a 54 km (56 mi) long, deep, and slow-moving portion of the river formed by the natural impoundment behind Willamette Falls. The tidally influenced reach below the falls is also relatively deep and has the longest estimated travel time: 241 hours (Rickert et al., 1975).

At its mouth the Willamette is the tenth largest river in the continental U.S. in terms of total discharge (Sedell and Frogatt, 1984), and the discharge per unit area is the highest of the large rivers in the U.S. due to the heavy winter rainfall at lower elevations in the basin during the winter months (Rickert and Hines, 1978). At higher elevations the winter precipitation occurs as snow which contributes to extended high flows as spring snowmelt runoff. The climate is temperate and characterized by wet, mild winters and dry, moderately warm summers. Most of the rainfall occurs in the fall, winter, and spring, with little rainfall during June, July, and August. The period of low river flow during the late summer coincides with the period of low rainfall and highest air temperatures.

Although the drainage basin contains most of the state's inhabitants, approximately half of the basin is forested. However, significant changes have occurred in the drainage basin since the arrival of European immigrants beginning in the early 1800s (Sedell and Frogatt, 1984; Gleeson, 1972). About one third of the basin is currently used for agriculture, and the forests have been exploited for timber products. Less than 10% has been urbanized or is in residential use. However, most of the urban and industrial areas occur along the banks of the main stem of the river, typically at or near the confluences with major tributaries, including the states three largest cities: Eugene, Salem, and Portland. As a result, the river receives direct inputs of treated municipal wastes and industrial effluents, primarily from pulp and paper processing facilities.

Concern for the water quality of the Willamette River was evident as early as the late 1920s when the first basinwide studies of the DO conditions in the river were begun (Rogers et al., 1930). Additional studies performed in 1944 indicated that further degradation of the river had occurred (Merryfield and Wilmot, 1945). By 1967 basinwide secondary treatment was mandated and then achieved for both municipal and industrial wastewater by 1972. In addition to basinwide secondary treatment, several other water-quality measures considerably improved DO conditions in the river. These measures included process changes at pulp and paper mills that reduced oxygen-demanding waste discharges, flow augmentation during late summer, and diversion of most of Portland's dry-weather combined sewer overflows to a treatment plant on the Columbia River (Velz, 1984).

However, concern continues for the maintenance of adequate water-quality conditions in a river basin faced with urban and industrial growth (Hines et al., 1977). This concern has led to additional water quality studies that included modeling of DO conditions of the river (Hines et al., 1977; Arnett and Waddel, 1975; Dunnette and Avedovech, 1983; HMS Environmental, 1991; HydroQual, 1990; McKenzie et al., 1979). However, the effects of nutrient discharges on phytoplankton growth and subsequent effects on DO concentrations in the Willamette have not previously been investigated. The interactions of nutrients, phytoplankton, and DO have historically been ignored based on the assumption that river phytoplankton have no net effect on DO concentrations when averaged over the course of a day.

MODEL SELECTION

The objective of the WRBWQS was to develop and calibrate a predictive DO model for the Willamette River to evaluate river basin management alternatives and meet regulatory mandates. In order to first identify appropriate predictive models, several DO models of varying complexity were identified and evaluated using several selection criteria: (1) Dimensionality — a one-dimensional model was considered adequate; (2) Temporal characteristics — a steady-state model was considered appropriate for the summer low-flow period of interest; (3) Consideration of relevant processes — including the capability to model phytoplankton growth and nutrient interactions; (4) Suitability for a range of applications — e.g., temperature or bacteria modeling; (5) Data requirements — the data required for model calibration had to be within the resources of the study; and (6) Ease of use — the selected model needed to be "user friendly" so that water-quality managers could easily use it as a decision-making tool.

Based on these selection criteria, the model QUAL2E-UNCAS (version 3.14) was selected. This model is supported by the U.S. Environmental Protection Agency-Center for Exposure Assessment Modeling (U.S. EPA-CEAM) (Brown and Barnwell, 1987). The QUAL2E model is a one-dimensional steady-state model that incorporates all of the relevant processes and has a menu-driven input and output system which facilitates the use of the model. The model also includes applications for component, sensitivity, first-order error analysis, and Monte Carlo simulations. The model and its components are more fully described by Brown and Barnwell (1987).

DATABASE DEVELOPMENT AND MODEL CALIBRATION

Reviews of historical water-quality data and previous DO modeling efforts were conducted to identify relevant data and modeling approaches that could be incorporated into the QUAL2E model calibration effort (Tetra Tech, 1992a,b). The historical data review also identified data gaps to support the design of a synoptic field sampling effort to provide a data set for calibration of the model. The field sampling effort was conducted in August 1992 and included diurnal DO and temperature measurements at 15 stations and measurements of nutrients and biochemical oxygen demand at 24 stations along the main stem (Tetra Tech, 1992c). Point source loading data were compiled for major municipal and industrial effluent discharges to the mainstem of the Willamette River using the permit-required monitoring reports submitted to the Oregon DEQ and additional data collected during the synoptic field study (Tetra Tech, 1992c,d).

The model was then calibrated to the 1992 synoptic water-quality survey data using a combination of visual best-fit and error minimization techniques (Tetra Tech, 1993b, c). The values of selected model rate constants used in the calibrated model are provided in Table 1. The calibrated-model fit to the synoptic survey DO and chlorophyll *a* data are shown in Figure 2. The location of minimum DO measured during the synoptic survey at RK 43 (RM 27) was also predicted by the model (7.3 mg/L). The model-predicted DO concentrations ranged up to 8.6% of the 24-hour average DO concentrations measured at 15 stations, with a mean and median relative difference of 2.5 and 1.7%, respectively. The model-predicted DO concentrations did not fit the concentrations measured using single grab samples collected at ten stations. In general, single grab samples for DO were considered inadequate for the calibration of a steady-state model, especially for the upper river reach where large diurnal fluctuations in DO occur (Rickert et al., 1975, 1976).

TABLE 1
Values of Selected Rate Constants Used in the Calibrated Model

Maximum specific algal growth rate, μ_{max}, 1/day	2.6
Linear algal self shading coefficient, 1/m μg Chl *a*	0.0049
Ammonia oxidation constant, β_1, 1/day	0.005 RK 0–88 (RM 0–55)
	0.743 RK 88–300 (RM 55–187)
Phytoplankton settling rate, σ_1, m/day	0.3
Sediment oxygen demand, K_4, g/m^2 day	0.32 RK 0–46 (RM 0–28.5)
	1.18 RK 46–80 (RM 28.5–50)
	0.27 RK 80–235 (RM 50–147)
	0.00 RK 235–300 (RM 147–187)

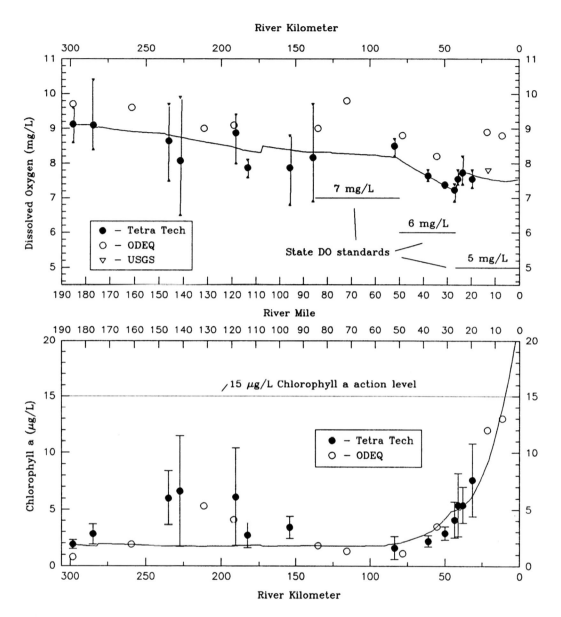

FIGURE 2. Comparison of the calibrated QUAL2E model output to the August 1992 field data for chlorophyll *a* and dissolved oxygen. The model has been calibrated to the solid circle data. Minimum and maximum dissolved oxygen concentrations measured over the 24-hr sampling period are shown. Error bars are the 95% confidence intervals for the solid circle chlorophyll *a* data (n = 3). Oregon DEQ (open circles) and U.S. Geological Survey data (open triangle) are also shown.

The maximum chlorophyll *a* concentrations measured in the lower river were also predicted well by the model, although the model prediction increased exponentially to much higher levels below RK 11 (RM 7). The model did not predict the relatively high chlorophyll *a* levels measured in the upper river. Suspended algal biomass in the upper river reach is considered to be derived from sloughing of periphyton in this relatively shallow stretch of river. Because the model does not consider the influence of periphyton growth, the model predictions for the upper river reach only reflect variation in the steady-state DO concentration due to point source inputs and reaeration.

MODEL VERIFICATION

Verification of a calibrated model with an independent data set is meant to substantiate the model's predictive power under environmental conditions similar to those under which the model was calibrated. With this goal in mind, the calibrated model was applied to August 1990 conditions using point source data provided by HydroQual (1990) and water-quality data available as part of Oregon DEQ's Ambient Monitoring Program. Although model-predicted and measured conditions generally agreed, the model was not considered to be validated because the DO concentrations reported by the Oregon DEQ are for single grab samples. These types of samples were not considered adequate for the calibration or verification of a steady-state model.

MODEL SENSITIVITY AND UNCERTAINTY ANALYSIS

Sensitivity and uncertainty analyses were conducted using the first-order error analysis and Monte Carlo simulation capabilities of the QUAL2E-UNCAS model. These analyses were conducted for the model output at RK 43 (RM 27) for DO and chlorophyll *a*.

Preliminary sensitivity analysis indicated that the model output for RK 43 for both DO and chlorophyll *a* was most sensitive to hydraulic variables (i.e., flow). Excluding hydraulic variables from the first-order error analysis, the model-predicted DO concentration at RK 43 was most sensitive to water temperature, sediment oxygen demand (SOD), and tributary DO load. Approximately 85% of the model output variance for DO was due to SOD (44%), water temperature (33%), and the atmospheric reaeration rate (9%). The model-predicted chlorophyll *a* concentration was sensitive to several model variables including temperature, algal maximum growth rate, incident photosynthetically active radiation, chlorophyll *a* to algal biomass ratio, the algal settling rate, and the light extinction coefficient. Approximately 40% of the model output variance for chlorophyll *a* was due to the algal maximum growth rate.

Using the model default values provided for the statistical distribution and variance associated with model input variables, 500 Monte Carlo simulations resulted in an estimated standard deviation and coefficient of variation of the model-predicted DO concentration at RK 43 of 0.25 mg/L and 3.5%, respectively. The model-predicted chlorophyll *a* concentration standard deviation and coefficient of variation at RK 43 was 2.9 ug/L and 52%, respectively. Because the actual statistical distribution of the model input variables is not presently known, the Monte Carlo results provide only a first-approximation of the uncertainty associated with the model output.

IMPLICATIONS FOR FLOW CONTROL OF DO
AND PHYTOPLANKTON BIOMASS

Although several case scenarios were explored to demonstrate the water-quality management capabilities of the model, the effect of variation in the Willamette River flow regime is the focus of this section. In general, variation in river flow had a noticeable effect on DO throughout the river and on chlorophyll *a* below RK 80 (RM 50) where the river enters the Newberg Pool reach (Figure 3). Flow augmentation of the Willamette River during the low-flow period of July through September has been recognized as an effective means of water-quality management (McKenzie et al., 1979). A minimum flow of 170 m³/s (6,000 cfs) at the Salem gauge [RK 134 (RM 84)] is maintained during these months to allow for navigation and also to maintain adequate DO levels in the river (Rickert et al., 1980). The relative effect of various flow regimes, ranging from 135 to 218 m³/s (4200 to 7700 cfs) measured at Salem, on the calibrated-model prediction of DO at RK 43 (using the August 1992 model inputs) and chlorophyll *a* at RK 16 (RM 10) are presented in Figure 4. The model-predicted DO concentration at RK 43 varies almost linearly from 7.0 to 7.5 mg/L over the range of flow regimes evaluated. The model-predicted effect of flow on chlorophyll *a* concentration was not linear. The model-predicted chlorophyll *a* concentration is predicted to

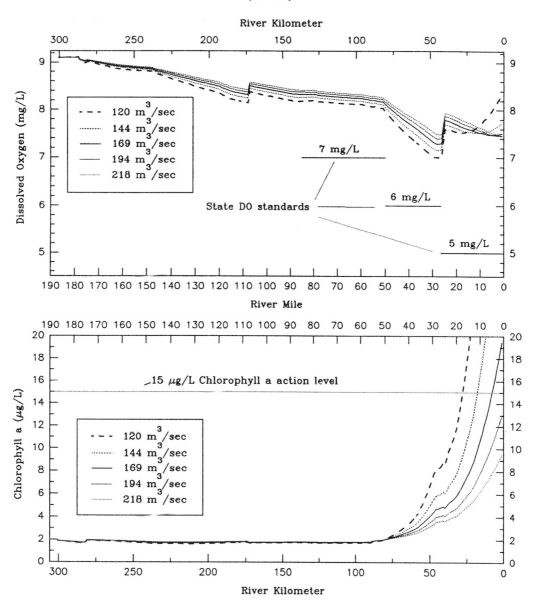

FIGURE 3. The simulation of the effect of variation in the flow regime of the Willamette River on the QUAL2E-predicted August 1992 DO and chlorophyll *a* concentrations.

increase rapidly when riverflow at Salem falls much below 150 m³/s (5300 cfs). These results support the hypothesis by Rickert et al. (1977) that phytoplankton biomass in the lower river is most strongly controlled by variation in the flow (i.e., the water residence time) and that control of the flow regime is not only an effective way to control DO levels, but also to control phytoplankton biomass.

CONCLUSIONS AND FUTURE STUDIES

The calibrated model contains several limitations which are outlined below:

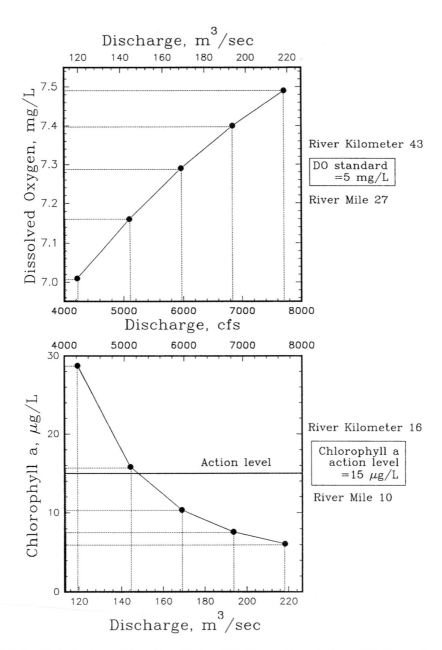

FIGURE 4.   Variation in model-predicted DO at RK 43 and chlorophyll *a* at RK 16 as a function of river discharge predicted at Salem.

1. The model does not incorporate the effect of periphyton production on DO. The effect of periphyton production on DO may be significant in the upper reach of the river above RK 80 (RM 50).
2. The model does not account for tidal mixing with the Columbia River. Therefore, the model output below RK 16 (RM 10) should be interpreted with caution.
3. The model does not consider nonpoint or diffuse sources of nutrients or oxygen-demanding substances. It has been assumed that these sources may be ignored during the low-flow period when heavy rainfall runoff does not frequently occur.
4. The model does not consider minor point sources of nutrients or oxygen-demanding substances. Preliminary estimates suggest that minor point sources could contribute as

much as an additional 10% to the estimated carbonaceous biochemical oxygen demand load to the mainstem of the river (Tetra Tech, 1992d).

5. The model-predicted DO concentration in the lower river was very sensitive to the model specified rate of SOD. However, no field data were collected to establish the reliability of the rates of SOD specified in the model.

As part of the Phase II WRBWQS, additional field work was conducted during the summer of 1994 by the Oregon DEQ and the U.S. Geological Survey, including measurements of SOD that allowed for further refinement of the QUAL2E-UNCAS DO model (Caldwell and Doyle, 1995; Tetra Tech, 1995). Phase II model improvements included updating the model to version 3.20, which allowed the incorporation of minor point sources and evaluation of the Phase I model-specified SOD rates. The Phase I model was recalibrated using the field measured SOD data. The validity of the recalibrated model was then assessed using the more recently collected field data (Tetra Tech, 1995).

ACKNOWLEDGMENTS

This work was performed under contract to the Oregon Department of Environmental Quality (Tetra Tech contract number 8983-03/04). The Phase I WRBWQS project managers for the Oregon DEQ were Mr. Don Yon and Ms. Barbara Priest.

REFERENCES

Arnett, R. C. and Waddel, W. W., An analysis of the waste load assimilation capacity of the Willamette River Basin, Prepared for Oregon Department of Environmental Quality, Portland, OR, Battelle, Pacific Northwest Laboratory, Richland, WA, 1975.

Brown, L. C. and Barnwell, T. O., The enhanced stream water quality models QUAL2E and QUAL2E-UNCAS: Documentation and users manual, EPA/600/3-87/007, U.S. Environmental Protection Agency, Office of Research and Development, Environmental Research Laboratory, Athens, GA, 1987.

Caldwell, J. M. and Doyle, M. C., Sediment oxygen demand in the lower Willamette River, Oregon, 1994. U.S. Geological Survey, Water Resources Investigations Report 95-4196, 1995.

Dunnette, D. A. and Avedovech, R. M., Effect of an industrial ammonia discharge on the dissolved oxygen regime of the Willamette River, Oregon, *Water Res.*, 17, 997, 1983.

Gleeson, G. W., The return of a river. The Willamette River, Oregon, The Willamette River Advisory Committee on Environmental Science and Technology and Water Resources Institute, Oregon State University, Corvallis, OR, 1972.

Hines, W. G., McKenzie, S. W., Rickert, D. A., and Rinella, F. A., Dissolved-oxygen regimen of the Willamette River, Oregon, under conditions of basinwide secondary treatment, U.S. Geological Survey Circular 715-I, 1977.

HMS Environmental, Impacts of pulp mill effluent discharge on the Willamette River: water quality modeling predictions. Part A: summer conditions, Prepared for Pope & Talbot, Inc., Halsey, OR, HMS Environmental, Inc., Beaverton, OR, 1991.

HydroQual, Dissolved oxygen data analysis and modeling for the Willamette River, Oregon, HydroQual, Inc., Mahwah, NJ, 1990.

Merryfield, F. and Wilmot, W. G., 1945 progress report on pollution in Oregon streams: Oregon State University, Corvallis, OR, Engineering Experiment Station Bulletin No. 19, 1945.

McKenzie, S. W., Hines, W. G., Rickert, D. A., and Rinella, F. A., Steady-state dissolved oxygen model of the Willamette River, Oregon, U.S. Geological Survey Circular 715-J, 1979.

Moffatt, R. L., Wellman, R. E., and Gordon, J. M., Statistical summaries of streamflow data in Oregon, Vol. 1 — Monthly and annual streamflow, and flow duration values, U.S. Geological Survey, Open-File Report 90-118, Prepared in cooperation with Oregon Water Resources Department, 1990.

Rickert, D. A., Hines, W. G., and McKenzie, S. W., Methods and data requirements for river-quality assessment, *Water Resources Bull.*, 11, 1013, 1975.

Rickert, D. A., Hines, W. G., and McKenzie, S. W., Methodology for river-quality assessment with application to the Willamette River Basin, Oregon, U.S. Geological Survey Circular 715-M, 1976.

Rickert, D. A., Petersen, R., McKenzie, S. W., Hines, W. G., and Wille, S. A., Algal conditions and the potential for future algal problems in the Willamette River, Oregon, U.S. Geological Survey Circular 715-G, 1977.

Rickert, D. A. and Hines, W. G., River quality assessment: implications of a prototype project, *Science*, 200, 1113, 1978.

Rickert, D. A., Rinella, F. A., Hines, W. G., and McKenzie, S. W., Evaluation of planning alternatives for maintaining desirable dissolved-oxygen concentrations in the Willamette River, Oregon, U.S. Geological Survey Circular 715-K, 1980.

Rogers, H. S., Mockmore, C. A., and Adams, C. D., A sanitary survey of the Willamette Valley, Oregon State University, Engineering Experiment Station, Bulletin Series No. 2, 1930.

Sedell, J. R. and Frogatt, J. L., Importance of streamside forests to large rivers: The isolation of the Willamette River, Oregon, U.S.A., from its floodplain by snagging and streamside forest removal, *Verh. Internat. Limnol.*, 22, 1828, 1984.

Tetra Tech, Willamette River Basin Water Quality Study. Component 3: Data review and summary for dissolved oxygen modeling on the Willamette River, Prepared for Oregon Department of Environmental Quality, Portland, OR, Tetra Tech, Inc., Redmond, WA, 1992a.

Tetra Tech, Willamette River Basin Water Quality Study. Component 4: Review and summary of nutrient and phytoplankton growth data for the Willamette River, Prepared for Oregon Department of Environmental Quality, Portland, OR, Tetra Tech, Inc., Redmond, WA, 1992b.

Tetra Tech, Willamette River Basin Water Quality Study. Component 11: Water quality survey data, Prepared for Oregon Department of Environmental Quality, Portland, OR, Tetra Tech, Inc., Redmond, WA, 1992c.

Tetra Tech, Willamette River Basin Water Quality Study. Component 7: Point source discharges and waste loading to the Willamette River basin during 1991, Prepared for Oregon Department of Environmental Quality, Portland, OR, Tetra Tech, Inc., Redmond, WA, 1992d.

Tetra Tech, Willamette River Basin Water Quality Study. Summary report, Prepared for Oregon Department of Environmental Quality, Portland, OR, Tetra Tech, Inc., Redmond, WA, 1993a.

Tetra Tech, Willamette River Basin Water Quality Study. Willamette River dissolved oxygen modeling component report. Volumes 1 and 2, Prepared for Oregon Department of Environmental Quality, Portland, OR, Tetra Tech, Inc., Redmond, WA, 1993b.

Tetra Tech, Willamette River Basin Water Quality Study. Willamette River nutrient and phytoplankton growth modeling component report. Volumes 1 and 2, Prepared for Oregon Department of Environmental Quality, Portland, OR, Tetra Tech, Inc., Redmond, WA, 1993c.

Tetra Tech, Willamette River Basin Water Quality Study. Phase II. Steady-state model refinement component: QUAL2E-UNCAS dissolved oxygen model calibration and verification, Prepared for Oregon Department of Environmental Quality, Portland, OR, Tetra Tech, Inc., Redmond, WA, 1995.

Velz, C. J., *Applied Stream Sanitation*, 2nd ed., John Wiley & Sons, New York, 1984, 540.

13 River Water-Quality Modeling in Poland

Marek J. Gromiec

BACKGROUND: WATER QUANTITY/QUALITY ANALYSIS

WATER RESOURCES AND WATER DEMANDS

The Republic of Poland is divided into 49 provinces (voivodships). The country is 312,520 km^2 in area, and its total water surface covers 5000 km^2 or 1.6% of the total area of the country. There are about 9300 lakes, covering an area of 3200 km^2. For example, the Masurian Lake District has 1063 lakes, the largest of which have areas of about 11,000 ha. Lakes and artificial reservoirs have a capacity of 33 km^3, and a large number of ponds hold an additional 1 km^3. The two most important rivers in the country are the Vistula River (basin area of 194,000 km^2), and the Odra River (basin area of 110,000 km^2).

The water balance during a normal annual cycle in Poland is presented below. The average annual amount of rainfall is 597 mm, equivalent to 186.6 km^3 of water per year over the whole country. Tributaries from outside Poland yield an additional 5.2 km^3 of water annually making a total input of water of 191.8 km^3. Underground water resources have been estimated at 33 km^3/yr for an area of 272,520 km^2 and the annual dynamic underground water resources have been evaluated at 9.2 km^3. Rivers and streams discharge about 58.6 km^3 of water into the Baltic Sea during a mean low-flow year, and about 34 km^3 in a mean dry-weather year. Obviously, only a portion of this volume is available for water use. About 10 km^3 is necessary as minimum flow to maintain biological life and for sanitary reasons; therefore, the available flow for water use is only 24 km^3.

Poland belongs to the group of European countries most deficient in water resources, ranking 22nd overall. Average annual water resources in Poland (estimated on the basis of atmospheric inputs and the number of population) amount to 1600 km^3 per inhabitant, compared with 2800 km^3 per inhabitant for Europe. In 1992 the total water consumption in Poland was 12.5 km^3, with 22.5%, 66.5%, and 11% for municipal, industrial, and agricultural purposes, respectively. Most water for agriculture is taken during the summer months. Water demands anticipated in the future, particularly during dry seasons, exceed current available water.

LEGISLATION AND ADMINISTRATIVE ASPECTS

The present basis for legal action in the field of water protection against pollution is the Water Law Act issued by the Polish Parliament in 1974. In 1975, on the basis of the Water Law, the Council of Ministers announced regulations concerning classification of waters and determination of effluent standards, as well as financial penalties for the effluent discharges that do not meet the requirements specified in the regulations. The following classes of surface water quality were established:

1. Class I waters are those used for municipal and food processing supply purposes, and for salmon fish growth.

2. Class II waters are intended for use as recreational waters, including water sports and swimming, and for growth of fish other than salmonids.

3. Class III waters (the lowest class) are only used as industrial water supplies and for irrigation purposes.

Water-quality standards are tailored to meet appropriate use of surface waters. In addition, the following provisions are laid down by the Water Law:

1. Industrial plants and other operations which discharge wastewaters to water or to land are obliged to construct, maintain, and utilize wastewater treatment facilities.

2. Without simultaneous operation of wastewater treatment systems, no industrial plant or any other plant from which wastewater is discharged can be started up.

3. A permit is required to maintain wastewater discharge.

In addition, the two most important decrees include:

1. Decree of Minister of Environmental Protection, Natural Resources and Forestry on classification of waters and on the conditions that must be fulfilled when the wastewater is discharged into waters or ground (1991.11.05).

2. Decree of Minister of Health and Social Welfare concerning conditions that must be met by drinking water and by water for industrial purposes (modified version of 1990.05.04).

In 1991, Polish Parliament passed a bill concerning ecological policy which determines the general rules, aims, and directions of future actions in Poland. Currently a new version of the Water Law Act is being prepared. The new system of water resources management introduces subdivision of the country's water system into river catchment based areas under responsibility of River Basin Water Authorities.

WATER-QUALITY PROBLEMS

Processes associated with intensive urbanization, growth of population, identification of agriculture, and the growth of industry have resulted in deterioration of surface water resources, despite the introduction of certain measures for water pollution control. Therefore, most of the major rivers have experienced serious degradation of water quality (Figure 1). In addition, lakes, especially in the northern region, are threatened by eutrophication with high phosphorus concentrations as the primary cause.

Water pollution control has become one of the most important environmental problems in Poland. Currently, about 70% of all wastewaters needing treatment are treated but with different degrees of treatment (Figure 2). A substantial part of the existing municipal wastewater treatment plants are overloaded. The majority of industry is situated in the south near the origins of the country's river systems. In addition, the main rivers — the Vistula and Odra — are heavily used for municipal and industrial water supply, agriculture irrigation, cooling purposes for power plants, and navigation. These rivers also receive discharges of wastewater with varying degrees of treatment and runoffs.

Multiple uses impose competing demands on waters, and water resources management must protect many desirable uses. The principal water quality problems in Poland are related to:

1. effects of municipal and industrial wastewater, including saline discharges from coal mines;

2. influence of nonpoint sources, such as agriculture and urban storm waters;

3. effects of dams and other water management structures resulting from phenomena associated with impounded water; and

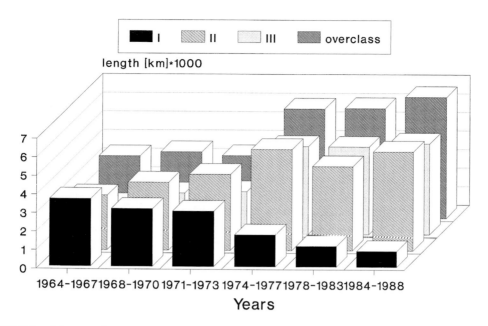

FIGURE 1.   Water-quality classification: physical and chemical criterion.

4. effects of power plants, since the discharge of heat from cooling operations is considered
    to be specific pollutant.

MONITORING SYSTEM AND DATA ANALYSIS

Poland has monitoring activities in rivers, lakes, coastal waters, and the Baltic Sea. The river water-quality surveillance system is composed mainly of conventional monitoring stations. The country is covered by a network of stations at established cross sections. The sampling frequency depends on the purpose for which data are recorded, ranging from a minimum of bimonthly sampling up

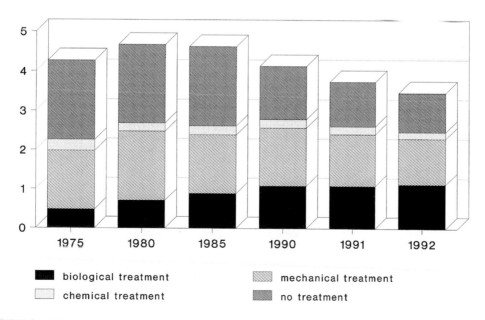

FIGURE 2.   Wastewater treatment.

to daily sampling at some points. The sampling of water is performed simultaneously with the rate of flow measurements.

The river monitoring network provides a large number of observed data. These data are analyzed by a statistical method based on the assumption that, at a given cross section, some correlation exists between the pollutant concentration and the rate of flow. The shape of the curve depends on many factors, such as the degree of water pollution, the type of pollutant, hydrological characteristics of the river, its self-purification capacity, the distance between monitoring stations, and others. From these relationships between stream flow and concentrations of water-quality constituents, indicative concentrations (IC) for a design flow are established. The mean low streamflow (MLQ) has been selected as the design streamflow at each site, based on the assumption that higher streamflows will result in higher DO concentrations and better water quality. In other countries, a similar approach has been taken. The IC values are plotted along the river for various water-quality constituents. Final interpretation is based on these hydrochemical profiles, and the overall river classification is performed after all measured water-quality constituents are compared with standards. A compendium of hydrochemical profiles for major rivers and streams is prepared each year by the Institute of Meteorology and Water Management (IMWM) which serves as an overall river classification system.

WATER-QUALITY MODELING

EARLY BOD-DO MODELS

Mathematical modeling of river systems in Poland has become an integral of water resources planning and water-quality management. These models can be used to aid in water-quality surveillance and to predict future water quantity/quality conditions (Gromiec et al., 1983). Various computerized models have been applied to water-quality simulations in the Odra and Vistula Rivers. As an example, a Streeter-Phelps model and QUAL-I model were used by the IMWM to evaluate concentrations of biochemical oxygen demand (BOD) in the Vistula River reaches. The first model is designed to simulate the spatial and temporal variations in BOD under various flow and temperature conditions. The second model is capable of routing BOD, DO, and temperature through a one-dimensional, completely mixed branching river system. These early BOD-DO models are representative of nonconservative coupled models. The predictions obtained from these models are only as reliable as the input data, proper measurement, and estimation of the various model parameters.

AN OVERVIEW OF THE QUAL2E AND QUAL2E-UNCAS MODELS

The Stream Water Quality Model QUAL2E (Brown and Barnwell, 1987) is a steady-state model for conventional pollutants in one-dimensional streams and well-mixed ecosystems. The conventional pollutants include conservative substances, temperature, bacteria, BOD, DO, nitrogen, phosphorus, and algae (Figure 3). The model is widely used for simulation of water quality and for waste load allocations and discharge permit determinations in the U.S., and it is a proven, effective analytical tool (Barnwell et al., 1987).

A major problem faced by the user when working with a complex model such as QUAL2E is model calibration and determination of the most efficient plan for collection and calibration data. This problem can be addressed by application of principles of uncertainty analysis. QUAL2E-UNCAS is a recent enhancement to QUAL2E which allows the user to perform uncertainty analysis on the steady-state water quality simulations. The above models are available from the U.S. Environmental Protection Agency's Environmental Research Laboratory (AERL) at Athens, GA.

THE U.S.-POLISH JOINT PROJECT

A variety of microcomputer models for water-quality simulation is supported by the U.S. EPA Center for Exposure Assessment Modeling (CEAM) at the Athens Laboratory. The Athens Envi-

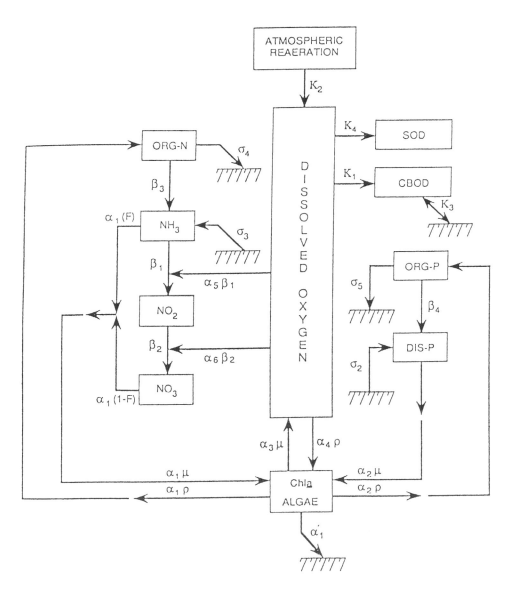

FIGURE 3. Major constituent interaction in QUAL2E.

ronmental Research Laboratory (AERL) and the IMWM have jointly selected a set of CEAM-supported computer models that address surface-water quality problems of mutual interest.

The QUAL2E and QUAL2E-UNCAS models have been applied to tributaries of the Vistula River in Poland (Gromiec et al., 1994). About 30 major rivers from the upper, mid, and lower part of the Vistula River Basin were chosen for this study. A summary of river characteristics is shown in Table 1.

For water-quality simulation of Polish rivers, the following water-quality constituents were chosen to be simulated in steady-state modeling: temperature, DO, BOD, nitrogen compounds (organic nitrogen, ammonia, nitrite, nitrate), phosphorus compounds (organic and dissolved), coliforms, chlorophyll a, conservative constituents (chlorates and sulfates), and arbitrary nonconservative (chemical oxygen demand).

The water-quality data base, which has been used for preparing input files, contains results from field measurements. Sampling points were located at important stream cross sections with respect to tributaries, point loads, and withdrawal points. They were located upstream and down-

silver filter in preparation for analysis of dissolved organic carbon (Wershaw et al., 1987); material retained on the filter was analyzed for suspended-organic carbon.

Trace element concentrations were determined by filtering part of the water remaining in the churn through a 0.45-mm pore-size filter into an acid washed, 200-mL Teflon* bottle for Cd, Cr, Cu, and Pb and into an acid washed, 250-mL bottle for Al, Sb, As, Ba, Be, B, Co, Fe, Li, Mn, Hg, Mo, Ni, Se, Ag, Sr, V, and Zn. To achieve lower method reporting limits for Cd, Cr, Cu, and Pb, filtered-water determinations were made by atomic adsorption in conjunction with a graphic furnace (AAGF); the remainder of the suite was analyzed by inductively coupled plasma (ICP) according to the methods of Fishman and Friedman (1989). At the time of sampling, water samples also were collected from each of the 10 sampling verticals and placed into 10-L polycarbonate containers for centrifugation used to isolate particles 0.45 μm or larger (assuming a density of 2.5 g/cm^3) for analysis of suspended Al, Sb, As, Be, Cd, Cr, Co, Cu, Fe, Pb, Mn, Mo, Ni, Th, Ti, V, and Zn. The suspended sediment samples were freeze-dried, and 50 mg aliquots were digested with a combination of HF/HClO$_4$/HNO$_3$ acids. Suspended trace-element determinations were made using inductively coupled plasma-optical emission spectroscopy (ICP-OES) and inductively coupled plasma-mass spectroscopy (ICP-MS) according to Arbogast (1990).

Aquatic Biota

Aquatic biota were sampled at 34 sites from 1989 to 1990; seven sites were located on the main stem (Figure 2). The aquatic biota medium included analyses of plant tissues from unidentified species of algae, curlyleaf pondweed (*Potamogeton crispus*), waterweed (*Elodea* sp.), and coontail (*Ceratophyllum demersum*); fish tissues from rainbow trout (*Oncorhynchus mykiss*), cutthroat trout (*O. clarki*), mountain whitefish (*Prosopium williamsoni*), sculpin (*Cottus* spp.), brook trout (*Salvelinus fontinalis*), bridgelip sucker (*Catostomus columbianus*), largescale sucker (*Catostomus macrocheilus*), chiselmouth (*Acrocheilus alutaceus*), and carp (*Cyprinus carpio*); asiatic clam (Corbiculidae *Corbicula fluminea*), and aquatic insects that included caddisflies (*Arctopsyche* spp., *cheumatopsyche* spp., *Parapsyche* spp., *Hydropsyche amblis*, *H. californica*, *H. cockerelli*, *H. occidentalis*), stoneflies (*Calineuria* spp., *Claassennia* spp., *Doroneuria* spp., *Hesperoperla* spp., *Isoperla* spp., *Megarcys* spp., *Perlinodes* spp., *Skwala* spp., *Pteronarcys* spp.), and mayflies (unidentified).

The apical portions of submerged aquatic plants were collected and prepared for analysis according to Crawford and Luoma (1993). Plant tissues analyzed for Al, As, Ba, Be, B, Cd, Cr, Cu, Fe, Pb, Mg, Mn, Mo, Ni, Se, Ag, Sr, Tl, V, and Zn were digested in a hot nitric-perchloric acid reflux; plants analyzed for Hg were digested in a hot nitric acid reflux, and diluted to 50 mL with 1% HCl prior to analysis. Hg was determined by cold-vapor AAS; As and Se by hydride-generation AAS; and the remaining elements by ICP, according to Fuhrer et al. (1994c).

Aquatic insects were collected from shallow riffles, placed in polyethylene bags with stream water for 6 to 8 hours in order to flush their stomach contents (purge), and frozen on dry ice. Prior to analysis, insects from each site were sorted by taxon to obtain samples with total dry weights of at least 100 mg and then analyzed by ICP. Asiatic clams were collected at five sites in the lower valley. At each site, three composite samples were collected, each of which consisted of approximately 20 clams. Clams were rinsed with stream water, purged, and prepared for analysis according to Crawford and Luoma (1993). Depending on stream size, fish were collected using backpack or boat-mounted electrofishing equipment. Fish livers were collected in most cases, except for sculpin, which were analyzed on a whole-body basis. All fish and aquatic insect tissues, except for those analyzed for As, Hg, and Se, were digested in hot-16 N HNO$_3$, reconstituted in 0.6 N HCl, and analyzed using a Thermo-Jarrel Ash ICP-61. Tissues analyzed for As and Se were digested using HF/HClO$_4$/HNO$_3$ and analyzed by hydride generation AAS. Tissues analyzed for Hg were digested using LeForte aqua regia and analyzed by cold vapor AAS.

---

* The use of brand names in this report is for identification purposes only and does not constitute endorsement by the U.S. Geological Survey.

RESULTS AND DISCUSSION

Most element enrichment in the Yakima River Basin results from natural geologic sources under-lying forest lands of the Kittitas and mid-Yakima Valleys — primarily in the Cle Elum, Upper Naches, Teanaway and Tieton Subbasins (Figure 1). These areas might be classified as "pristine" by the casual observer; however, they are geologic sources of Sb, As, Cr, Cu, Hg, Ni, Se, and Zn. For example, in the Kittitas Valley, As, Cr, and Ni concentrations in streambed sediment affected by geologic sources were as high as 61, 1700, and 1900 µg/g, respectively. These elevated con-centrations ranged from 13 to 74 times higher than the respective median concentrations in stre-ambed sediment in agricultural areas unaffected by geologic sources in the lower Yakima Valley. The pre-Tertiary metamorphic and intrusive rocks geologic unit is the source of the high trace element concentrations in the Kittitas Valley; high trace element concentrations in the pre-Tertiary Rocks geologic unit (Figure 3) were typical for As, Co, and Ni as well. The pre-Tertiary Rocks geologic unit covers portions of the Cle Elum and Teanaway Subbasins, and several sites in these subbasins have high Cr concentrations that exceed the 95% range of concentrations in Western United States soils (Table 1).

As a result of geologic sources, several of these elements, including As, Cr, Cu, and Ni, leave chemical signatures that are measurable in streambed sediment and suspended sediment of higher order streams, including the main stem. Chromium concentrations in streambed sediment of the main stem, for example, exceeded 200 µg/g at the Yakima River at Cle Elum — a fixed site located near the geologic source (Figure 4). Downstream, however, concentrations decreased to just 64 µg/g at the Yakima River at Umtanum, which receives Cr-poor sediment transported to the main stem from Wilson Creek. Sediment formed in the Wilson Creek drainage originates from the Quaternary deposits and loess geologic unit where the median Cr concentration was only 56 µg/g. The distribution of suspended-Cr concentrations was similar to that in streambed sediment. Con-centrations of Cr in suspended sediment, measured monthly and during several storms for the period of 1987 to 1990, ranged from 28 to 160 µg/g (Table 2). On the basis of median suspended-Cr concentrations, the highest concentrations were measured at the Cle Elum and Umtanum sites in the Kittitas Valley — in the vicinity of the geologic source (Figure 5).

The large variation in monthly suspended-Cr concentrations in the Yakima River at Umtanum (Figure 6) is indicative of seasonal changes in the source of Cr. The Cr-rich sediment of the pre-Tertiary Metamorphic and Intrusive Rock geologic unit in the Cle Elum Sub-basin contributes large concentrations of suspended Cr during both the nonirrigation season and the early and midirrigation season at Umtanum (Figure 6). The concentrations of suspended Cr decrease sharply in the late irrigation season (September and October); however, these decreases coincide with the curtailment of reservoir releases upstream of Umtanum — including releases from Cle Elum Lake. The cur-tailment of reservoir releases indirectly increases the proportion of irrigation-return flow at Umtanum — an important factor because the agriculturally affected sediment entering the main stem in irrigation-return flow upstream of Umtanum is formed in the Cr-poor Quaternary deposits and loess geologic unit. The net result of an increase in the proportion of Cr-poor sediment entering the main stem is a dampening or dilution of suspended Cr concentrations during the September and October time period. Similar temporal patterns also existed for suspended As and Ni concentrations.

In addition to measurable concentrations in streambed sediment and suspended sediment, some of the geologically derived elements, including Cr, Ni, and Se, can be detected in aquatic biota of higher-order streams. For example, Cr concentrations in aquatic insects in the North Fork of the Teanaway River ranged from 2.2 to 33 µg/g and are 4 to 52 times higher (depending on insect species) than the minimum concentrations measured in the basin. Enrichment from the Teanaway River also affects biota in the main stem at the Yakima River at Umtanum — although to a lesser extent. Similarly, chromium enrichment in the main stem at Umtanum is evident in caddisflies, stoneflies, and curlyleaf pondweed (Table 3), in addition to streambed sediment and suspended sediment. The sharp contrast between the high Cr concentrations in the main stem at Umtanum and the low Cr concentrations in Umtanum Creek (a reference site) underscores the effect of geology

CHROMIUM IN STREAMBED SEDIMENT

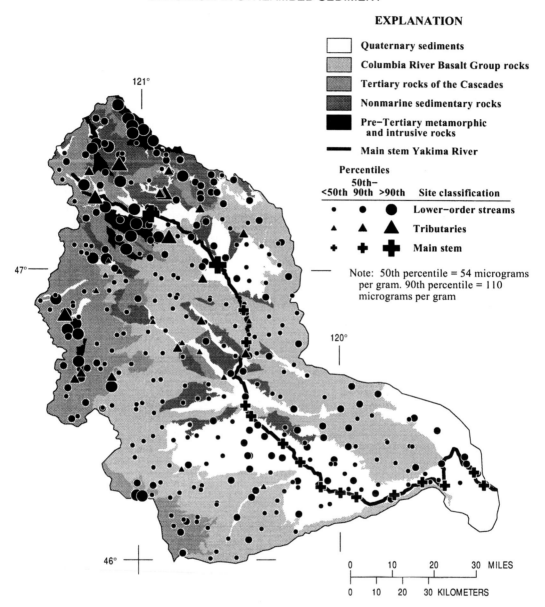

FIGURE 3. Chromium concentrations in streambed sediment, Yakima River Basin, WA, 1987.

on the distribution of trace elements among media. Umtanum Creek is located in the Cr poor Columbia River Basalt Group Rock geologic unit where the median Cr concentration was only 48 mg/g. Low Cr concentrations also are found in Rattlesnake Creek in the Naches Subbasin and are generally typical of the Naches Subbasin. Concentrations are also low in the main stem at RM 72 and are low to medium in several other agriculturally affected tributaries (not listed in this abbreviated table). The lower concentrations among media reflect the Cr-poor nature of the Quaternary deposits and loess geologic unit. Although geologic sources affect aquatic biota in some lower- and higher-order streams, the influences of these sources are offset, overall, by the diluting effect of the other Cr-poor geologic units. Consequently, the median concentration of Cr (1.7 mg/g) in benthic insects in the Yakima River Basin is less than the 4.9 to 14.2 mg/g reported for uncontaminated to minimally contaminated basins in other studies (Elwood et al., 1976; Lynch et al., 1988;

TABLE 1
Concentrations of Selected Elements in Soils of the Western U.S. and in Fine-Grained Stream Sediment of the Yakima River Basin, Washington, 1987

Element	No. sites	Min. value	Value at indicated percentile						Max. value	Baseline conc.	Western U.S. soils		
			10	25	50	75	90	98			No. observations	Expected 95% range, baseline soils	Maximum
Antimony	404	0.1	0.2	0.3	0.4	0.5	0.8	2.0	4.8	0.7	na	na	na
Arsenic	404	.7	1.6	2.3	3.9	5.9	11	66	310	8.5	730	1.2–22	97
Beryllium	407	<1	<1	<1	<1	2	2	2	4	3	778	0.13–3.6	15
Cadmium[a]	407	<2.0	<2.0	<2.0	<2.0	<2.0	<2.0	<2.0	<2.0	na	na	na	na
Cadmium[b]	5	0.2	0.2	0.2	0.2	0.2	.5	.6	0.8	na	na	na	na
Cerium	407	5	33	39	45	56	69	100	120	57	683	22–190	300
Chromium	407	14	32	44	54	74	110	554	1800	320	778	8.5–200	2000
Copper	407	13	20	24	28	34	46	96	190	40	778	4.9–90	300
Lead	407	2	9	11	13	17	25	130	890	20	778	5.2–55	700
Mercury	406	<.02	<.02	<.02	.02	.08	.16	.48	3.1	.30	733	.008–.25	4.6
Nickel	407	4	13	17	21	32	53	690	1900	120	778	3.4–66	700
Selenium	99	<.1	<.1	.2	.4	.6	1	1.3	14	.7	733	.04–1.4	4.3
Zinc	407	32	71	80	93	110	130	220	710	120	766	17–180	2100

Note: To avoid statistical bias that may be associated with elements analyzed in duplicate or triplicate at a site, only one element concentration per site was statistically summarized; concentrations are in units of micrograms per gram (µg/g), dry weight; data statistically summarized in this table are from Fuhrer et al. (1994a); element percentile concentrations shown in bold print equal or exceed baseline concentrations for the Yakima River Basin (Fuhrer et al., 1994a).

[a] Analyzed by inductively coupled plasma-atomic emission spectroscopy (ICP-AES); limit of determination = 2.0 mg/g.
[b] Analyzed by ICP-AES with organometallic-halide extraction; limit of determination = 0.05 mg/g.

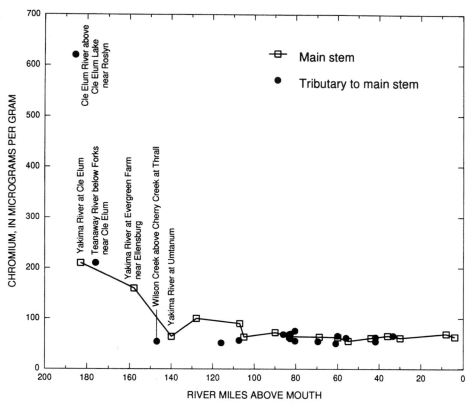

FIGURE 4. Chromium concentrations in streambed sediment of the mainstem and selected tributaries, Yakima River Basin, WA, 1987.

TABLE 2
Summary of Trace Element Concentrations in Suspended Sediment, Yakima River Basin, Washington (1987 to 1990)

Element	No. samples	Min. value	Value at indicated percentile						Max. value
			10	25	50	75	90	95	
Antimony	211	.3	.5	.5	.6	.7	.8	.9	3.1
Arsenic	211	2.8	4.7	5.4	6.6	8.2	11	14	20
Beryllium	211	<2	<2	<2	<2	<2	2	2	3
Cadmium	211	<.1	.2	.3	.5	.7	1.4	1.7	32.6
Chromium	184	28	46	55	60	83	110	120	160
Copper	211	21	33	39	44	55	74	96	680
Lead	211	6	12	15	19	24	27	30	410
Nickel	184	12	22	29	37	55	82	105	170
Silver	211	<.1	.2	.2	.4	.5	.9	1.3	7.7
Zinc	184	88.0	112	123	142	172	202	231	521

Note: Data statistically summarized in this table are from monthly and selected hydrologic-event samplings from the Yakima River at Cle Elum, Yakima River at Umtanum, Naches River near North Yakima, Yakima River above Ahtanum Creek at Union Gap, Sulphur Creek Wasteway near Sunnyside, Yakima River at Euclid Bridge at RM 55 near Grandview, and Yakima River at Kiona; to avoid statistical bias that may be associated with constituents analyzed in duplicate or triplicate at a site, only one element concentration per visit was statistically summarized; concentrations of trace elements are in units of micrograms per gram.

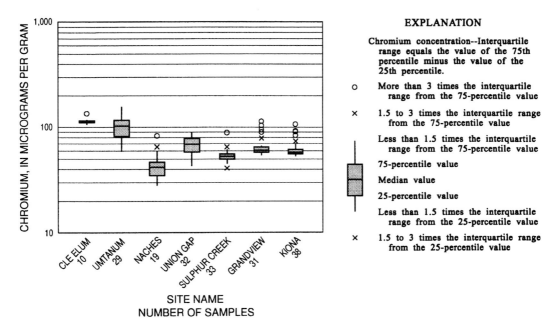

FIGURE 5. Distribution of chromium concentrations in suspended sediment at fixed sites, Yakima River Basin, WA (1987–1990).

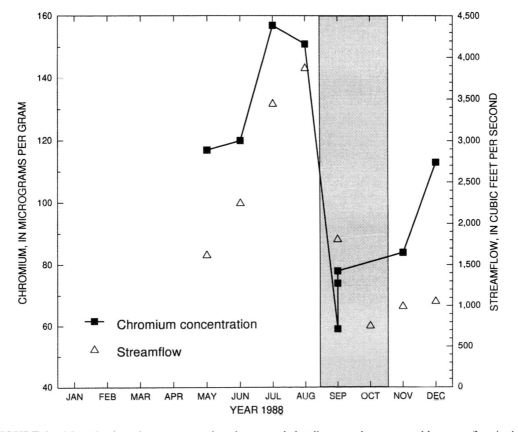

FIGURE 6. Monthly chromium concentrations in suspended sediment and mean-monthly streamflow in the Yakima River at Umtanum, WA, 1988.

TABLE 3

Comparison of Low, Medium, and High Chromium Concentrations in Water, Sediment, and Aquatic Biota for Selected Sites, Yakima River Basin, WA, 1987 to 1991

Site ref. no.	Site name	Filtered water	Sediment Streambed	Sediment Suspended	Largescale-sucker liver	Insects Caddisfly	Insects Stonefly	Asiatic clam	Curlyleaf pondweed
3	Jungle Creek near mouth near Cle Elum	—	H	—	—	—	—	—	—
4	North Fork Teanaway River	—	—	—	—	—	H	—	—
5	Teanaway River below Forks near Cle Elum	—	H	—	—	—	—	—	—
6	Yakima River at Cle Elum	H	H	H	—	*	*	—	—
8	Taneum Creek at Taneum Meadow near Thorp	—	H	—	—	—	*	—	—
12	South Fork Manastash Creek near Ellensburg	—	H	—	—	—	*	—	—
19	Yakima River at Umtanum	L	—	H	—	H	H	—	H
20	Umtanum Creek near mouth at Umtanum	—	L	—	—	L	—	—	—
21	Rattlesnake Creek above Little Rattlesnake near Nile	—	L	—	—	L	L	—	—
27	Wide Hollow Creek near Ahtanum	—	L	—	—	L	—	—	—
48	Yakima River at river mile 72 above Satus Creek	*	*	—	—	—	—	L	L
50	Yakima River at Kiona	H	*	*	*	*	—	*	*

Aquatic biota

Note: This table is a partial listing of table 23 in Fuhrer et al., 1996. For *filtered water,* the low- and high-concentration assignments are based on a percentile distribution of the 75th-percentile values for each fixed site. For *streambed sediment and aquatic biota,* the low- and high-concentration assignments are based on a percentile distribution of the mean concentrations for each fixed site. For *suspended sediment,* the low- and high-concentration assignments are based on a percentile distribution of the 50th-percentile values (median) for each fixed site. High concentrations (H) represent that portion of the distribution which is greater than or equal to the 75th-percentile value. Low concentrations (L) represent that portion of the distribution which is less than or equal to 25th-percentile value. Concentrations greater than 25th, but less than 75th-percentile value are denoted with an *. The term *filtered water* is an operational definition referring to the chemical analysis of that portion of a water-suspended sediment sample that passes through a nominal 0.45-μm filter; sample species; largescale sucker (*Catostomus macrocheilus*), caddisfly (*Hydropsyche* spp.), stonefly (*Hesperoperla* spp.), Asiatic clam (Veneroida: Corbiculidae *Corbicula fluminea*), and curlyleaf pondweed (*Potamogeton crispus*); data statistically summarized for fixed sites are from monthly and selected hydrologic-even samplings from the Yakima River at Cle Elum, Yakima River at Umtanum, Naches River near North Yakima, Yakima River above Ahtanum Creek at Union Gap, Sulphur Creek Wasteway near Sunnyside, Yakima River at Euclid Bridge at river mile 55 near Grandview, and Yakima River at Kiona; to avoid statistical bias that may be associated with constituents analyzed in duplicate or triplicate at a site, only one element concentration per visit was statistically summarized; — = no data.

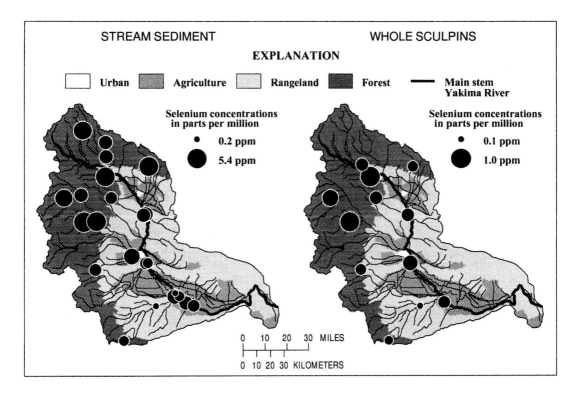

FIGURE 7.   Selenium concentrations in whole sculpins and streambed sediment, Yakima River Basin, WA (1987–1990).

Smock, 1983). Additionally, in the lower Yakima Valley and in tributaries of the mid-Yakima Valley, Cr concentrations in fish and Asiatic clams were typically small (<2 µg/g) and varied little among sites.

The distribution of Se concentrations is affected principally by the Miocene and older volcanic rocks, marine sedimentary rocks, and pre-Tertiary metamorphic and intrusive rocks geologic units located in forests of the Cascades (Figure 7). Selenium concentrations at several sites in these geologic units are two to three times the median concentration (0.4 µg/g) in the basin. The geologically affected Se concentrations in streambed sediment of forest lands results in high Se concentrations in whole sculpin (*Cottus* spp.). Selenium concentrations in sculpin ranged from 0.2 to 5.4 µg/g (dry weight) and were high in Rattlesnake Creek, which is located in the Naches Subbasin; the concentrations were also high in nearby Taneum Creek. Concentrations of Se in sculpin were also higher at sites located near the mouths of creeks carrying irrigation-return flow than at sites located upstream from agricultural activity. For example, 2.6 µg/g of Se was measured in whole sculpin from Ahtanum Creek at Union Gap. This site is located near the mouth of Ahtanum Creek and receives irrigation-return flow from the Ahtanum Subbasin. Upstream of agricultural activity on Ahtanum Creek, concentrations of Se in sculpin were only 1.0 µg/g. Similar concentration gradients also exist in the Satus Creek drainage. Satus Creek at gauge at Satus is located near the mouth and receives irrigation-return flow. Similarly, the concentration of Se in rainbow trout liver samples from Wide Hollow Creek (31 µg/g), a creek affected by agriculture and urban activities, was four to eight times higher than concentrations in rainbow trout from other sites in the basin.

In some parts of the basin, human activities (such as farming) can reduce element concentrations, in other parts, they can increase element concentrations. Elements whose distributions increase in areas affected by human activities include Sb, Cd, Cu, Pb, Hg, Se, and Zn. Concentrations of those elements frequently are highest in the Wide Hollow Subbasin, which drains urbanized and lightly industrialized lowland in addition to agricultural land in the upper reaches of Wide Hollow Subbasin. Concentrations of Pb in streambed sediment of Wide Hollow Creek, for example, were more than twice that expected in streambed sediment from geologic sources in Wide Hollow Subbasin and also exceeded the 5.2 to 55 µg/g range of concentrations found in 95% of Western United States soils (Table 1). In addition to urban runoff, past applications of lead arsenate in apple orchards, including those in the upper reaches of Wide Hollow Subbasin, may be a source of Pb. Concentrations of Pb in soils of former apple orchards in the mid-Yakima Valley (historically treated with lead arsenate) were as high as 890 µg/g; Sb concentrations in treated soils were as high as 2.9 µg/g. The relation between elevated Pb and Sb concentrations is probably a result of pesticide formulation — Pb produced from domestic sources contains residual Sb (U.S. Geological Survey, 1969). In eastern Washington, beginning in 1908, the pesticide lead arsenate was applied to control codling moths in apples, and this practice continued until the introduction of DDT in 1947. From 1908 to 1947, lead-arsenate applications increased from 50 lb of Pb and 18 lb of As to 192 lb of Pb and 71 lb of As per acre (Peryea, 1989). Benthic insects also reflect Pb enrichment in Wide Hollow Creek. Lead concentrations in caddisflies (*Hydropsyche* spp.) in Wide Hollow Creek represented the basin maximum (9 µg/g) for caddisflies and were 15 times higher than concentrations in caddisflies from Umtanum Creek — a reference site not affected directly by human activities. Lead concentrations in caddisflies in Wide Hollow Creek, however, are only slightly enriched compared to concentrations of Pb, which ranged from 0.1 to 1.8 µg/g, in benthic insects in other uncontaminated or minimally contaminated systems (Cain et al., 1992; Lynch et al., 1988).

Although the irrigation-season discharge of suspended sediment indirectly attenuates geologically derived As concentrations in streambed sediment of the Kittitas Valley, the discharge may increase As concentrations in streambed sediment of lower order streams that receive agricultural return flow in the mid- and lower Yakima Valley. This situation is likely a result of historical applications of lead-arsenate pesticides to apple orchards in the mid- and lower Yakima Valley. Approximately 3000 acres of apple orchards existed (primarily in the mid- and lower Yakima Valley) prior to 1955 (U.S. Department of Agriculture, 1986). Two geologic units that cover a large proportion of the mid- and lower Yakima Valley are the Quaternary deposits and loess and the nonmarine sedimentary rocks geologic units. Taking into consideration the prevailing land-use scheme and the fact that a proportion of the land use designated as agriculture represents orchards historically treated with lead-arsenate pesticides, concentrations of As in streambed sediment formed from these two geologic units were larger for land use designated as agriculture than for land use designated as nonagriculture (Figure 8). Additionally, a strong Kendall's Tau-B correlation (r = 0.03; n = 77) between Pb and As further supports the presence of a historical lead-arsenate source in agricultural lands of the lower Yakima Valley.

In the mid- and lower Yakima Valley, human activities also augment As concentrations in filtered water, suspended sediment, and aquatic biota. Consequently, agricultural drains are useful as indicators of past As use. Suspended-As concentrations in Sulphur Creek Wasteway, an agricultural drain, ranged from 4.9 to 20 µg/g and were the highest in the basin. Between the Kittitas Valley and the mid-Yakima Valley, the annual suspended-As loads from 1987 to 1990 increased by as much as threefold. During the irrigation season in particular, about 2.2 lb of suspended As per day enter the mid-Yakima Valley over a 9.4 RM reach that receives irrigation return flow from Moxee Subbasin and Wide Hollow Subbasin. This As load represents about half the irrigation season's, daily mean load in the Yakima River above Ahtanum Creek at Union Gap. Irrigation drains are likely large contributors of suspended As in the mid-Yakima Valley — Moxee Drain alone is estimated to contribute nearly 1 lb of suspended As per day during the irrigation season. During the irrigation season in the lower Yakima Valley, the June contributions of suspended As from Sulphur Creek Wasteway (2 lb/day) typically accounted for most of the suspended-As load in the Yakima River at Euclid Bridge at RM 55 near Grandview — located 6 miles downstream from Sulphur Creek Wasteway.

Filtered water samples, collected monthly for 1987 to 1990, had As concentrations ranging from <1 to 9 µg/L (Table 4). As concentrations were higher in the mid- and lower Yakima Valley,

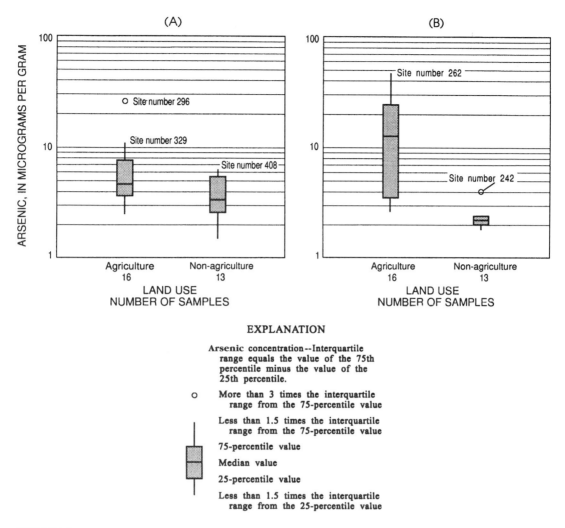

FIGURE 8.   Arsenic concentrations in agricultural and nonagricultural land uses for streambed sediment of lower-order streams in Quaternary deposits and loess geologic unit (A), and in the nonmarine sedimentary rocks geologic unit (B), Yakima River Basin, WA, 1987.

where waters are affected primarily by agricultural-return flow (Table 5). Arsenic concentrations in Sulphur Creek Wasteway and in the main stem in the lower Yakima Valley exceeded the 90th percentile (3 μg/L) for the basin. Concentrations of As in Sulphur Creek Wasteway ranged from 2 to 9 μg/L and were the highest in the basin. As concentrations also varied seasonally in agriculturally affected waters. In Sulphur Creek Wasteway, As concentrations generally ranged from 2 to 4 μg/L during the irrigation season and from 7 to 9 μg/L during the nonirrigation season. Seasonal variations in As concentration in the main stem, however, occur in an opposite pattern to those in Sulphur Creek. Variations in concentrations are the result of dilution processes. During the irrigation season, As concentrations in Sulphur Creek are diluted by excess canal water from Rosa Canal and by irrigation-return flow from the Sunnyside Subbasin. In the main stem, however, mean stream flows are smaller during the irrigation season than during the nonirrigation season. As a consequence, less water is available, and the main stem takes on the water-quality character of the agricultural drains.

In addition to higher concentrations of As in filtered-water samples from agriculturally affected parts of the basin, the load of As in agricultural drains represents a large proportion of the As load passing the Yakima River at Kiona, the terminus of the basin. Sulphur Creek Wasteway, for example,

TABLE 4
Summary of Trace Element Concentrations in Filtered-Water Samples, Yakima River Basin, Washington (1987–1990)

Element	No. samples	Min. value	Value at indicated percentile						Max. value
			10	25	50	75	90	95	
Antimony	18	<1	<1	<1	<1	<1	<1	1	1
Arsenic	106	<1	<1	<1	<1	2	3	7	9
Beryllium	36	<.5	<.5	<.5	<.5	<.5	<.5	<.5	<.5
Cadmium	279	<.2	<.2	<.2	<.2	<.2	.3	.5	2.2
Chromium	26	<.5	<.5	<.5	<.5	.6	1.0	1.1	1.1
Copper	280	<.5	<.5	.6	.9	1.3	1.9	3.0	20
Lead	279	<.5	<.5	<.5	<.5	<.5	<.5	.6	1.9
Mercury	283	<.1	<.1	<.1	<.1	<.1	<.1	.1	.6
Nickel	36	<10	<10	<10	<10	<10	<10	<10	<10
Selenium	22	<1	<1	<1	<1	<1	<1	2	2
Zinc	36	<3	<3	<3	5	12	18	29	30

Note: Data statistically summarized in this table are from monthly and selected hydrologic-event samplings from the Yakima River at Cle Elum, Yakima River at Umtanum, Naches River near North Yakima, Yakima River above Ahtanum Creek at Union Gap, Sulphur Creek Wasteway near Sunnyside, Yakima River at Euclid Bridge at RM 55 near Grandview, and Yakima River at Kiona; to avoid statistical bias that may be associated with constituents analyzed in duplicate or triplicate at a site, only one element concentration per visit was statistically summarized; concentrations of trace elements are in units of micrograms per liter.

TABLE 5
Distribution of Arsenic Concentrations in Filtered-Water Samples at Fixed Sites, Yakima River Basin, Washington (1987–1990)

Site reference number	Site name	No. samples	Min. value	Value at indicated percentile						Max. value
				10	25	50	75	90	95	
6	Cle Elum	16	<1	<1	<1	<1	<1	<1	<1	<1
19	Umtanum	11	<1	<1	<1	<1	<1	<1	<1	<1
26	Naches	15	<1	<1	<1	<1	<1	<1	1	1
32	Union Gap	23	<1	<1	<1	<1	<1	<1	1	1
50	Kiona	25	<1	<1	1	1	2	3	4	4
52	Sulphur Cr.	15	2	2	2	3	7	8	9	9
56	Grandview	14	<1	<1	<1	1	2	2	3	3

Note: To avoid statistical bias that may be associated with constituents analyzed in duplicate or triplicate at a site, only one element concentration per visit was statistically summarized; concentrations are in micrograms per liter. Cle Elum, Yakima River at Cle Elum; Umtanum, Yakima River at Umtanum; Naches, Naches River near North Yakima; Union Gap, Yakima River above Ahtanum Creek at Union Gap; Sulphur Creek, Sulphur Creek Wasteway near Sunnyside; Grandview, Yakima River at Euclid Bridge at RM 55 near Grandview; Kiona, Yakima River at Kiona.

has an annual streamflow representing only about 8% of the annual streamflow at the Kiona fixed site, yet it accounted for nearly 20% of the filtered-As load at Kiona. Comparisons between loads determined from filtered-water samples (an operational approximation of dissolved load) and loads determined from As in suspended sediment, show that most of the As load in the basin is in a dissolved form. For example, the annual dissolved-As loads in the lower Yakima Valley at Sulphur Creek Wasteway, Grandview, and Kiona sites were from four to nine times higher than their respective suspended loads.

Arsenic is present in the aquatic biota of the basin. In curlyleaf pondweed, an aquatic plant, concentrations ranged from 0.48 to 1.5 μg/g and were threefold higher in the main stem in the lower Yakima Valley than in the main stem in the Kittitas Valley. Concentrations of As in caddisflies collected from agricultural drains in 1989 in the lower Yakima Valley were as large as 5.4 μg/g and exceeded the 85th percentile concentration for the basin. Asiatic clams were collected only from the lower Yakima Valley, and As concentrations varied little (3.6 to 4.6 μg/g) among sites. Compared to other studies, however, As concentrations in Asiatic clams of the lower Yakima Valley were an order of magnitude higher than concentrations reported in the Apalachicola River in Florida (Elder and Mattraw, 1984) and at least three times higher than in the Sacramento River Basin of California (McCleneghan et al., 1981), but were similar to As concentrations in Asiatic clams of the San Joaquin River in California, which are considered to be affected by minor anthropogenic sources (Johns and Luoma, 1990; Leland and Scudder, 1990).

SUMMARY

The multimedia sampling approach used to spatially define the distribution of trace elements has been a successful method for identifying sources of trace elements in the Yakima River Basin. Additionally, the method has provided multiple lines of evidence for assessing the effects of geologically derived and anthropologically derived trace elements. Streambed sediment is a long-term integrator of trace element concentrations; it is, therefore, ideal for assessing potential sources of trace elements. The significance of these trace-element sources, however, was better assessed by temporal measurements of suspended and dissolved trace elements at fixed location sampling sites. In the Yakima River Basin, temporal measurements equated to monthly samplings at fixed sites with additional samplings for selected storms. This sampling frequency provided data necessary to assess temporal variations in trace-element concentrations as well as magnitude and flux of daily mean and annual trace-element loads. Synoptic samplings of aquatic biota, in addition to depicting spatial distributions of trace elements, provided a measure of the tendency of trace elements to move up the food chain.

REFERENCES

Arbogast, B. F., Ed., Quality assurance manual for the U.S. Geological Survey Branch of Geochemistry, U.S. Geological Survey Open-File Report 90-668, 1990, 184.

Baker, R. A., Ed., *Contaminants and Sediments, Vol. 1 — Fate and Transport, Case Studies, Modeling, Toxicity,* Ann Arbor Science, Ann Arbor, MI, 1980, 558.

Cain, D. J., Luoma, S. N., Carter, J. L., and Fend, S. V., Aquatic insects as bioindicators of trace element contamination in cobble-bottom rivers and streams, *Can. J. Fish. Aquatic Sci.,* 49 (10), 2141, 1992.

Columbia Basin Inter-Agency Committee, River-mile index, Yakima River, Columbia River Basin Hydrology Subcommittee Report 15, 1964, 39.

Crawford, T. K. and Luoma, S. N., Guidelines for studies of contaminants in biological tissues for the National Water-Quality Assessment program, U.S. Geological Survey Open-File Report 92-492, 1993, 69.

Edwards, T. K. and Glysson, G. D., Field methods for measurement of fluvial sediment, U.S. Geological Survey Open-File Report 86-531, 1988, 118.

Elder, J. F. and Mattraw, H. C., Jr., Accumulation of trace elements, pesticides, and polychlorinated biphenyls in sediments and the clam *Corbicula manilensis* of the Apalachicola River, Florida, *Arch. Environ. Contamin. Toxicol.,* 13, 453, 1984.

Elwood, J. W. and Hildebrand, J. J., Contribution of gut contents to the concentration and body burden of elements in *Tipula* spp. from a spring-fed stream, *J. Fish. Res. Board Canada,* 33, 1930, 1976.

Fishman, M. J. and Friedman, L. C., Eds., Methods for the determination of inorganic substances in water and fluvial sediments, 3rd ed., U.S. Geological Survey Techniques of Water-Resources Investigations, 1989, 545.

Forstner, U. and Wittmann, G. T. W., *Metal Pollution in the Aquatic Environment,* Springer-Verlag, New York, 1979, 486.

Fuhrer, G. J., McKenzie, S. W., Rinella, J. F., Sanzolone, R. F., and Skach, K. A., Surface-water quality assessment of the Yakima River Basin in Washington: analysis of major and minor elements in fine-grained streambed sediment, 1987, with sections on geologic overview and geologic classification of low-order stream sampling sites, by Marshall W. Gannett, U.S. Geological Survey Open-File Report 93-30, 1994a, 226. [Pending publication as U.S. Geological Survey Water-Supply Paper 2354-C].

Fuhrer, G. J., Fluter, S. L., McKenzie, S. W., Rinella, J. F., Crawford, J. K., Cain, D. J., and Skach, K. A., Surface-water-quality assessment of the Yakima River Basin in Washington — Major- and minor-element data for water, sediment, and aquatic biota, 1987– 91, U.S. Geological Survey Open-File Report 94-308, 1994b, 223.

Fuhrer, G. J., Cain, D. J., McKenzie, S. W., Rinella, J. F., Crawford, J. K., Skach, K. A., and Hornberger, M. I., Surface-water-quality assessment of the Yakima River Basin in Washington — Spatial and temporal distribution of trace elements in water, sediment, and aquatic biota, 1987–91, U.S. Geological Survey Open-File Report 95-440, 1996.

Horowitz, A. J., *A Primer on Sediment-Trace Element Chemistry,* 2nd ed., CRC/Lewis, Boca Raton, FL, 1991, 136.

Horton, R.E., Erosional development of streams and their drainage basins; hydrophysical approach to quantitative morphology, *Geol. Soc. Am. Bull.,* 56, 275, 1945.

Johns, C. and Luoma, S. N., Arsenic in benthic bivalves of San Francisco Bay and the Sacramento/San Joaquin River Delta, *Sci. Total Environ.,* 97/98, 673, 1990.

Leland, H. V. and Scudder, B. C., Trace elements in *Corbicula fluminea* from the San Joaquin River, California, *Sci. Total Environ.,* 97/98, 641, 1990.

Levinson, A. A., *Introduction to Exploration Geochemistry,* 2nd ed., Applies Publishing, Wilmette, IL, 1980, 924.

Lynch, T. R., Popp, C. J., and Jacobi, G. Z., Aquatic insects as environmental monitors of trace metal contamination: Red River, New Mexico, *Water Air Soil Pollut.,* 42, 19, 1988.

McCleneghan, K., Meinz, T. M., Crane, D., Setu, W., and Lew, T., Toxic substances monitoring program 1980, State of California, State Water Resources Control Board Water Quality Monitoring Report No. 81-8TS, 1981, 54.

Peryea, F. J., Leaching of lead and arsenic in soils contaminated with lead arsenate pesticide residues, Washington State Water Research Center Project No. A-158-WASH, 1989, 59.

Rankama, K. and Sahama, T. G., *Geochemistry,* University of Chicago Press, Chicago, IL, 1950, 911.

Ryder, J. L., Sanzolone, R. F., Fuhrer, G. J., and Mosier, E. L., Surface-water-quality assessment of the Yakima River Basin in Washington: Chemical analyses of major, minor, and trace elements in fine-grained streambed sediment, U.S. Geological Survey Open-File Report 92-520, 1992, 60.

Smock, L. A., The influence of feeding habits on whole-body metal concentrations in aquatic insects, *Freshwater Biol.,* 13, 301, 1983.

U.S. Department of Agriculture, Washington fruit survey, 1986, Olympia, Washington Agricultural Statistics Service, 1986, 73.

U.S. Geological Survey, Mineral and water resources of Oregon, U.S. Geological Survey for Committee on Interior and Insular Affairs, United States Senate, 90th Congress, 2nd session, 1969, 462.

Wershaw, R. L., Fishman, M. J., Grabbe, R. R., and Lowe, L. E., Eds., Methods for the determination of organic substances in water and fluvial sediments, U.S. Geological Survey Techniques of Water-Resources Investigations, 1987, 80.

15 Colloidal Iron and Its Effect on Phosphorous Dynamics in the Tualatin River Basin, Oregon

T. D. Mayer and W. M. Jarrell

INTRODUCTION

An important consequence of the oxidation of Fe(II) in aquatic systems is the formation of colloid-sized particles. Colloids are particles with at least one linear dimension of 0.001 to 1.0 μm (Hiemenz, 1986). For particles of this size, the effects of gravitational forces are small compared to the effects of Brownian motion. As a result, they are very slow to settle out of solution. In addition, as particle dimensions are reduced, the surface area per unit mass of material increases. This is an important consideration for surface-reactive species, such as P, that may interact with Fe oxides.

In general, colloid-associated elements or compounds will behave differently from species that are truly dissolved or associated with larger, settleable particles. Colloid-associated material may be transported much further than expected or predicted due to slower settling characteristics (O'Melia, 1980; McCarthy and Zachara, 1989). Longer residence times in the water column may increase the bioavailability of colloid-associated material (Williams et al., 1980; Kuwabara et al., 1986). Management strategies, such as settling ponds or constructed wetlands, may not be effective at removing colloidal sediments or colloid-associated pollutants.

Colloidal Fe oxides have been described in several systems including lakes, rivers, estuaries, and groundwaters (Tipping et al., 1989; Laxen and Chandler, 1983; Gschwend and Reynolds, 1987; Leppard et al., 1988; Fox, 1989). In general, the colloids ranged from 0.05 to 0.5 μm in diameter and contained P, Ca, Si, and Al in addition to Fe. In several of the lakes, the Fe oxide colloids were found in redox transition zones, suggesting that they resulted from the oxidation of ferrous Fe.

FILTRATION ARTIFACTS

Colloids may have been ignored or underestimated in the past due to analytical difficulties in their measurement and identification. Filtration, the most commonly used method for separation of soluble and solid phases (Laxen and Chandler, 1982), can underestimate the extent of colloidal constituents in two ways. First, colloids can be small enough to pass through the filter pores and, if detected analytically, be included in the "dissolved" phase (Kennedy et al., 1974; Danielsson, 1982; Tarapchak et al., 1982). Second, filters can retain colloids that are much smaller than the filter pores due to high flow rates during filtration, low permeability of the filters, and clogging at high filter loads (Buffle et al., 1992; Horowitz et al., 1992). Our research indicates that both of these artifacts occur simultaneously during filtration and subsequent filtrate analysis.

In the field of water science, a 0.4 or 0.45 μm pore size filter is frequently used to separate soluble and dissolved phases. But this is larger than the lower diameter limit (0.05 μm) of Fe oxide colloids described in the literature. We compared P and Fe concentrations in the filtrates of 0.4 and 0.05 μm pore size membrane filters for several surface water samples. Phosphorus and Fe concentrations were mostly higher in the 0.4 μm filtrate (Figure 1), especially in the case of Fe, indicating

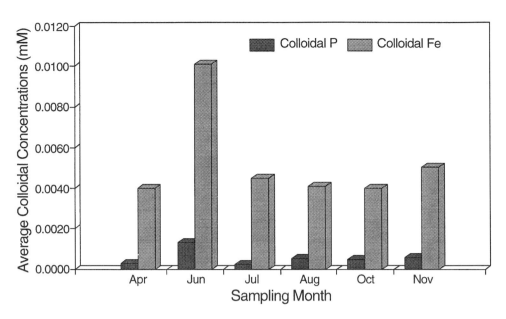

FIGURE 2. Average colloidal concentrations of P and Fe in the Tualatin River Basin for five of the six sites sampled.

The lower portion of the mainstem of the river is characterized by low flow and high biological activity. Colloidal concentrations of P and Fe in this section were very low during July, August, and October, essentially going to zero in the case of colloidal P. Chlorophyll *a* concentrations increased dramatically during the same period at this site (Unified Sewerage Agency, unpublished data). This suggests that colloidal P and Fe may have been removed by biological activity, either due to nutrient uptake, dissolution, or desorption under the higher pH from photosynthesis, or settling of attached algal masses.

Colloidal P and colloidal Fe correlated fairly well ($r^2 = 0.56$; $p < 0.001$) in the Tualatin Basin samples. Including data from a preliminary study collected on a single tributary in 1992 increased the number of data points in the higher concentration range and improved the correlation ($r^2 = 0.83$; $p < 0.001$). Figure 3 shows the correlation of colloidal P and colloidal Fe for both studies.

Significantly, there was no correlation between P (total, colloidal, or soluble) and suspended sediment. If surface runoff was a significant source of P, as has been documented in other studies (Sharpley, 1980), then such a correlation should have existed. In this system, sediment and sub-surface sources of P may be just as important as surface sources, and control of P inputs may require more than just the elimination of surface runoff. Data from a preliminary study in 1992 indicated that colloidal P and Fe were more prevalent during low flow and low dissolved O_2 periods, confirming that surface runoff was not the dominant source of colloids.

Total P and total Fe were poorly correlated, despite the correlation between colloidal P and colloidal Fe. While P and Fe in the larger size fraction are undoubtedly associated with each other to some extent, they probably occur as other forms as well.

LABORATORY EXPERIMENTS

Much more P was associated with the Fe solid phase when the P was added initially ("coprecipitation" experiments) rather than added after solid formation ("adsorption" experiments). However, much of this solid phase P was released back into solution as the Fe oxide recrystallized. This release continued for 200 hr or more. Other studies have described the release of coprecipitated species during the recrystallization of Fe oxides (Fuller et al., 1993; Schwertmann and Thalmann,

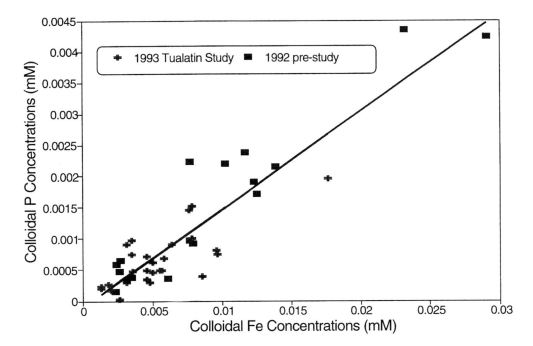

FIGURE 3.   Correlation of colloidal P and Fe. Data from the Tualatin study and a preliminary study of one tributary in the Tualatin Basin.

1976). In contrast, equilibrium was reached in the adsorption experiments after 24 hr; there was no change in soluble P concentrations after this time.

The presence of silicate during the coprecipitation of P and Fe slowed the release of P substantially. Silicate may inhibit the recrystallization of the oxide (Schwertmann and Thalmann, 1976). This effect of silicate is important because many systems, including the Tualatin River, have high background concentrations of silicate. In such systems, the coprecipitation of P and Fe will occur in the presence of silicate.

Data from several coprecipitation experiments, with silicate, and the adsorption experiment with the highest observed P sorption density are presented in Figure 4. The maximum sorption densities observed in the adsorption experiments are very similar to values reported by other studies at similar concentrations and pH (Dzombak and Morel, 1990). At a given concentration of soluble P, the coprecipitation experiments prescribe a much higher P/Fe molar ratio than the adsorption experiments.

When the colloid data from the Tualatin study are included in the same graph, the data clearly fall in the range of the coprecipitation experiments (Figure 4). This is evidence that colloidal P and Fe in this system formed through the oxidation of Fe(II) and coprecipitation of Fe(III) and P. Silicate plays an important role in inhibiting the recrystallization of the Fe oxides and stabilizing the colloids at high P/Fe molar ratios.

CONCLUSIONS

A significant fraction of the P and Fe oxides in suspended sediments of the Tualatin River system are in the colloid size fraction (0.05 to 1.0 μm). The mass percent of P and Fe in colloidal form ranged from 2 to 77% and 0 to 48% of the total P and Fe. Phosphorus correlated well with Fe concentrations in the colloidal size fraction but not in the larger size fraction. Laboratory experiments with synthetic solutions suggest that the association of colloidal P and Fe occurs through a

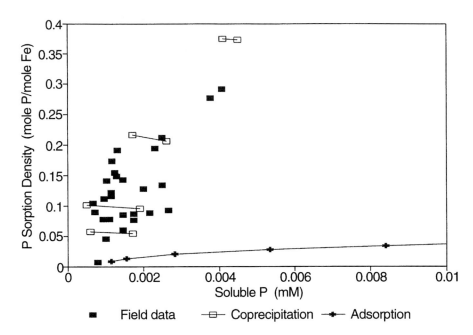

FIGURE 4. Comparison of laboratory results from adsorption and coprecipitation experiments (terms defined in text) and field samples from the Tualatin River Basin. Since sorption densities and soluble P values changed over time in the coprecipitation experiments, individual experiments are represented by a line joining the values at 24 hr and 73–165 hr into each experiment.

coprecipitation process. Silicate may be important in stabilizing the colloids and inhibiting the release of solid phase P during recrystallization of the Fe oxides.

REFERENCES

Abrams, M. M. and Jarrell, W. M., Soil phosphorus as a potential nonpoint source for elevated stream phosphorus levels, *J. Environ. Qual.*, 24, 132, 1995.

APHA, *Standard Methods for the Examination of Water and Wastewater,* Amer. Public Health Assoc., New York, 1989, chap. 4.

Buffle, J., Perret, D., and Newman, M., The use of filtration and ultrafiltration for size fractionation of aquatic particles, colloids, and macromolecules, in *Environmental Particles*, Buffle, J. and van Leeuwen, H., Eds., Lewis Publishers, Boca Raton, FL, 1992, chap. 5.

Crosby, S. A., Butler, E. I., Turner, D. R., Whitfield, M., Glasson, D. R., and Millward, G. E., Phosphate adsorption onto iron hydroxides at natural concentrations, *Environ. Technol. Lett.*, 2, 371, 1981.

Danielsson, L. G., On the use of filters to distinguish between dissolved and particulate fractions in natural waters, *Water Res.*, 16, 179, 1982.

Dzombak, D. A. and Morel, F. M., *Surface Complex Modeling: Hydrous Ferric Oxide,* John Wiley & Sons, New York, 1990, chap. 7.

Fox, L. E., A model for inorganic control of phosphate concentrations in river waters, *Geochim. Cosmochim. Acta*, 53, 417, 1989.

Fuller, C. C., Davis, J. A., and Waychunas, G. A., Surface chemistry of ferrihydrite: Part 2. Kinetics of arsenate adsorption and coprecipitation, *Geochim. Cosmochim. Acta*, 57, 2271, 1993.

Gschwend, P. M. and Reynolds, M. D., Monodisperse ferrous phosphate colloids in an anoxic groundwater plume, *J. Contam. Hydrol.*, 1, 309, 1987.

Hiemenz, P. C., *Principles of Colloid and Surface Chemistry,* Marcel Dekker, New York, 1986, chap. 1.

Honeyman, B. D. and Santschi, P. H., The role of particles and colloids in the transport of radionuclides and trace metals in the oceans. in *Environmental Particles*, Buffle, J. and van Leeuwen, H., Eds., Lewis Publishers, Boca Raton, FL, 1992, chap. 10.

Horowitz, A. J., Elrick, K. A., and Colberg, M. R., The effect of membrane filtration on dissolved trace element concentrations, *Water Res.*, 26, 753, 1992.

Kennedy, V. C., Zellweger, G. W., and Jones, B. F., Filter pore size effects on the analysis of Al, Fe, Mn, and Ti in water, *Water Resour. Res.*, 10, 785, 1974.

Kuwabara, J. S., Davis J. A., and Chang, C. C. Y., Algal growth response to particle-bound orthophosphate and zinc, *Limnol. Oceanogr.*, 31, 503, 1986.

Laxen, D. P. H. and Chandler, I. M., Comparison of filtration techniques for size distribution in freshwaters, *Anal. Chem.*, 54, 1350, 1982.

Laxen, D. P. H. and Chandler, I. M., Size distribution of iron and manganese species in freshwaters, *Geochim. Cosmochim. Acta*, 7, 731, 1983.

Leppard, G. G., Buffle, F., DeVitre, F. F., and Perret, D., The ultrastructure and physical characteristics of a distinctive colloidal iron particulate isolated from a small eutrophic lake, *Arch. Hydrobiol.*, 113, 405, 1988.

Mayer, T. D. and Jarrell, W. M., Formation and stability of iron (II) oxidation products under natural concentrations of dissolved silica, *Water Res.*, 30, 1208, 1996.

McCarthy, J. F. and Zachara J. M., Subsurface transport of contaminants, *Environ. Sci. Technol.*, 23, 496, 1989.

O'Melia, C., Aquasols: The behavior of small particles in aquatic systems, *Environ. Sci. Technol.*, 14, 1052, 1980.

Schwertmann, U. and Thalmann, H., The influence of [Fe(II)], [Si], and pH on the formation of lepidocrocite and ferrihydrite during oxidation of $FeCl_2$ solutions, *Clay Minerals*, 11, 189, 1976.

Sharpley, A. N., The enrichment of soil phosphorus in runoff sediments. *J. Environ. Qual.*, 9, 520, 1980.

Tarapchak, S. J., Bigelow, S. M., and Rubitschun, C., Soluble reactive phosphorus measurements in Lake Michigan: Filtration artifacts, *J. Great Lakes Res.*, 8, 550, 1982.

Tipping, E., Woof, C., and Cooke, D., Iron oxide from a seasonally anoxic lake, *Geochim. Cosmochim. Acta*, 45, 1411, 1989.

Williams, J. D. H., Shear, H., and Thomas, R. L., Availability to *Scenedesmus quadricanda* of different forms of phosphorus in sedimentary materials from the Great Lakes, *Limnol. Oceanogr.*, 25, 1, 1980.

16 Chemical Characteristics of a Seep at the St. Johns Landfill in Portland, Oregon

William Fish, William Romanelli, and Cheryl Martin

INTRODUCTION

From about 1940 until its closure in 1991, the St. Johns Landfill (SJL) was the primary municipal waste disposal site for the Portland, OR metropolitan area. Because of the long use of this site, it is important that the closed landfill be managed in a way that minimizes the risk of materials escaping from the site and affecting adjacent ecosystems, i.e., the waterways and wetlands of the Columbia Slough system. Through the slough, the entire system is connected hydrologically to the Willamette and Columbia Rivers (Figure 1).

Recently, the landfill has been under the jurisdiction of Metro, a regional governmental agency. Since closure, Metro has made extensive efforts to ensure the long-term environmental integrity of the site. For example, the surface of the landfill is being sealed with a multilayer cap that should cut off >95% of the water input from precipitation. These sensible postclosure measures will sharply diminish movement of leachate from the landfill (Metro, 1995). To monitor for potential releases to the groundwater, some 30 wells in and around the landfill are routinely checked for indications of contamination (Fish, 1993).

An important component of this postclosure work is the modeling of the fate and transport of known and suspected leachate-borne materials in both surface and groundwater. The accuracy and utility of such modeling depends critically on chemical transport parameters such as adsorption coefficients, retardation factors, and source conditions. All potential sources must be considered.

Seeps at the margins of the landfill are a potential connection between the landfill and the surrounding sloughs. The seepage varies spatially and appears to focus at specific locations around the perimeter (Metro, 1995). Previous surveys visually identified the apparent seeps and documented their locations and surficial characteristics. In those surveys a limited number of grab samples of seep water were analyzed, revealing various levels of constituents typical of leachate.

These isolated data are of little practical use because such point samples give no indication of the spatial variability across a seepage zone; a single point rarely is representative of an entire seep. Furthermore, these data do not capture essential information about temporal variability. Seep water composition is likely to fluctuate widely with time due to the direct influence of rainfall and near-surface baseflow. This chapter describes a detailed examination of the chemistry of a seepage zone at the site.

LEACHATE CATIONS AND SOIL EXCHANGER PROPERTIES

Ammonia (NH_3) is a common by-product of the anaerobic breakdown of nitrogenous wastes in a municipal landfill (Antonopoulos, 1984). The chemical speciation of ammonia is important to its transport properties and is sensitive to pH. Leachate at the SJL site appears to be well buffered in

17 Determination of Organochlorine Compounds in Water from the Vistula River and Seawater from the Gulf of Gdańsk

Jacek Czerwiński, Marek Biziuk, Jacek Namieśnik and Pat Sandra

INTRODUCTION

At present organohalogen compounds are among the most dangerous environmental pollutants, due to their toxic, mutagenic, and carcinogenic properties (Dojlido, 1987; Helman, 1987; de Kruijf and Kool, 1985; Fresenius et al., 1988; Moore and Ramamoorthy, 1984; Nemerow, 1985). The problem also concerns volatile organohalogen compounds such as pesticides used in agriculture, forestry, and home use. These compounds do not usually occur in nature, hence their content is a measure of anthropogenic environmental pollution. Volatile organohalogen compounds (the industrial products and by-products of water disinfection by chlorination) enter rivers and seas through the municipal and industrial sewage. Pesticides are washed out from soil by rain and groundwater and flow into rivers and seas. The Baltic Sea is especially threatened in this respect. The threat results from the topography of the Baltic Sea, which is an inland sea that is characterized by poor water exchange with the North Sea.

EXPERIMENTAL

Due to its simplicity, speed of preconcentration, minimal water adsorption, and the absence of a solvent background, adsorption on solid sorbents is used for preconcentration of organic compounds from water. The compounds sorbed can be released from the sorbent layer by thermal desorption or solvent extraction. There are two approaches to the determination of organohalogen compounds: determination of group parameters such as VOX (Volatile Organic Halogen) or TOX (Total Organic Halogen) (Namieśnik et al., 1990; Biziuk and Polkowska, 1990; Biziuk et al., 1991, 1993), or determination of individual compounds. We have used thermal desorption for the determination of the group parameter VOX (expressed as chlorine), which can serve as indicator of anthropogenic water pollution by organohalogen compounds (Biziuk and Polkowska, 1990; Biziuk et al., 1991, 1993). The organohalogen compounds to be analyzed were sorbed on an Amberlite XAD-4 (Aldrich, USA) layer (3.5 g of XAD-4 in a glass tube). The sample (maximum volume 10 L) was pumped through sorbent layer by a peristaltic pump at a flow rate of 50 mL/min. The trapped compounds were desorbed from the sorbent bed at 200°C for 40 min in a stream of purified argon at a flow rate of 20 mL/min. The desorbed organohalogen compounds (in a stream of argon) were mixed

0-56670-138-4/97/$0.00+$.50
© 1997 by CRC Press, Inc.

TABLE 1
Results of VOX Determination in the Surface Water from the Gdańsk District

Sampling site	Sampling date	VOX (µg Cl/L)
Kacza River in Orłowo	90.04.19	12
Borowo Lake	90.06.25	ND
	90.07.23	ND
	90.10.12	0.6
	90.11.28	1
Straszyn Reservoir	90.06.25	2.4
	90.07.25	2
	90.10.19	3.1
	90.12.07	1.8
Vistula River in Kiezmark	88.05.26	16.3
	89.05.24	11.6
	90.04.04	40.2
	90.06.29	18.6
	90.07.27	13.6
	90.10.17	7.9
	90.12.05	4.6
Baltic Sea, Gulf of Gdańsk	89.06.01	1.8
Baltic Sea, Pomerania Bay	89.06.02	2.0
Baltic Sea, Depth of Gdańsk	89.06.01	0.2
Baltic Sea, open sea	89.06.03	ND
Gdynia Orłowo, quay	88.08.08	0.6
	89.05.11	4.5
	90.04.19	4.3
	90.05.17	7.2
Sopot, quay	88.07.01	0.7
	89.10.25	1.3
	90.03.27	1.3
	90.05.10	3.2
	90.05.31	2.9

Note: ND, not detected.

---

with oxygen and combusted in an empty quartz tube heated at 900°C. The products of mineralization were washed out from the tube outlet and all organic halogen was determined as halogenide ions by a coulometric argentometric titration to a preset potential. In order to determine VOX in natural samples, the sorbent layer was additionally washed after sorption, with 500 ml of distilled and purified water to remove inorganic salts. Preconcentration of the analytes on the sorbent bed was carried out directly at the sampling site. The final determination was performed in the laboratory.

Table 1 presents the results of VOX determinations of water samples collected from the Vistula River near Kiezmark, the Kacza River which enters the Baltic Sea, Borowo Lake situated in an agricultural region, the Baltic Sea (determinations were carried out during a research cruise of the yacht r/v "Oceania" by the Baltic Sea), and sea water from Sopot and Orłowo quays. We have found a relatively high anthropogenic pollution of the Vistula River (c_{VOX} = 11 to 45 µg Cl/L). Depending on the season, the instantaneous VOX concentration can be very high. We have found a very high VOX concentration in the water samples from the Kacza River. This small river is highly polluted due to municipal and industrial sewages. The Borowo Lake, situated in an agricultural region, is not very polluted by volatile organohalogen compounds, which probably originated from rain and small camps around this lake. Very high concentrations of VOX in sea water have been found near outlets of big rivers such as Vistula or Odra or even small rivers such as Kacza in Gdynia Orłowo.

VOX is a very good and often used group indicator, but generally we like to know the concentration of individual compounds. This is possible after sorption on a solid sorbent, solvent extraction and chromatographic determination (Namieśnik et al., 1990; Biziuk et al., 1993). In our case, the compounds were concentrated from 0.5 L of water sample at a flow rate of 15 mL/min on a sorbent layer (0.8 g of XAD-4) (Biziuk et al., 1993). The trapped compounds were extracted with 10 mL of nanograde pentane (J. T. Baker, Germany) at a flow rate of 0.5 mL/min. The extracts were evaporated to a volume of 0.5 mL. Extracts were analyzed by the GC-ECD. Due to the possibility of multiple injections of one extract to a chromatograph, this method enables the study of extracts using various capillary columns to determine volatile or nonvolatile (pesticides) organohalogen compounds.

The determination of volatile organohalogen compounds were performed using a VEGA 6180 gas chromatograph (Carlo Erba, Italy), equipped with ECD-40 (electron capture detector), with ^{63}Ni as source, operated in the constant current mode at 350°C. One or two ml samples were injected cold on-column onto a 30 mL × 0.32-mm I.D. column coated with a 5 mm d_f film of the apolar stationary phase PS-255, through a deactivated 2 mL × 0.32-mm I.D. retention gap. Samples were analyzed isothermally at 85°C using hydrogen as carrier gas at a linear velocity of 40 cm/s. Nitrogen (99.999%) at a flow of 50 mL/min was used as the make-up gas.

Table 2 lists the results of the volatile organohalogen compounds in the water samples taken from Vistula River, Borowo Lake, and Straszyn Reservoir (water intake of Gdańsk). We have found trichloromethane ($CHCl_3$), trichloroethylene (C_2HCO_3), and tetrachloromethane (CCl_4) in these samples. The highest concentrations of trichloromethane were in the Vistula River and in Straszyn Reservoir. The Straszyn Reservoir also had the highest concentration of trichloroethylene.

Using the same extracts, we performed determinations of pesticides using different GC columns and different conditions, and different instruments in Belgium (Department of Organic Chemistry, University of Ghent).

Determination of organochlorine pesticides were performed using GC-8000 gas chromatograph (Fisons, Italy), equipped with ECD-80 (electron capture detector), with ^{63}Ni as source, operated in the constant current mode at 350°C. One or two ml samples were injected cold on-column onto a 60 mL × 0.32-mm I.D. column coated with a 0.25 μm d_f film of the apolar stationary phase SE-

---

TABLE 2
Volatile Organochlorine Compounds in the Vistula River, Borowo Lake, and Impoundment Reservoir in Straszyn

Sampling site	Sampling date	$CHCl_3$	C_2HCl_3	CCl_4	Σ
Vistula River in Kiezmark	90.10.17	0.41	0.07	0.05	0.53
	90.06.25	0.57	0.24	0.03	0.84
Borowo Lake	90.06.28	0.15	0.13	0.05	0.33
	90.07.23	0.22	0.44	0.14	0.8
	90.10.12	0.16	0.55	0.05	0.76
	90.11.28	0.18	0.11	0.02	0.31
Straszyn Reservoir	90.06.25	0.36	1.38	0.08	1.82
	90.07.25	0.62	1.16	0.38	2.16
	90.10.19	0.15	0.5	0.02	0.67
	90.12.07	0.35	1.31	0.17	1.38

Note: Measurements are in micrograms per liter.

TABLE 3
Organochlorine Pesticides in Surface Water and Drinking Water Intakes from the Gdańsk District

Sample site	Sampling date	Lindane	p,p' DDT	Methoxychlor
Surface water intakes				
Vistula River in Kiezmark	92.05.03	19	ND	21
	92.09.17	11	ND	13
Seawater from Orłowo (quay)	92.05.03	6.5	0.7	0.3
	92.09.17	4	0.6	ND
Seawater from Sopot (quay)	92.05.03	3.2	ND	0.2
	92.09.17	3	ND	ND
Kacza River in Orłowo	92.05.03	1.2	ND	1.1
	92.09.17	0.7	ND	1.2
Drinking water intakes				
Bitwy pod Płowcami	93.07.08	27	80	ND
	93.12.21	ND	ND	ND
Dolina Radości	93.07.08	ND	ND	ND
	93.12.21	ND	ND	ND
Zaspa	93.07.08	ND	90	ND
	93.12.21	0.1	ND	ND
	94.06.07	2	4	ND
Czarny Dwór	93.07.08	ND	ND	ND
	93.12.21	ND	ND	ND
	94.06.07	11	ND	ND
Pręgowo	93.12.21	76	ND	ND
	94.06.06	ND	ND	ND
Straszyn: reservoir	93.07.08	ND	ND	ND
	93.12.22	77	34	ND
	94.06.07	19	ND	8
Straszyn: treated water	93.07.08	ND	ND	ND
	93.12.22	ND	ND	ND
	94.06.07	2	ND	ND

Note: ND, not detected. Measurements are expressed in nanograms per liter.

54, through a deactivated 2 mL × 0.32-mm I.D. retention gap. Samples were analyzed at the following temperature conditions:

$$50°C \rightarrow 15°C/min \rightarrow 170°C \rightarrow 5°C/min \rightarrow 260°C \ (15 \ min)$$

using hydrogen as carrier gas at a linear velocity of 40 cm/s. Argon with 5% methane (99.999%) at a flow of 50 mL/min was used as the make-up gas.

The pesticide concentration results are listed in Table 3. If the high concentrations of lindane and methoxychlor in the Vistula River are believable, the relatively high concentrations of these analytes in the sea water in Gdynia Orłowo are striking.

Table 4 presents results obtained for the surface waters from the Gdańsk District. The analyses have been carried out after sorption on homemade columns packed with Amberlite XAD-4 and on commercially available Supelclean LC C-18 cartridges. The analytes were extracted with pentane and analyzed by GC-ECD as previously described. The results obtained for XAD and C-18 solid

TABLE 4

Comparison in Results of Organochlorine Pesticides Determination in Surface Water from the Gdańsk District Using Different Sorbents

Sampling site	Sampling date	Sorbent	Lindane	p,p′ DDT	Methoxychlor
Vistula River in Kiezmark	0.7.03.94	C-18	43	ND	29
		XAD-4	48	<10	24
	14.03.94	C-18	60	16	30
		XAD-4	56	<10	32
	09.07.94	C-18	24	13	ND
Motława River in Gdańsk	09.07.94	C-18	18	8	4
Radunia River in Straszyn	09.07.94	C-18	17	ND	4
Seawater Orłowo (quay)	07.03.94	C-18	31	ND	20
		XAD-4	27	12	17
	14.03.94	C-18	35	12	19
		XAD-4	33	<10	21

Note: ND, not detected. Measurements are expressed in nanograms per liter.

sorbents are comparable. In the analyzed waters, the presence of lindane, p,p′ DDT, and metoxychlor were detected. These pesticides were found in the surface-water intake in Straszyn Lake, the drainage-water intake in Pręgowo, and even in the underground water intake from the street Bitwy pod Płowcami. This is the same intake where we have also found a relatively high concentration of trichloroethylene.

CONCLUSIONS

The studied surface waters, rivers, and lakes were found to be very polluted by anthropogenic compounds originating from municipal and industrial sewage and from agriculture. This results in the potential pollution of sea water and even underground and tap waters. These compounds are all very dangerous for human health and for the environment. We need permanent control of the water pollution and an improvement of low-cost technologies to reduce this pollution.

ACKNOWLEDGMENT

Jacek Czerwiński thanks the European Community for a study grant (TEMPUS project JEP 0379-92/3).

REFERENCES

Biziuk, M. and Polkowska, Ż., Determination of volatile organic halogens in water after sorption on XAD-4 and thermal desorption, *Analyst,* 115, 393, 1990.

Biziuk, M., Kozłowski, E., and Błasiak, A., Determination of volatile halogenated compounds in tap and surface waters from the Gdańsk District, *Int. J. Environ. Anal. Chem.*, 44, 147, 1991.

Biziuk, M., Czerwiński, J., and Kozłowski, E., Identification and determination of organohalogen compounds in swimming pool water, *Int. J. Environ. Anal. Chem.*, 50, 109, 1993.

de Kruijf, H. A. M. and Kool, J. H., Eds., *Organic Micropollutants in Drinking Water and Health*, Elsevier, Amsterdam, 1985.

Dojlido, J., *Chemia Wody*, Arkady, Warsaw, 1987.

18 Ambient Water-Quality Monitoring in the Willamette River Basin

Gregory A. Pettit

INTRODUCTION

The beautiful Willamette River Basin, located in the fertile northwest of Oregon, is vitally important to the state. The basin supports over two thirds of Oregon's population and supplies water to much of the state's industry. The river is also a valuable resource for drinking water, fishing, irrigation, recreation, wildlife habitat, and commerce.

The Willamette Basin drains an area of 31,000 km². Oregon's three largest cities (Portland, Eugene, and Salem) are located in the basin. The Willamette joins the Columbia River just below Portland. The Willamette Valley, extending from Eugene to Portland (160 km), was the destination of most of the early settlers that came to Oregon. Land use in the basin is 60% forests, 35% agricultural, and 5% urban.

On the east side of the Willamette Valley are the Cascade Mountains with elevations ranging from 3400 to 1000 m. These mountains receive up to 5 m of precipitation a year, much of which falls as snow in the winter. Snowpacks are commonly 1 to 2 m and can reach depths of 6 m or more in the higher mountains. Spring and summer runoff from snowmelt is an important source of water for the Willamette River. On the west side of the Willamette Valley are the Coastal Mountains. These mountains are lower (500 to 1000 m) and receive most of their precipitation in the winter as rainfall (2 to 3 m/yr). Annual precipitation in the valley floor is usually around 1 m and falls primarily as rain during fall, winter, and spring.

HISTORY

As the population and industrial activity in the early part of this century increased, the quality of the Willamette River declined. In the 1930s the discharge of untreated sewage and wastes from industries, canneries, and slaughterhouses reached such a level that the river was not fit for swimming. Fish were sometimes killed within minutes of contact. The people of Oregon responded in 1938 by passing an initiative that created the State Sanitary Authority to restore the Willamette River and manage water quality throughout the state. By the early 1970s the Willamette's water was once again fit for recreation. Steelhead and salmon returned to the river in large numbers.

WATER-QUALITY MONITORING

The Oregon Department of Environmental Quality is responsible for monitoring and maintaining Oregon's water quality. Monitoring activities are conducted by the Water Quality Monitoring Section of the Laboratory Division of the agency. The Water Quality Monitoring Section is composed of three units: (1) Surface Water Monitoring, (2) Groundwater Monitoring, and (3) Biological Monitoring.

The Surface Water Unit conducts ambient river monitoring, estuary monitoring, total maximum daily load (TMDL) studies, and other special studies. This chapter focuses on ambient river monitoring and TMDL studies in the Willamette River Basin.

WHY DO WE MONITOR WATER QUALITY?

Determine Beneficial Use Protection

Water quality management in Oregon is based on the concept of protecting the beneficial uses of the water. Standards are set based on the level of water quality necessary to provide for a designated beneficial use. Monitoring provides data that can be compared to established standards and thereby determine if water quality is adequate to support designated beneficial uses of the water.

The beneficial uses for which water quality shall be managed are defined in the Oregon Revised Statute 468.710:

Water supply
  Domestic
  Public
  Industrial
  Municipal
Propagation
  Wildlife
  Aquatic Life
  Fish
Recreation
  Fishing
  Boating
  Water Contact Recreation

Long-Term Trending

A consistent ambient water quality monitoring program is essential to conducting long-term water quality trending. Trending is important because it answers questions regarding the long-term effectiveness of the overall water quality management plan. Trending can indicate progress, or lack of progress, in improving water quality. It can also identify deteriorating water quality that may be caused by population growth or other factors.

Problem Identification

A water quality monitoring program can help identify sources of contamination and quantify the impacts of particular point or nonpoint sources. This information can help establish water quality management priorities that will result in the most effective water-quality restoration and protection activities. Problems identified may be addressed through individual permits or specific management strategies.

Develop Models

Conceptual and mathematical models are used to determine what levels of treatment are necessary to meet water quality standards. These models are then used to establish specific load allocations for point and nonpoint sources of pollution in a basin. In order for these models to be properly

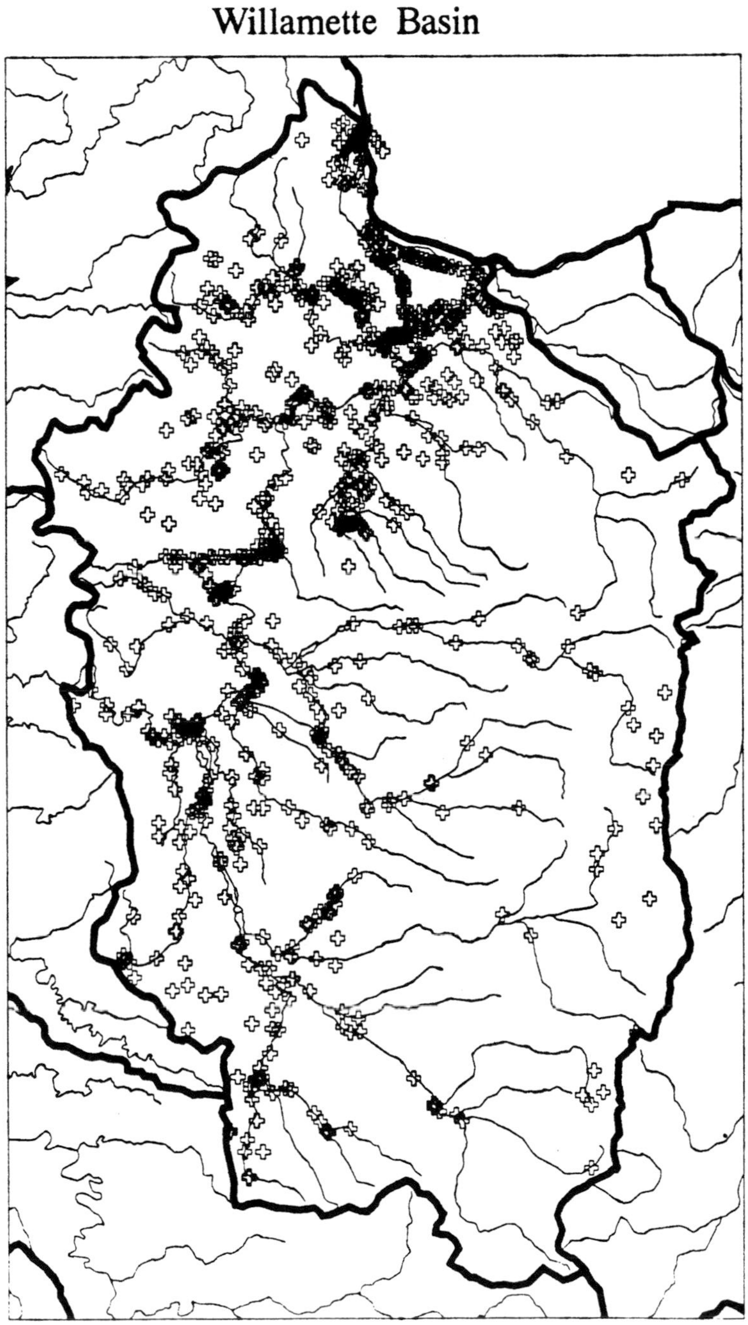

Willamette Basin

FIGURE 1.   Water quality monitoring sites.

calibrated and to work reliably it is essential that accurate data be obtained through monitoring for the various parameters that are being used in the model.

AMBIENT MONITORING NETWORK

There are 837 sampling surface water sampling sites in the Willamette Valley that have been sampled at least once (Figure 1). The current ambient monitoring network consists of 15 key sites that are

Willamette Basin

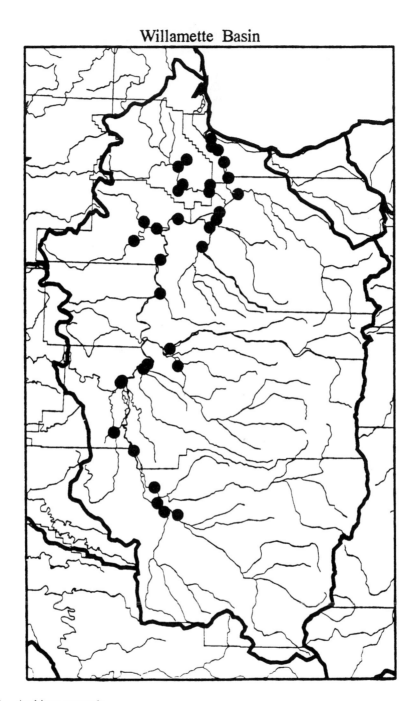

FIGURE 2. Ambient network.

sampled monthly and 19 additional secondary sites that are sampled quarterly (Figure 2). These sites are part of a statewide ambient water-quality monitoring network of 150 sites.

Table 1 contains parameters that are measured for all samples.

SAMPLING METHODOLOGIES

Samples are collected using protocols prescribed in the Water Quality Monitoring Section Mode of Operations Manual. Samples are analyzed by the DEQ lab using EPA-approved methods.

TABLE 1
Ambient Network Parameters

Field measurements	Temperature, specific conductivity, dissolved oxygen (DO), dissolved oxygen % saturation, pH, total alkalinity
Laboratory measurements	BOD_5, chorophyll *a*, COD, fecal coliform, Enterococcus, ammonia, nitrate, TKN, total phosphorous, dissolved phosphorous, suspended solids, total solids, TOC, turbidity, calcium, color, hardness, iron, magnesium, manganese, potassium, sodium, TOX, aluminum, lithium

Samples are collected primarily from road bridges and then gathered in specially designed sampling buckets. These weighted buckets are made of stainless steel and hold the sample bottles. Bottles are placed into the bucket for nonacidified parameters, acidified parameters, bacteria, dissolved oxygen (DO), biochemical oxygen demand (BOD), and TOX. The buckets are designed with filling tubes so that the sample bottles are flushed with sample then filled. The sampling buckets are rinsed with distilled water between sites. These buckets are lowered by rope from the bridges at mid-channel and samples are collected at a depth of 1 m below the surface.

Duplicate samples are taken at 10% of the sites or once per day whichever is more frequent. An equipment blank is collected once per trip. Conductivity and pH are duplicated in the laboratory for 10% of the samples. Chemists maintain historical files for all sites and examine data for outliers before data is released. Quality assurance and control data is reviewed by the Lab Quality Assurance Section before data is released.

Samples are analyzed immediately after collection for the field parameters listed in Table 1. Calibration log books are maintained for all field instruments. Instruments are checked and calibrated prior to leaving the lab and at least once daily while in the field. Dissolved oxygen is determined by Winkler Titration. Chlorophyll *a* samples are filtered in the field and frozen on dry ice immediately. Samples for dissolved analyses are field filtered. Samples are preserved according to analytical protocols, put on ice, and then shipped to the lab.

It typically takes 20 to 30 min at a site to collect the samples and conduct the field work. Five to ten sites a day are usually sampled by a sampling crew of two people.

DATA STORAGE

Once data is released from the lab, it is entered into the U.S. EPA national water quality database STORET (STOrage and RETrieval). This database was created by the U.S. EPA in the 1970s to store water-quality data. The idea behind STORET is that all agencies collecting water-quality data should enter their data into a single database. As a result, anyone who needs to gain access to water-quality data will only need to go to one source to get all available data. The DEQ Water Quality Monitoring Section provides STORET data retrievals free of charge to whoever requests it. Typically, data is provided to requestors three or four times per week. Requestors include: industry; federal, state, and local government agencies; consulting and engineering firms; and public interest and environmental organizations.

DATA EVALUATION

Every 2 years, the DEQ evaluates all of the data that is collected in its ambient water quality monitoring program. This evaluation is published in a report known as the *Water Quality Status Assessment, 305(b) Report.*

Data from the previous ten years for each site is compared to water-quality standards. Appendix A of the report contains this evaluation. Because many water-quality problems are seasonal the evaluation is done on a seasonal basis of summer, fall, winter, and spring. For each sampling location,

TABLE 2
Degree of Beneficial Use Support Monitored
Willamette Basin River Miles

Miles monitored	Miles fully supported	Miles partially supported	Miles not supported
1407	63	424	920

and for each parameter, Appendix A contains data on: number of data points, % of data points exceeding standard, degree of beneficial use support (fully, partial, not), and whether or not the stream has been designated as water quality limited, which means water quality is limiting beneficial uses. Beneficial use support is determined by looking at the percent violation. If the percent of violation is 10% or less, then the site is considered fully supported for that parameter; if the percent violation is 11 to 25% exceedance then the site is considered partially supported for that parameter; and if the percent exceedance is greater than 25% then the site is considered not supported for that parameter. Remember, each parameter relates to one or more specific beneficial uses.

A particular monitoring site or group of sites is used to determine beneficial use support for a specified reach of river. These reaches were determined by looking at the locations of tributaries and sources of pollution. The condition of these reaches is then used to determine beneficial use support in terms of miles of rivers. If any site within a designated reach is not supported for a given parameter, then the entire reach is classified as not supported for that parameter as well. Table 2 provides data on degree of beneficial use support within the Willamette Basin by river miles assessed. Most of the small streams and tributaries in the Willamette Basin are not monitored. Monitoring is concentrated on the mainstem and the lower portions of the major tributaries. Virtually all waters that receive major effluent discharges are monitored.

As can be seen in Table 2, only 4.5% of the reaches monitored in the Willamette Basin are considered fully supported for all parameters at all sites.

Table 3 breaks down the causes of beneficial uses not being supported by parameter. As shown, the most commonly exceeded standard in the Willamette Basin is bacteria. The fecal coliform standard for the Willamette Basin is 400 for a single sample and 200 for a geometric mean of at least five samples collected in a 30-day period. The next most commonly exceeded standard is for DO, and nutrient and pH criteria exceedances are even less frequent. Under the criteria used in the 305b report, a river reach cannot be categorized as "not supported" solely on the basis of nutrient exceedances, only as "partially supported," regardless of the number of exceedances.

Table 4 contains a list of streams in the Willamette River Basin that have been designated as "water quality limited" for pH and dissolved oxygen. Figures 3, 4, and 5 show the locations of streams in the Willamette Basin that are water-quality limited for pH, DO, and bacteria.

TABLE 3
Percentages of Monitored Willamette Basin
River Miles Supported, Partially Supported,
Not Supported by Parameter

Parameter	Fully supported	Partially supported	Not supported
Bacteria	24%	13%	63%
DO	29%	33%	38%
pH	94%	11%	5%
Nutrients	92%	8%	0%

TABLE 4
Willamette Basin Water Quality Limited Streams

Dissolved oxygen	pH
Tualatin	Columbia Slough
Pudding	Johnson Creek
Yamhill	Yamhill River
N. Fork Yamhill	Coast Fork Willamette
S. Fork Yamhill	
Rickreall Creek	
Bashaw Creek	
Luckiamute	
Calapooia	
Santiam	
Coast Fork Willamette	

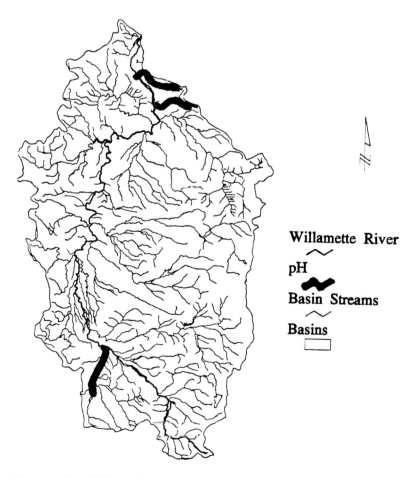

Willamette River

pH

Basin Streams

Basins

FIGURE 3. Water quality, pH limited.

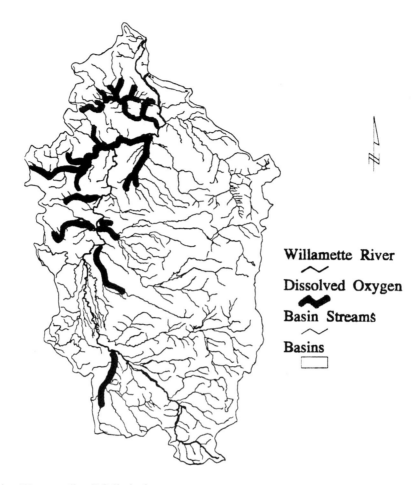

FIGURE 4.   Water quality, DO limited.

TRENDING

METHODS

One of the most important functions of an ambient water-quality monitoring program is to provide data to conduct long-term trending. The objective of most water-quality management plans is to maintain or improve water quality. Progress toward this goal can only be determined by a statistically valid trending analysis of data. Because trending is a fairly rigorous exercise and resources are limited, trending analysis is conducted only on selected sites and parameters.

Three sites in the lower Willamette River [Salem at river mile (RM) 83.9, Newberg Pool at RM 48.6, and Portland Harbor at RM 7.0] were selected for assessment. One site on the Tualatin River, a major tributary of the Willamette, at RM 16.0 was also selected.

In addition to routine monitoring, intensive surveys have also been conducted. At stations which had multiple monthly observations, a statistical process was used to select one value to represent the data for the month.

Seven key water quality parameters were selected for trend analysis: DO, BOD, temperature, ortho and total phosphorus, ammonia nitrogen as nitrogen (N), and inorganic nitrogen. The trend tests were run on the annual data sets and on seasonal subsets (May to October and November to April). For time periods less than 10 years the trending technique used for this assessment is the *Seasonal Kendall Test* (Hirsch et al., 1982). For time series greater than 10 years the *Seasonal Kendall Test* with correction for serial correlation (Hirsch and Slack, 1984) was used.

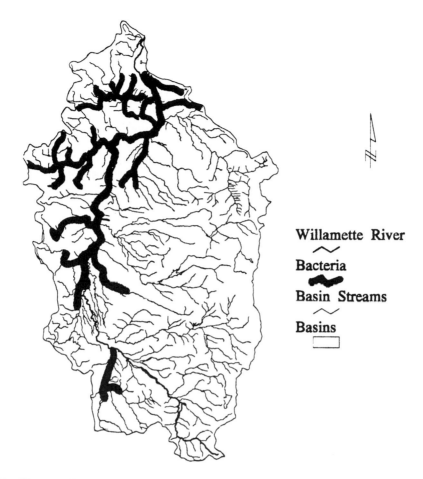

Willamette River
~

Bacteria
━

Basin Streams
~

Basins
▭

FIGURE 5. Water quality, bacteria limited.

A seasonal version of Sen's nonparametric method (Sen, 1986) was used to estimate the magnitude of the slope. These nonparametric (distribution free) statistical tests are most appropriate for many water-quality parameters which do not follow a normal (bell-shaped) statistical distribution curve.

Certain parameters tend to be strongly correlated to stream flow and/or hour of collection. A data-smoothing procedure (*LOWESS* — Cleveland, 1979) was used to adjust the data for all parameters, except DO, to account for differences in streamflow at the time of sampling at the Portland and Salem Willamette River sites. The differences between the observed values and the values presented by the smoothing procedure (referred to as residuals) were tested for trends.

The time of day at which the samples are collected can affect water-quality trends for diurnally influenced parameters such as DO. DO data was adjusted to account for the hour of collection, using the same technique used to adjust for streamflow.

Willamette and Tualatin Rivers

A complete description and results of the trend analysis, including plots, can be found in *Oregon's 1992 Water Quality Status Assessment Report.*

Willamette River

Increases in DO and decreases in BOD, ammonia, and nitrogen are considered to be beneficial to the health of a stream. There was significant decreasing trends in ammonia in the lower Willamette

River at Portland Harbor and Newberg Pool. At the Newberg Pool and Portland Harbor, DO levels have generally increased over time while ammonia levels have decreased. It appears, however, that in the mid-1980s, improvements in concentrations of DO and ammonia have leveled off.

Tualatin River

The analysis indicated increases in temperature, levels of total phosphorus, and inorganic nitrogen from 1981 to 1990. Increases in temperature and in levels of phosphorus and nitrogen in the Tualatin can contribute to excessive growth of algae. The Tualatin has been designated as water-quality limited and studies conducted in the process of establishing TMDLs have identified the Unified Sewage Agency wastewater treatment plants as the major source of pollutant loading. Compliance schedules for reducing pollutant loads have been established and water-quality improvement efforts are in progress.

TMDL STUDIES

Waters (predominantly streams) are designated "water-quality limited" when water-quality standards are not being met and there is a need for increased treatment of wastes (beyond secondary treatment) prior to discharge to the stream. EPA and DEQ are required to set TMDLs on the water-quality limited streams or stream segments.

A TMDL is the total amount of a pollutant that can enter a water body without causing it to violate a water-quality standard for that pollutant. Once a TMDL is established, the "load" is divided into load allocations (that part of the load that is either from natural background sources or point sources) and waste load allocations (that part of the load that is allocated to point sources of pollution).

In order to determine the assimilative capacity of the stream and establish TMDLs, a conceptual and mathematical model of the stream needs to be developed. This model needs to account for all of the significant processes that affect stream-water quality, including hydrology, biological processes, physical processes, and chemical processes. Once this model is established limiting factors can be identified and TMDLs developed.

The DEQ undertakes comprehensive water quality studies (TMDL studies) in order to provide data necessary to develop a TMDL. Table 5 identifies the assessment components that are included in most TMDL studies, and Table 6 identifies the status of TMDL studies that have been conducted in the Willamette River Basin.

CONCLUSION

In the Willamette Basin, as anywhere, an effective monitoring program is an essential component of the water-quality protection strategy. Monitoring provides the basic information on the status of water quality. Identifying problems, trends in water quality, compliance with standards, water-quality limited streams, potential health risks, and impact of individual contaminant sources are

TABLE 5
TMDL Study Elements

Hydrology	Water chemistry	Diurnal effects	Biological processes
Flow studies	Longitudinal sampling	Continuous monitoring	Sediment oxygen demand studies
Time of travel (dye studies)	Synoptic sampling		Biological oxygen demand studies
Groundwater studies	Source studies		Periphyton studies
			Phytoplankton studies

TABLE 6
TMDL Studies in Willamette Basin

River	Parameters of concern	TMDL adopted
Tualatin	DO, ammonia, chlorine	Yes: phosphorus, ammonia, nitrogen
Yamhill	Bacteria, DO, pH	Yes: Phosphorus
Rickereall Creek	DO	Yes: BOD
Coast Fork Willamette	DO, pH, bacteria	No: in development
Pudding	DO, bacteria	Yes: BOD

Note: DO, dissolved oxygen; BOD, biochemical oxygen demand.

---

only a few of the many uses of water-quality data in a water-quality management program. Water-quality data is essential for valid regulatory decision making. As the DEQ has shifted from a technology-based water-quality management strategy, which was developed to address more obvious water-quality problems, to a water-quality based management strategy, the importance of water-quality information has become even greater. In summary, an effective water-quality management program starts and ends with reliable water-quality data. These data guide the agency and the public in their efforts to maintain and improve water quality.

REFERENCES

Cleveland, W. S., Robust locally weighted regression and smoothing scatterplots, *J. Am. Stat. Assoc.*, 74, 829, 1979.
Hirsch, R. M., Slack, J. R., and Smith, R. A., Techniques of trend analysis for monthly water quality data, *Water Resour. Res.*, 18 (1), 107, 1982.
Hirsch, R. M. and Slack, J. R., A nonparametric test for seasonal data with serial dependence, *Water Resour. Res.*, 20 (6), 727, 1984.
Oregon Department of Environmental Quality, Oregon's 1992 Water Quality Status Assessment Report (305b Report). Water Quality Division, Oregon Department of Environmental Quality, Portland, OR, 1992.
Sen, P. K., Estimates of regression coefficient based on Kendall's Tau, *J. Am. Stat. Assoc.*, 63, 1379, 1986.
U.S. EPA STORET. Water Quality Monitoring Database.
Wentz, D. A. and McKenzie, S. W., National Water-Quality Assessment Program — The Willamette Basin, Oregon. Water Fact Sheet, U.S. Geological Survey, Department of the Interior Open-File Report 91-167, 1991.

19 State Environmental Monitoring Program in Poland

Adam Mierzwiński and Elżbieta Niemirycz

DEVELOPMENT OF MONITORING OF SURFACE WATERS IN POLAND

The water quality of rivers has been the subject of interest as far back as the last century. In 1876, Mendelejew, according to his project of new water supply system and sewage system for Warsaw analyzed the Vistula water quality (Dojlido and Wojciechowska, 1974). He determined the elements that were responsible for water hardness, but the results of his measurements cannot constitute the base for the assessment of the Vistula water quality in the last century.

The first complex investigation on the Vistula water quality was made by Kirkor from 1923 to 1924 (Kirkor, 1928). He determined concentrations every 2 weeks of 14 basic parameters in the river water. He found that the Vistula water above Warsaw was satisfactorily clean despite its opacity and yellowish color. The mean values of pollutants for the investigation period were rather low, e.g., BOD_5 — 2.2 mg O_2/L; iron — 0.24 mg/L; NH_4N — 0.1 mg/L, etc. Only the amount of suspended matter was very high. It was also observed that the increase in suspended matter was closely dependent on increase in water flow.

The establishment of the National Inspection for Water Protection in the mid-1950s and the water testing laboratories in provinces was the first step in organized monitoring of surface water. Inspection frequency depended on the economic importance of the river and varied from a few inspections in 5 years to 12 in one year. The range of inspection was at first narrowed to oxygen conditions, BOD_5, COD, suspended matter, and salinity. Based on the archival data of the Provincial Inspectorate of Environmental Protection in Gdańsk (Previous Centre of Environmental Investigation and Control), it was found that significant increase in pollution of the Vistula occurred in the 1960s.

Apart from the routine monitoring, special investigations were undertaken to determine specific substances, migrations of chosen elements, and similar phenomena that cannot be characterized by routine monitoring. Examples of such investigations were the monitoring carried out in 1972 to 1973 of polychlorinated pesticides (Taylor and Bogacka, 1979) and nutrients (Szarejko and Bogacka, 1980) made twice a week, based on large populations of data.

Monitoring of the Vistula mouth sections has been performed regularly twice a week since 1975 in a wide range of measured parameters according to this special program (Niemirycz et al., 1980; Januszkiewicz et al., 1974). This monitoring covers the following indicators: temperature, color, pH, dissolved oxygen (DO), BOD_5, COD, organic carbon, nitrogen according to Kjeldahl, ammonia, nitrate, nitrite, phosphorus (total and suspended), phosphate phosphorus (total and dissolved), chlorophyll *a*, ether extract, nonpolar substances of the extract, volatile phenols, anion detergents, PCB, DDT and metabolites, DMDT, gamma-HCH, iron, manganese, zinc, copper, lead, cadmium, mercury, sodium, potassium, magnesium, calcium, total hardness, chlorides, sulfates, dry residue, dissolved substances, and NPL *coli* fecal.

FIGURE 6.   Classification of riverine water quality based on nutrients.

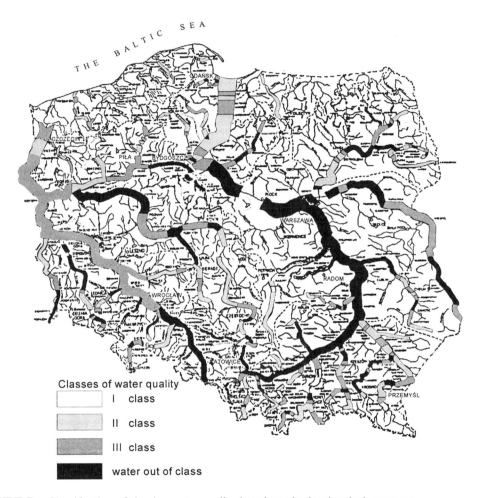

FIGURE 7. Classification of riverine water quality based on physicochemical parameters.

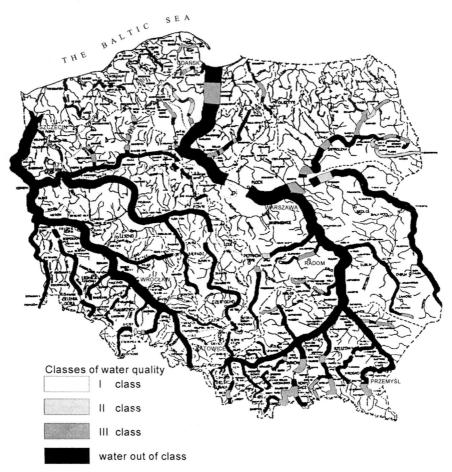

FIGURE 8.   Classification of riverine water quality based on physicochemical parameters and bacteriological criteria.

TABLE 1
Variation of Water Quality of the Vistula River from 1974–1991 Using the Polish Classification Method

Period	Parameters considered	River length studied (km × 1000)	Percent of river in water-quality class			
			I	II	III	Out of class
1947–1977	Physicochemical	17.8	9.6	30.7	26.5	33.0
1978–1983	Physicochemical	16.2	6.8	27.8	29.0	36.4
1984–1988	Physicochemical	17.6	4.8	30.3	27.8	37.1
	Bacteriological	16.5	0.0	3.9	20.3	75.8
1990	Physicochemical	10.1	6.0	27.9	30.3	35.8
	Bacteriological	10.1	0.0	3.0	16.8	80.2
1991	Physicochemical	10.5	2.3	32.7	30.0	35.0
	Bacteriological	10.5	0.0	3.7	18.4	77.9
	Physicochemical and bacteriological (combined)	10.5	0.0	3.3	14.5	82.2

Note: Class I is for drinking water supply; class II is for recreation and fish breeding; class III is for industrial supply and irrigation. Out of class means very polluted water.

From Walewski, 1992.

REFERENCES

Dojlido, J. and Wojciechowska J., Physical-chemical changes of water parameters in the Vistula River at Warsaw, 1945–1970, 1974.

Dz.U.Nr 116. Regulation of the Ministry of Environmental Protection, Natural Resources, and Forestry regarding water-quality classification, dated Nov. 11, 1991.

Januszkiewicz, T., Kowalewska, K., Szarejko, N., and Żygowski, B., Impact of agriculture (fertilizer) on the Vistula River, Mat. IMGW, 1974.

Kirkor, T., Results of systematic measurements of water from the Vistula River, Przemysł chemiczny, Volume XII, z. 6, Lwów, 1928.

Niemirycz, E., Rybiński, J., and Korzec, E., Transport of specific pollutants in the Vistula River with special considerations to substances of agricultural origin, Mat. IMGW Gdańsk, 1980–1984.

Niemirycz, E. and Borkowski, T., Riverine input of pollutants in 1992. Environmental conditions, in *The Polish Zone of the Southern Baltic Sea during 1992*, Maritime Branch Materials, Institute of Meterologic Water Management, 205 and 247, 1987.

PIOŚ, State of the purity of rivers, lakes, and the Baltic Sea based on measurements made within the framework of national environmental monitoring from 1991 to 1992, Biblioteka Monitoringu Środowiska, 1993.

Szarejko, N. and Bogacka, T., Measurement of the Vistula River water quality from Tczew to Kiezmark, Mat. IMGW, Gdańsk, 1980.

Taylor, R. and Bogacka, T., Transport of pesticides to the sea by the Vistula River, *Oceanology*, 11, 129, 1979.

Walewski, A., State environmental monitoring program, Warsaw, State Inspectorate of Environmental Protection, 1992.

20 Concentration of Nutrients in the Vistula River, Poland

Jacek Rulewski, Karin Sundblad, and Andrzej Tonderski

INTRODUCTION

Two of the most important nutritional elements in the biosphere are nitrogen and phosphorus. They are both essential for flora and fauna, but excess concentrations destroy the biotic structure of a water body. The natural cycling of nitrogen and phosphorus is increasingly disturbed by human activities. This causes severe environmental problems, including contamination of ground- and surface waters with nutrients. The enrichment of the litho- and hydrosphere with N and P compounds has severe consequences. For example, water may become unsuitable for drinking or industrial purposes.

The Baltic Sea is surrounded by 10 countries with relatively high populations and considerable concentrations of industry. Poland is one of the major countries in the region — both with regard to population and amount of pollutants discharged into the sea. A substantial part of the nutrient load is carried by the Vistula River, which is the largest river in Poland with a catchment area of 200,000 km^2 and a total length of 1038 km. A large part of the Polish population, agriculture, and industry is situated within the basin. The greater part of the population and also the industry is located in the southern part of the basin (Bieruń-Tyniec section) and in Warsaw. Moreover, the Vistula River receives a considerable load of pollutants from the Ukraine via the Bug River.

An important part of a strategy to improve the water-quality conditions in the region must be the identification and reduction of emissions of nutrients from point and nonpoint sources into the Vistula Basin. To do this, it is essential to have information about the concentration and load changes along the river course. This chapter presents the concentration variations in the river water from July 1992 to February 1993 from its sources to the mouth. In order to get a rough estimation of the geographical distribution of point sources, statistical data were collected from national statistic (GUS, 1992).

SAMPLING AND CHEMICAL ANALYSES

Water was collected once a month from 11 sites along the river course, according to flow direction on the Vistula (Wisła), as follows: at Bieruń, Tyniec, Sandomierz, Puławy, Warsaw, Modlin, Płock, Włocławek, Fordon, and Kiezmark. Samples were also collected at the mouth of two tributaries: the Przemsza upstream from Cracow (Kraków), and Narew downstream from Warsaw. Samples were taken from bridges at nine points within a lateral cross-section and combined into one sample for each site (Tonderski et al., 1994). They were stored in a freezer until analyzed.

Concentrations of NO_3^--N, NH_4^+-N, TOT-N, and TOT-P were obtained by means of Flow Injection Analysis technique (FIA Star 5020 Tecator AB, Hoganas, Sweden). TOT-N was analyzed as NO_3^--N, after oxidation with alkaline peroxidisulfate solution and TOT-P as PO_4^{-3}-P after acidic oxidation with peroxidisulfate according to Swedish standard methods.

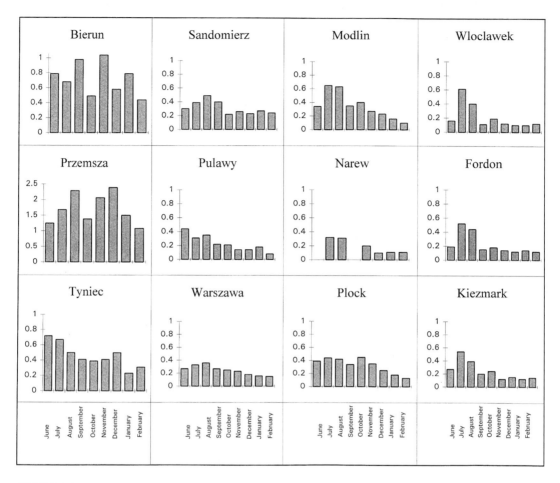

FIGURE 1.   Temporal distribution of total phosphorus concentration (in mg/L) in the Vistula River in 1992–1993.

VARIATIONS OF TOTAL PHOSPHORUS

The phosphorus contamination in Bieruń and Tyniec varied seasonally (Figure 1). In this zone the concentrations were relatively high, attaining a maximum of about 1.0 mg/L. This can be explained by the high population density, which results in the discharge of a large amount of wastewater (GUS, 1992). In addition, the water flow here was relatively low. Extremely high concentrations, up to 2.4 mg/L, were observed in Przemsza, which drains the densely populated Silesia region. A substantial decrease of the concentrations of P between Tyniec and Sandomierz is due to a relatively low discharge of wastewater in this section. The main source is Kraków's municipal and industrial sewage discharge. A slight drop in the phosphorus concentration was observed between Sandomierz and Puławy where the large and relatively clean tributary San joins the river. A moderate concentration increase toward Warsaw is probably related to the discharge of the Wieprz tributary, which drains an intensively cultivated area. Some of the phosphorus may originate from soil runoff also in the section between Tyniec and Warsaw, since there is a considerable amount of agriculture in this region. However, the concentrations of P generally decreased when the water flow increased (e.g., in the Sandomierz-Warsaw section; Figure 2), which indicates a dominant input of point sources with relatively constant flow.

The second maximum of P concentrations appeared in the Modlin and Płock section, where the maximum concentration exceeded 0.6 mg/L. The reason for this increase is the discharge of

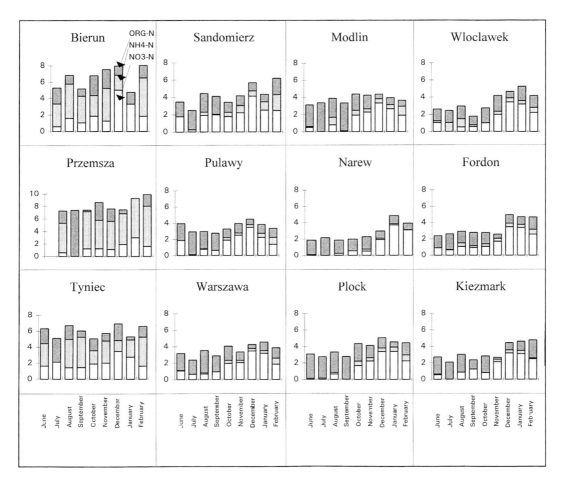

FIGURE 2. Temporal distribution of nitrate–nitrogen, ammonia–nitrogen, and organic nitrogen concentrations (in mg/L) in the Vistula River in 1992–1993.

large amounts of municipal and industrial wastewater from Warsaw and from the Bzura River. The total discharge of P from Warsaw during the study period (calculated as the difference of loads between Modlin and Warsaw) was 90 tons/month. In Płock, the concentration of P was higher than in Modlin during high flow periods (October to February) and lower during low-flow months (July to September). This can possibly be ascribed to seasonal sedimentation and resuspension of phosphorus from the sediments in this part of the river. Such phenomenon are common in the Włocławek reservoir between Płock and Włocławek (Giziński et al., 1989; Sundblad et al., 1994).

The TOT-P concentrations downstream Włocławek were on average 44% lower than in Płock. This difference is due to consumption of phosphorus by plants, sedimentation, precipitation, and adsorption reactions, which occur in the reservoir. No retention, however, was observed in July and August. A possible explanation for this finding is that P was released from the sediment during anaerobic conditions, which likely occurred in the bottom layers of the reservoir during the low-flow period in the summer. A similar effect has been observed by others (Sundblad et al., 1994). The increasing concentrations of P downstream Włocławek may be related to point-source discharges since a negative correlation with flow was observed in Kiezmark.

In summary, concentrations of phosphorus in the Vistula River usually exceeded those allowed for the lowest water-quality class (Dojlido and Woyciechowska, 1987). In general, high concentrations of TOT-P were observed during the summer months at low water flow, and concentrations decreased when the flow increased (Figure 2). This suggests that the major part of phosphorus in

Vistula comes from point sources since such discharge is constant during the year and pollutants are diluted when the water flow increases.

VARIATIONS OF DIFFERENT FORMS OF NITROGEN

Nitrogen concentrations were very high and varied randomly in the southern part of the river until they were diluted by cleaner water downstream Tyniec (Figure 3). The high concentrations are caused by the same factors as for phosphorus, i.e., low flows, heavy industrialization, and high population density in the Silesia region. This resulted in TOT-N concentrations reaching 8.06 mg/L in Bieruń, and 8.66 mg/L in the Przemsza River. ORG-N concentrations were also relatively high (maximum 2.4 mg/L) in this section. The predominance of point source pollution was clearly shown in Bieruń and Tyniec, where concentrations of ammonium reached 4 mg/L. The high ammonium concentrations means that the wastewater discharge exceeded the nitrification capacity of the river. It is also possible that, since the discharge of organic matter is high as well, low content of dissolved oxygen-inhibited nitrification processes.

In the section Sandomierz-Warsaw, the ammonium concentration decreased substantially, while the nitrate concentration was often higher than in the preceding section. This implies that substantial nitrification occurred. Some ammonium could also have been lost by volatilization. The wastewater discharge from Warsaw affected the nitrogen concentrations in the river much less than the phosphorus concentrations, since the N concentrations upstream Warsaw were already high (Figure 3).

A local maximum of the N concentrations appeared in the Modlin-Płock section and Włocławek reservoir, which has a more eutrophic character than the rest of the river. A decrease of the N concentration downstream from the Włocławek reservoir during June to October was observed (Figure 3). This suggests that denitrification occurred in the reservoir. Consumption by plants followed by sedimentation could also temporarily decrease the content of N. This is supported by the fact that retention in the reservoir does not occur in winter when low temperatures inhibit biological activity. When anaerobic conditions prevail in the deeper zones, the mineralized nitrogen remain as NH_4^+-N in the deep water explaining the higher concentrations of this compound at the outlet of the reservoir.

TOT-N concentrations slowly increased downstream from the Włocławek reservoir in some months. This was due to increases in either NO_3^--N or organic N. The total amount of nitrogen transported in the main stream of the Vistula delta varied from 2000 tons N/month in summer to 14,000 tons N/month in winter.

Vistula water oscillated between second and third water quality class, with regard to the nitrogen content which is in agreement with previous studies and Dojlido et al. (1987).

The general conclusion is that nonpoint sources constitute the major input of nitrogen (Figure 2). Along the whole length of the river, nitrate concentrations are often positively correlated with flow. In addition, NO_3^--N was the main component of the increasing concentrations in autumn. At the maximum in December, concentrations of NO_3^--N were about 70% of the TOT-N in the lower part of the river. With no plants growing in the fields, mineralization and nitrification processes could result in large amounts of NO_3^--N being available in the soil for potential leaching.

N:P RATIO IN THE VISTULA RIVER

The ratio of concentrations of nitrogen to phosphorus in the cases investigated was, on average, equal to 4.1 in the Przemsza, 8.3 in Bieruń and varied from 12 to 16.2 for sites downstream. The primary low ratio of N:P (Bieruń, Tyniec) increases in the neighborhood of Puławy and then after P-enriching at Modlin and Płock increases rapidly at Włocławek. These values indicate that nitrogen was the limiting factor for biological activities in the upper parts of the Vistula drainage basin, and phosphorus restricted the biomass growth in the remaining part of the river water.

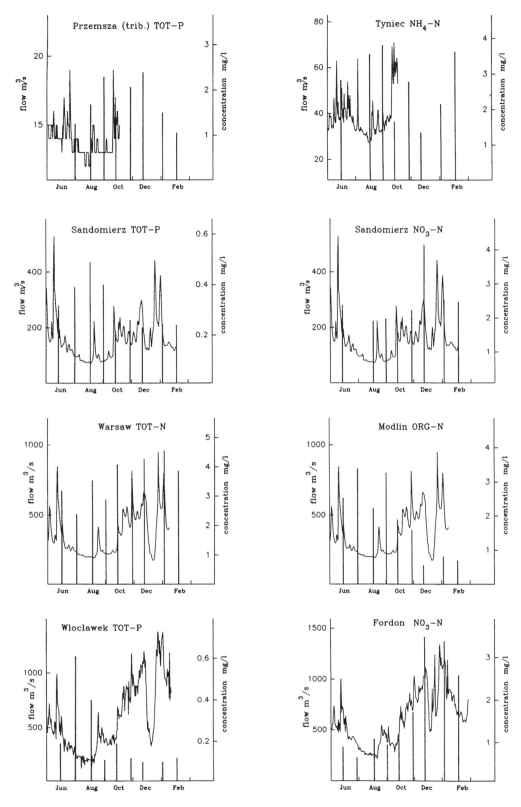

FIGURE 3. The correlation of concentrations of parameters examined (bars) vs. water flows (lines) for chosen sampling sites along the Vistula River in 1992–1993.

TABLE 1
Concentration of TCDD (pg/g) and TCDF (pg/g) in Whitefish and Carp Collected from the Willamette River Upstream and Downstream of the Bleach Kraft Pulp Mill Discharge Located at RM 147

Location	TCDD 1990	TCDD 1991	TCDF 1990	TCDF 1991
Whitefish US, RM 147	0.64 (0.11)	0.27	2.98 (0.52)	1.70
Whitefish DS, RM 147	4.77[a] (1.59)	1.93 (0.32)	20.17[b] (5.08)	6.50 (1.07)
Carp US, RM 147	ns	0.41	ns	0.41
Carp DS, RM 147	ns	0.49 (0.04)	ns	0.52 (0.03)

Note: US, upstream; DS, downstream; RM, river mile; ns, no sample.

[a] Significantly different from whitefish collected upstream of RM 147 ($p < 0.05$).
[b] Significantly different from whitefish collected in 1991 ($p < 0.05$).

upstream (Table 1). There was no statistical difference between 1990 and 1991 downstream whitefish average TCDD concentrations but TCDF was higher in whitefish collected in 1990 than in 1991. Downstream average TCDD carp concentrations were significantly lower than downstream whitefish (Figure 1). Correction of TCDD and TCDF values for lipid content did not affect the downstream to upstream or the species to species comparisons (data not shown).

TOX concentrations increased from near the detection level of 0.005 mg/L at upstream locations to a maximum of 0.112 mg/L and 0.122 mg/L at river mile (RM) 131 and 119, respectively. TOX concentrations measured at all downstream locations, including RM 7 (140 mi downstream from the discharge), remained above the levels measured upstream of the bleach kraft pulp mill discharge (Figure 2).

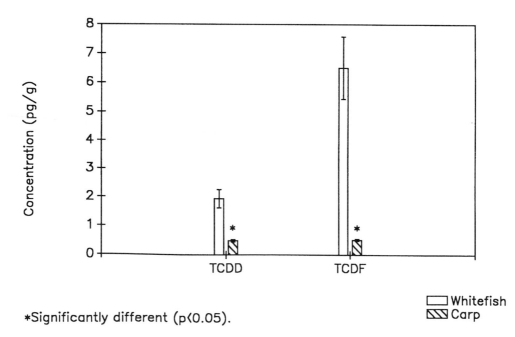

*Significantly different (p<0.05).

FIGURE 1.   Concentrations of TCDD and TCDF in whitefish and carp collected in 1991 downstream of the bleach kraft pulp mill discharge located at RM 147.

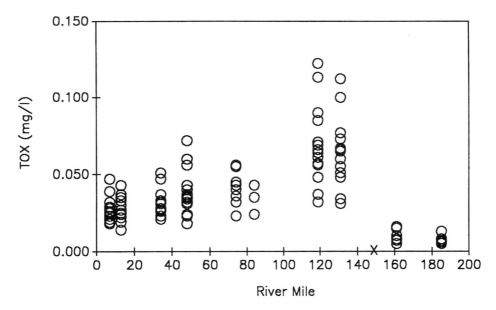

FIGURE 2. TOX concentration in water samples collected from the Willamette River in 1991 and 1992.

DISCUSSION

There are 175 dioxin and furan congeners with 18 congeners that are considered to have toxico-logical significance. The most toxic of the dioxins and furans, TCDD, has been described as the most potent anthropogenic carcinogenic compound tested (U.S. EPA, 1987). Dioxins and furans exhibit similar physical and chemical properties: hydrophobicity, lipophilicity, low volatility, resis-tance to thermal destruction, biological stability, and susceptibility to photolysis. Because of the physical and chemical properties, TCDD and TCDF behavior in the aquatic environment is expected to be adsorption to dissolved and suspended solid particles, particularly organic matter and subse-quent availability for bioaccumulation and magnification. Bioaccumulation of TCDD has been documented in laboratory and field studies (Batterman et al., 1989; Mehrle et al., 1988; Short et al., 1990).

A water-quality standard for TCDD of 0.013 pg/L was adopted by the Oregon Environmental Quality Commission in 1987. Effluent monitoring data and dilution analysis using ten year low-flow data for the Willamette River indicated that the bleach kraft pulp mill discharge at RM 147 violated the TCDD water quality standard. The study was designed to determine if the bleach kraft pulp mill was having an effect on TCDD and TCDF levels in fish near the discharge and far field effect on water quality.

Whitefish showed an increase in TCDD and TCDF concentrations downstream from the bleach kraft pulp mill discharge, while carp collected downstream compared to whitefish collected upstream from the discharge exhibited no change in TCDD or TCDF concentrations. This was surprising as laboratory and field studies reported TCDD and TCDF accumulation in carp which was attributed to individual longevity, high lipid content, and association with sediments (Kuehl et al., 1987; U.S. EPA, 1987). However, the carp were collected from the sloughs and backwater areas of the river. Although these areas were hydrologically connected to the main stem, effluent may not have been completely mixed with river water at the time of diversion into the sloughs, resulting in a lower exposure to TCDD and TCDF. Whitefish, on the other hand, can be found year-round in the main stem Willamette River.

TOX levels increased below RM 147 and did not return to upstream concentrations. TOX, a measure of total organic halides, would indicate the organic chlorine discharge from the bleach kraft pulp mill which was not detected by TCDD or TCDF analysis. Other sources of TOX would include sewage treatment plant and industrial discharges. Sewage treatment plants and industrial dischargers are located upstream and downstream of RM 147. The increased levels of TOX downstream from RM 147, which remain elevated to RM 7, indicate that the bleach kraft pulp mill discharge can affect water quality to the confluence of the Willamette and Columbia rivers.

The data collected during these surveys indicate that the bleach kraft pulp mill discharge was having both near and far field effects on water quality. The in-stream biological effects are unknown.

REFERENCES

Batterman, A. R., Cook, P. M., Lodge, K. B., Lothenbach, D. B., and Butterworth, B. C., Methodology used for a laboratory determination of relative contributions of water, sediment, and food chain routes of uptake for 2,3,7,8-TCDD bioaccumulation by lake trout in Lake Ontario, *Chemosphere*, 19, 451, 1989.

Kuehl, D. W., Cook, P. M., Batterman, A. R., Lothenbach, D., and Butterworth, B. C., Bioavailability of polychlorinated dibenzo-*p*-dioxins and dibenzofurans from contaminated Wisconsin River sediment to carp, *Chemosphere*, 16, 667, 1987.

Mehrle, P. M., Buckler, D. R., Little, E. E., Smith, L. M., Petty, J. D., Peterman, P. H., and Stalling, D. L., Toxicity and bioconcentration of 2,3,7,8 tetrachlorodibenzo-*p*-dioxin and 2,3,7,8 tetrachlorodibenzo-*p*-furan in rainbow trout, *Environ. Toxicol. Chem.*, 7, 47, 1988.

Short, R. A., Aungst, N. J., Yagley, T. J., and Preddice, T. L., Results and discussion of Lake Ontario fish sampling and analysis. in Lake Ontario TCDD Bioaccumulation Study. Final Report, U.S. EPA, 1990.

U.S. EPA, Ambient water quality criteria for 2,3,7,8-tetrachlorodibenzo-*p*-furan, EPA-440/5-84-007, 1984.

U.S. EPA, National Dioxin Study, EPA 530-SW-87-025, 1987.

U.S. EPA, Interim procedures for estimating risks associated with exposures to mixtures of chlorinated dibenzo-*p*-dioxins and dibenzofurans (CDDs and CDFs), EPA/625/3-89/016, 1989.

22 Assessment Programs for Detecting Changes in Water Quality

Margaret A. House

The utility of physicochemical indices of water quality for detecting changes in water quality has been assessed. Advantages and disadvantages of specific indices to water quality assessment have also been highlighted.

INTRODUCTION

Rivers are dynamic systems possessing large, temporal and spatial variations in water quality. This can be the result of inherent variability, episodic pollution events, or a consequence of sampling strategies or analytical techniques. This chapter evaluates the relative merits of existing U.K. approaches to the chemical assessment of river water quality and explores the potential advantages offered by the use of physicochemical indices of water quality. The use of these indices will be demonstrated via their application to river reaches within England and Wales that are subject to transient urban runoff.

CHEMICAL WATER-QUALITY MONITORING

Long-term trends in water quality are monitored in the U.K. using the National Water Council Classification (NWC, 1977). In 1995 this was replaced by a new form of classification; a General Quality Assessment (GQA) scheme (Department of the Environment and Welsh Office, 1992). This consists of six broad categories of water quality which are assessed with reference to four separate classifications: (1) a chemical classification; (2) a biological classification; (3) an aesthetic classification; and (4) a nutrient status classification. The six water quality categories for the chemical classification are based on the recorded 90 percentile concentrations for biochemical oxygen demand (BOD) and ammonia, and the 10 percentile concentration for dissolved oxygen (DO). This type of classification provides a "broad-brush" approach to the identification of long-term changes in river water quality, but provides little information on short-term fluctuations.

A more precise and rigorous method of utilizing physicochemical water quality data for the detection of both long- and short-term changes in river water quality is provided by the use of Water Quality Indices (WQIs).

A Water Quality Index has been defined as:

a form of average derived from relating a group of variables to a common scale and combining them into a single number. The group should contain the most significant parameters of the data set, so that the index can describe the overall position and reflect change in a representative manner (Scottish Development Department, 1976).

0-56670-138-4/97/$0.00+$.50

A series of four independent WQIs have been developed by House (1986). Two of these — the General Water Quality Index (WQI) and the Aquatic Toxicity Index (ATI) — shall be discussed here. The former is based on eight routinely monitored physicochemical determinants. It has a scale of 10 to 100 and reflects water quality in terms of a range of potential water uses. The latter, an index of aquatic toxicity, is an example of a use-related index which is based on less frequently monitored toxics such as heavy metals, hydrocarbons, and pesticides. The ATI reflects water quality in terms of its ability to support healthy fish and wildlife populations. The lower end of the ATI scale has been reduced to zero to reflect the fact that as a use-related index it must reflect water that is unacceptable for use.

The aforementioned indices were developed in four stages:

1. Determinant selection
2. Determinant transformations
3. Determinant weighting
4. Determinant aggregation

Detailed information on their formation can be obtained from House (1986, 1989). Each index has been developed in relation to specific use-related European Commission (now Union) Directives (EC, 1975, 1978). These were used extensively to develop the determinant transformations (rating curves) such that the WQI and ATI scales can be interpreted in terms of these EU Directives and can indicate river reaches where one or more determinant concentration is in excess of these legal standards.

APPLICATION OF THE WQI

The WQI Program operates in two stages. WQI scores are initially calculated for all of the river samples collected for an annual or longer time period. The output (Table 1) provides information on the determinant concentrations, the Water Quality Ratings (WQRs) achieved by each determinant (the determinant transformation scores), the weighting, and the determinant WQI scores. The sample WQI score is derived using Equation 1 which is a modified arithmetic formula first used by the Scottish Development Department (SDD, 1976) in the development of the SDD Index.

$$\frac{1}{100} \cdot \left(\sum_{i=1}^{n} q_i w_i \right)^2$$

where: q_i represents the rating for the ith determinant, w_i represents the weighting for the ith determinant, and n represents the number of determinants.

The provision of the WQRs for each determinant makes it possible to highlight not only a deterioration in water quality, but also the specific determinant(s) responsible for this deterioration. From the example in Table 1 it can be seen that in this instance a sample WQI score of 37 was recorded and that total ammonia and BOD concentrations were responsible for the low WQI score. The WQRs have an additional value in that they can highlight situations in which one or more determinant concentration has exceeded the EC Directive for the protection of fisheries (EC, 1978) or potable water abstractions (EC, 1975) as the definition of the WQI scale and the development of each determinant rating curve was undertaken with reference to the standards contained therein. Thus, additional information of value to the operational management of water quality is provided by using the WQI.

Secondly, a summary WQI score is calculated for the time series of data for each site. This is calculated as a parametric 5 percentile of the individual sample WQI scores for the period. By using all of the sample data sets it is additionally possible to calculate 90% upper and lower confidence limits around the summary WQI scores. These were calculated using a statistical package

TABLE 1
Sample Output from Water Quality Index (WQI) Program

Determinant	Concentration	Water quality rating	Weighting	Determinant WQI
pH	7.4	99.90	0.10	10.11
Temperature	12.0	100.00	0.02	2.25
DO	40.5	47.87	0.22	10.76
BOD	10.0	36.67	0.20	7.41
NH_4	3.4	31.80	0.18	5.72
NO_3	1.5	92.75	0.10	9.38
SS	12.0	92.00	0.12	11.37
Cl	60.0	88.00	0.05	3.96

Sample WQI = 37.1

Annual summary data

Maximum sample WQI	Mean sample WQI	Minimum sample WQI	No. of samples
67.3	44.4	22.3	21

Lower 90% confidence	5 percentile WQI	Upper 90% confidence
14.06	21.65	29.24

Note: DO, dissolved oxygen; BOD, biochemical oxygen demand; SS, suspended solids.

developed by Kinley (1988). This provides an indication of the resultant water quality for the period which, for the example in Table 1, is 22 ± 7.

These indices provide a more rigorous approach to the assessment of both long- and short-term changes in river water quality, enabling the impact of intermittent discharges to be highlighted. The continuous scale provided by the use of an indexing system allows the quantification of "good" and "bad" water quality. All the samples within an annual or longer time series are used in the final assessment of water quality thus allowing the detection of secular cycles and trends in water quality.

Despite the advantages afforded by the use of these indices they are only as good as the database available. Therefore, sampling frequencies that fail to monitor intermittent, episodic events will still be left undetected.

The WQI has been extensively validated through its application to data covering a number of years and geographical locations within the U.K. (House and Ellis, 1987; House and Newsome, 1989; Tyson and House, 1989), and it is used by Southwest and Northwest NRA Regions for water-quality monitoring and operational management. In addition, it is a key variable in the "Mersey Measure" used by the Mersey Basin Campaign to monitor improvements in the River Mersey associated with an intensive cleanup campaign (Mersey Basin Campaign, 1991). The ATI is still within the validation phase of development.

CASE STUDIES

In order to provide a more detailed demonstration of the information provided by the WQI, the results obtained from the application of the WQI to data for a number of river reaches are outlined below.

TABLE 2
5 Percentile WQI Scores for Selected River Reaches
within Southwest NRA Region

	5 Percentile WQI score	WQI range	No. data sets
Hollocombe Water			
1985	85 ± 2	81–93	15
1986	85 ± 2	81–93	17
1987	68 ± 6	49–93	18
1988	66 ± 6	49–92	17
1989	68 ± 6	49–92	19
River Trewince			
1985	46 ± 11	44–87	16
1986	62 ± 5	44–88	21
1987	62 ± 4	44–90	32
1988	70 ± 3	62–90	33
1989	68 ± 3	62–90	30
River Fal			
1985	62 ± 2	57–89	78
1986	61 ± 2	46–89	84
1987	57 ± 3	46–87	69
1988	57 ± 3	46–86	52
1989	54 ± 5	50–90	39
The Cranny			
1985	45 ± 6	36–80	19
1986	47 ± 6	37–77	18
1987	27 ± 11	14–77	16
1988	29 ± 8	14–77	22
1989	24 ± 9	14–83	30

The results in Tables 2 and 3 are for four river reaches in the Southwest NRA Region. These results are for triennial data which covers 7 years (1983 to 1989), thus allowing the detection of both spatial and temporal variations in river water quality.

USE OF SUMMARY WQI SCORES

The results from the calculation of the triennial 5 percentile WQI scores (Table 2) not only provide an indication of the water quality for each of the four reaches over the preceding 3-year period, but also highlight improvements or deterioration in water quality that have occurred over the seven year period. In addition, the use of the WQI allows the river reaches to be ranked in order of water quality.

The results for Hollocombe Water and the River Trewince indicate waters of very different quality for the 1983 to 1985 period (85 ± 2 and 46 ± 11). By 1989, however, the water quality at Hollocombe Water had deteriorated to 68 ± 6, while that of the River Trewince had improved to 68 ± 3. The use of an indexing system with a continuous scale allowed these changes in water quality to be monitored, investigated, and managed.

It is apparent from these triennial 5 percentile WQI scores that the WQI is sufficiently sensitive to fluctuations in water quality to be of value in detecting long-term changes in water quality and as such is useful to the operational management of surface water quality.

TABLE 3
Sample WQI Scores and Lowest
Determinant Ratings for the River Trewince

Year	Sample WQI score	Determinant	Lowest water quality rating
1985	84	NO_3	69
	87	NO_3	80
	76	NH_4	68
	44	NH_4	29
1986	84	NO_3	72
	86	NH_4	74
	79	NH_4	74
	85	NO_3	80
	88	NO_3	80
	84	NO_3	82
	84	NO_3	82
	85	NO_3	78
	85	NO_3	75
	86	NO_3	78
	81	NO_3	75
	81	NO_3	76
	81	NO_3	80
1987	89	NO_3	73
	62	SS	38
	85	NO_3	79
	62	SS	36
	87	NO_3	74
	79	NO_3	78
	86	NO_3	77
	90	NO_3	76
	86	NO_3	75
	77	NH_4	80
	74	NH_4	71
	77	NO_3	76
	87	NO_3	79
	83	NO_3	77
	85	NO_3	75
1988	86	NO_3	78
	88	NO_3	73
	84	NO_3	78
	83	NO_3	76
	88	NO_3	72
1989	74	SS	66
	89	NO_3	72
	85	NO_3	77
	81	NO_3	80
	85	NO_3	81
	81	NO_3	79
	81	NO_0	81
	85	NO_3	80
	88	NO_3	71
	84	NO_3	75

ATI is reflecting waters that are toxic and likely to support only sporadic populations of tolerant fish and wildlife species (ATI scores of 30 to 44 and a 5 percentile of 27 ± 6).

Finally, the WQI results for site D, which is directly below a storm sewer outfall, produce a 5 percentile score of 37 ± 4. Although this value is not dramatically different to that of site B, an analysis of the WQI range recorded for each site (Table 4) indicates that, for site D, this is the result of consistently poor water quality rather than the product of two probable pollution events as appears to be the case for site B. This conclusion is confirmed by the narrower confidence limits calculated for this site. The results provided by the use of the ATI indicate that site D has the greatest variations in heavy metal toxicity and that this reach is unlikely to support any fish or wildlife populations (House, 1986).

The combined use of the WQI and ATI scoring systems provide a more accurate and informative assessment of the water quality of the Salmon's Brook than the use of any one by itself. The WQI provides information on the presence and impact of organic pollutants while the ATI provides an assessment of the presence and impact of urban/industrial discharges.

The Mersey Measure (Mersey Basin Campaign, 1991) combines the use of chemical (the WQI) and biological indices (Biological Monitoring Working Party Score and Average Score per Taxa; Bascombe et al., 1989), along with flow to monitor the water quality of the Mersey Basin as part of the cleanup campaign for this urban catchment. This additional inclusion of biological data would complete the overall picture for receiving water quality.

CONCLUSIONS

The results from these index applications to the Salmon's Brook, the Irwell catchment, and the rivers in the Southwest NRA Region indicate the importance of reporting tools that can provide a continuous scale when dealing with aquatic environments subject to intermittent and toxic discharges.

REFERENCES

Bascombe, A. D., House, M. A., and Ellis, J. B., *The Utility of Chemical and Biological Monitoring Techniques for the Assessment of Urban Pollution. River Basin Management,* Vol. 5, Advances in Water Pollution Control, Pergamon Press, Oxford, 1989, 59.

Commission For European Communities, Council directive of 16 June 1975 concerning the quality required of surface water intended for the abstraction of drinking water in member states, Official Journal No. L194/26, 1975.

Commission For European Communities, Council directive of 18 July 1978 on the quality of freshwaters needing protection or improvement in order to support fish life 78/659/EEC, Official Journal, Series L222, 1978.

Department of the Environment and Welsh Office, River quality. The government's proposals: a consultation paper, Department of the Environment Publication, London, U.K., 1992.

House, M. A., Water Quality Indices. Unpublished Ph.D. thesis, Middlesex Polytechnic, Queensway, Enfield, Middlesex, U.K., 1986.

House, M. A. and Ellis, J. B., The development of water quality indices for operational management, *Water Sci. Technol.,* 19, 145, 1987.

House, M. A. and Newsome, D. H., Water quality indices for the management of surface water quality, *Water Sci. Technol.,* 21, 1989.

House, M. A., A water quality index for the classification and operational management of rivers, in *River Basin Management, Vol. 5, Advances in Water Pollution Control,* Pergamon Press, Oxford, 37, 1989.

Kinley, R. D., Genstat Program NQUANT, Northwest Water Authority, New Town House, Warrington, U.K., 1988.

Mersey Basin Campaign, How the Mersey Measure System monitors river water quality throughout the Mersey Basin, Department of the Environment, Manchester, U.K., 1991.

National Water Council, Final report of the working party on consent conditions for effluent discharges to freshwater streams, National Water Council, London, U.K., 1977.

Scottish Development Department, Development of a water quality index. Report ARD 3. Applied Research and Development Engineering Division, HMSO Edinburgh, 1976.

Tyson, J. M. and House, M. A., The application of a water quality index to river management, *Water Sci. Technol.*, 21, 1989.

23 Developing Indicators of Ecological Condition in the Willamette Basin: An Overview of the Oregon Prepilot Study for EPA's EMAP Program

Alan Herlihy, Phillip Kaufmann, Lou Reynolds, Judith Li, and George Robison

In order to develop robust (quantifiable, reproducible) fish, macroinvertebrate, and physical habitat indicators of ecological condition, a 4-year prepilot survey was conducted on wadeable streams in the mid-Willamette River basin in western Oregon. This chapter presents an overview of the prepilot survey and a summary of preliminary findings.

The first year of sampling focused on quantifying the field sampling effort needed to develop robust indicators using 18 streams in the Willamette Valley and Cascade Mountain ecoregions. For fish assemblages, preliminary results show that a one-pass electrofishing sample over a reach that is 40 times as long as the mean wetted width yields adequate information on indicators of species richness and proportionate abundance. Quantitative measures of physical habitat condition also yielded robust indicators within this 40-channel-width interval. For macroinvertebrates, multiple surber samples were collected from each stream. Their cumulative species richness increased in a direct, but diminishing relationship to the number of samples counted; even though more than 50 surber samples were collected in some streams. The compositing of 11 samples taken over reaches 40 times their baseflow wetted width stabilized the variance of diversity measures and the proportional abundance metrics for the macroinvertebrate assemblage.

The second year of sampling investigated the response of the fish and macroinvertebrate assemblages along gradients of disturbance represented by 35 stream samples in the study area. Preliminary results indicate that biotic assemblage metrics are related to habitat and human disturbances. Data collected during the last 2 years of the study are being assembled to evaluate the among-year and within-sample period variability in the biological and physicochemical indicators of ecological condition.

INTRODUCTION

In response to the need for better assessments of the condition of the nation's ecological resources, the EPA's Office of Research and Development has begun planning and designing the Environmental Monitoring and Assessment Program (EMAP). EMAP is a strategic approach designed to quantify the ecological condition of surface waters and to estimate the extent and magnitude of surface water degradation (Whittier and Paulsen, 1992; Paulsen and Linthurst, 1994). Another major objective of EMAP is to detect improvement or degradation in environmental condition over time by repeat sampling of the populations of interest.

In order to develop indicator concepts and field methods for use by EMAP to monitor physicochemical habitat, basin and riparian condition, and fish and macroinvertebrate assemblages, a 4-year prepilot study was conducted on wadeable streams in the mid-Willamette River basin of western Oregon. The overall goal of this Oregon prepilot study was not to monitor streams, but to *develop* indicators of ecological condition for streams. These indicators must be: (1) biologically relevant, (2) implementable on national and regional scales, (3) robust (repeatable and quantitative), and (4) sensitive to anthropogenic disturbance. The major objectives of the prepilot were in four areas of study:

Adequate Sampling Effort — To determine the field sampling design, including the minimum sampling reach length and level of sampling effort required for robust characterization of channel habitat and assemblages of fish and macroinvertebrates in wadeable streams of various sizes and ecoregional settings in the Mid-Cascade and Willamette Valley region of Oregon.

Landscape Stress Characterization — To develop and test approaches for quantifying potential stress on stream ecosystems due to alterations of land cover and land use. Based on remote imagery, these landscape stress indicators will incorporate the qualities of areal cover, disturbance intensity, and proximity to stream channels.

Metric Scaling and Responsiveness — On a regionally extensive set of sample sites, to use efficient, one-day physical, chemical, and biological assessment field protocols that: (1) test the response of physical and biological indicator metrics along disturbance gradients, and (2) develop approaches for scaling physical and biological metric expectations for stream size, gradient, and ecoregional setting.

Metric Precision — To assess the regional, among-year, and between-visit components of physical, chemical, and biological indicator variability.

This chapter will give an overview of the prepilot survey, present details of site selection, and present a summary of preliminary findings. Details about any of the methodology or data analyses for any of the individual metrics will not be covered. Detailed results are either in review or in preparation.

SURVEY DESIGN

Sample stream reaches were selected in four strata in a 2×2 design with two stream size classes (small and large) and two ecoregion classes (Cascade Mountain and Willamette Valley). Ecoregion boundaries were determined from Omernik's (1987) ecoregion map. The stream network represented on 1:100,000 scale USGS topographic maps was used as the sample frame from which to select sample sites. Small streams were defined as first order streams on the 1:100,000 scale maps. Large streams were defined as second- or third-order streams on the same maps. At this map scale, fourth- and larger-order streams are often unwadeable and were not considered for this study.

In the first year of the study, a probability sample of 16 stream reaches (four in each strata) and two hand-picked reaches were greatly oversampled to determine the minimum sampling effort required to obtain robust indicators of condition. In the second year, 17 of the 18 streams from year 1 were revisited along with an additional 18 streams. Streams were sampled using the physical and biological assessment protocols developed in year 1 in order to derive metrics of stream biotic condition along gradients of human disturbance. In the third and fourth year of the study, eight of the streams sampled in the first 2 years of the study were revisited twice during each of the summer sampling periods to quantify the within-sample season and among-year variability in physicochemical and biological indicators of stream conditions in the Willamette Basin.

Year-1 Site Selection

Year 1 sampling was initiated in summer 1992. For logistical (manpower, transportation) reasons, only streams between 44 and 45 degrees north latitude (roughly between the cities of Salem and Eugene, OR) in the two study ecoregions were considered for sampling (Figure 1). Stream segments were defined as the length of stream between "blue line" confluences or between a confluence and the headward extent of the blue line. The stream segments to be sampled were chosen in two steps. The first step involved obtaining a randomized systematic sample of stream segments using the approach developed for the National Stream Survey (Kaufmann et al., 1988, 1991). A clear acetate sheet dotted with a grid of rectangular points was randomly dropped on top of the 100,000 scale USGS topographic maps for each ecoregion. The stream segment "hit" by following the topographic fall line down from each of the acetate grid point was selected as a potential field sampling segment. This procedure resulted in a set of about 200 potential Cascade and Valley stream segments for sampling. The second step involved reduction of the sample size to a workable field-study size. Initial segments were assigned to either the small or large size class based on stream order and randomized within each of the four ecoregion/size class strata. The first four wadeable streams on the randomized list in each stratum, for which site access could be obtained, were selected for field sampling. The exact sampling location on each stream segment was chosen using a random number table. In addition to the 16 randomly selected streams, two hand-picked stream sites (Lookout and Mack Creeks) in the H.J. Andrews Experimental Forest/Cascade Mt. Long-term Ecological Research area were also sampled to allow comparison to a large database available for these sites. Mack Creek is a first-order stream and Lookout Creek is a relatively large third-order stream.

Year-2 Site Selection

The 18 sites sampled in year 1 were all scheduled for revisits in year 2 (summer 1993); however, access was denied to one of the year 1 sample sites (a small valley stream), so only 17 of the sites were actually revisited. To increase the sample size for the "year 2" disturbance gradient study, an additional 18 sites were selected for sampling. A desired attribute of the year 2 sample sites was that they represent the range of disturbance types and intensities in each ecoregion. Because data from the "year 1" sites did a good job of representing the expected gradient of conditions in the Cascade Mountains, an additional eight stream sites were randomly sampled in that ecoregion for a total of 18 Cascade sites. In the Willamette Valley ecoregion, the extremes in disturbance were not well represented and weren't likely to be represented in this relatively small random sample due to their rarity in the stream population. Consequently, five streams were purposefully chosen to reflect "very good" and "very poor" conditions in the valley. Additionally, 5 stream sample sites were randomly selected for a grand total of 17 Willamette Valley sites.

In year 2, the random site selection was accomplished using a geographic information system (GIS) and the digitized version of the 1:100,000 stream network following the newly established EMAP stream survey protocol. To better fit the new sampling design, hydrologic unit boundaries rather than map edge boundaries were used to define the study area. Streams within the Willamette Valley ecoregion in USGS hydrologic units 17090003 (Willamette River mainstem), 17090004 (McKenzie River), 17090005 (North Santiam River), 17090006 (South Santiam River), 17090007 (Willamette River mainstem), and 170900009 (Mollala/Pudding River) were used as the sample frame. As a result, the year-2 Willamette Valley sample area extended northward to near Oregon City (south of the Clackamas and Tualatin River subbasins). The Cascades sample area was restricted to the McKenzie, Calapooia, and Santiam drainage basins and thus remained roughly the same as that used in the year-1 sample. The location of all 35 sites sampled in year 2 is shown in Figure 1.

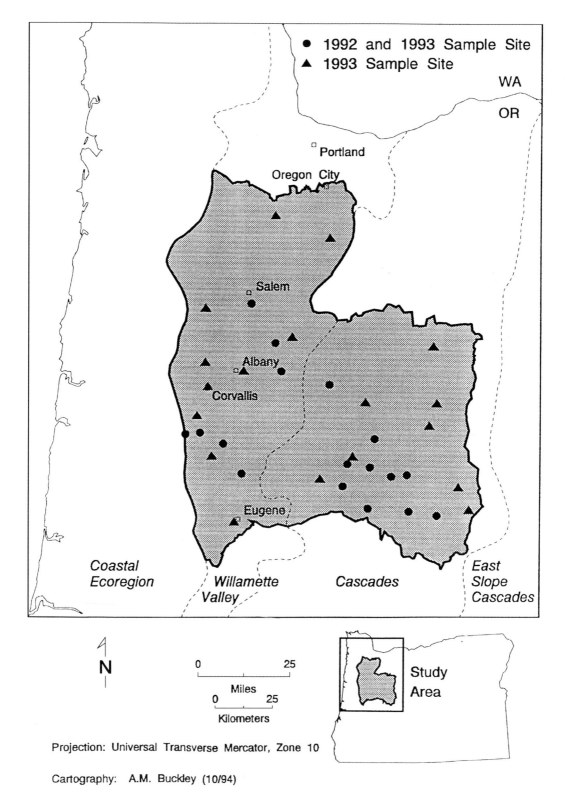

FIGURE 1. Location of Oregon prepilot sample sites. The year-2 (1993) survey area is shaded on the map. The year-1 (1992) study area has the same eastern and western boundaries but the northern and southern boundaries were 45 degrees/44 degrees north latitude (roughly between Salem and Eugene on the map).

FIELD SAMPLING

In year 1, field measurements were made on stream reaches longer than anticipated for future measurement. Stream lengths of approximately 100 wetted-channel widths were used to examine the gain in information with the increase in sampling effort and reach length. This procedure resulted in sample lengths of about 120 to 350 m in small streams and 250 to 600 m in large streams. The average total field time at each site was 0.5, 1.0, and up to 3 days, respectively for macroinvertebrates, physical habitat, and fish sampling. Field work was done from June to September 1992.

In year 2, a new field protocol was established using Year-1 experience. For each of the 35 sample sites, a 3 to 4 person crew combined physical, chemical, and biological (fish and macro-invertebrate) measurements into a 1-day sampling protocol. For six stream sites, crews made a second field visit to repeat the physical habitat measurements. On the six largest streams, the capture efficiency of the rapid fish sampling protocol was evaluated by doing a second intensive three-pass electrofishing effort 2 to 3 weeks later. Field work was done from June to September 1993.

In years 3 and 4, a stratified random subset of eight stream sites were sampled to quantify temporal and sampling variability. Two sites in each of the four study strata were randomly selected from the 17 sites sampled in both year 1 and year 2. Field crews collected benthic macroinvertebrate samples three times at these sites between May and August 1994 (year 3). From the same sites, habitat, chemistry, and fish assemblage data were collected in two field visits between July and August 1994. Habitat, chemistry, macroinvertebrate, and fish assemblage data were also collected in two sample visits between July and August 1995 (year 4). From this information, the index variability (combined measurement plus within-season variability), among-year variability, and the interaction between year and index variability of the sampling protocols can be evaluated. This will allow for the evaluation of the utility of various habitat and biotic assemblage metrics for determining status and trends in the condition of stream resources.

In order to begin defining the environmental stress gradients, remote data on drainage basin and riparian characteristics are being assembled and processed for all 35 sample reaches.

STREAM CHARACTERIZATION

In the Willamette Valley, agriculture and urbanization have influenced low-gradient, fine substrate streams for over a century, while logging and human settlement have altered the higher gradient, coarse-bedded Cascade streams over a similar period. Our study streams are a probability sample from the considerable diversity of wadeable streams in the region. They include low gradient, meandering streams, agricultural ditches, and cobble-bedded mountain cascades. Dominant substrates in the sample streams range from silt to boulders. Riparian vegetation includes old-growth conifers, deciduous trees (alder, ash, or maple), grasses and herbs, or agricultural stubble. Nutrient concentrations, clarity, and water temperature vary markedly among these streams.

Results of this study are representative of a wide variety of stream types. Streams in the small-size class typically had watersheds between 0.3 and 17 km^2 in size and wetted widths between 1 and 5 m, whereas streams in the large-size class typically had areas between 12 and 75 km^2 and widths between 4 and 8 m (Table 1). Streams in the Cascades had much steeper gradients and their basins were predominantly forest cover in a patchwork of varying ages. Streams in the Valley ecoregion were low gradient (< 1%), were dominated by fine substrate, and had primarily agricultural and urban land use/land cover. Chemically, there was a clear distinction between Cascade and Valley streams (Table 1). Valley streams had higher conductivity, chloride, and nutrient concentrations; all indicative of longer groundwater contact times and/or anthropogenic additions of chemical substances. The pH of streams in the study area were almost all in the 7.0 to 7.5 range.

TABLE 1
Characteristics of Sample Streams within Each of the Four Study Strata

Parameter	Small cascade streams	Large cascade streams	Small valley streams	Large valley streams
Watershed area (km²)	3.6 (2.6–5.1)	26.7 (12–55)	2.2 (0.3–17)	45.5 (27–75)
Elevation (feet)	2520 (2080–3280)	2750 (1320–3460)	340 (240–410)	260 (220–280)
Mean wetted width (m)	2.7 (2.1–3.0)	6.5 (3.9–8.3)	1.9 (0.7–4.8)	4.9 (4.1–6.4)
Mean depth (cm)	22 (16–24)	41 (34–45)	21 (16–25)	48 (31–62)
Stream gradient (%)	8.6 (7.8–11)	3.6 (3.0–4.4)	0.42 (0.28–1.0)	0.21 (0.07–0.60)
Substrate — % sand/fines	4 (2–4)	2 (0–6)	67 (38–82)	72 (37–87)
Watershed % forested[a]	100 (100–100)	100 (99–100)	1 (0–83)	41 (0–60)
Watershed % agriculture + urban[a]	0 (0–0)	0 (0–0)	77 (17–100)	59 (38–98)
Road density(km/km²)	1.6 (0–1.8)	0.80 (0–1.2)	1.9 (0.5–9.3)	1.8 (1.6–2.0)
pH	7.3 (7.3–7.4)	7.4 (7.4–7.5)	7.1 (6.9–7.4)	7.3 (7.2–7.7)
Conductivity (µS/cm)	35 (33–40)	42 (34–50)	82 (54–156)	133 (64–248)
Chloride (µeq/L)	22 (21–24)	21 (17–27)	116 (98–199)	112 (64–188)
Total N (µg/L)	41 (26–69)	43 (17–52)	410 (197–1500)	516 (249–1040)
Total P (µg/L)	7 (6–11)	17 (6–23)	9 (4–26)	69 (41–84)
Sample size	9	9	7	10

Note: Values are medians with first and third quartiles in parentheses.

[a] Land-use data from 1:250,000 scale USGS (1990) land-use and land-cover digital data.

PRELIMINARY FINDINGS

Data interpretations from each of the various physicochemical and biological indicators are being prepared for publication and are in varying stages of submission. The preliminary findings from these studies are presented in this section. Details of the results and supporting analyses will appear over the next few years. The study shows that combined biological, chemical, and physical habitat characterizations can be made rapidly (1 day) and economically over sample reaches 40 channel widths in length by a crew of 3 to 4 people.

Physical habitat indicators were evaluated by calculating signal-to-noise ratios based on sample variance. The "noise" was calculated as the variance in indicator values from repeated sample visits to the same sample site. It consists of both temporal variability within the sample period and measurement variability among different crew members. The "signal" was calculated as the variance in indicator values among all the sample streams. Thus, the signal-to-noise ratio is a measure of indicator precision; or how repeatable an indicator metric value is compared with the variability of the metric across the streams in the region.

Quantitative physical habitat measurements and design-based visual estimates of habitat characteristics were more robust than qualitative observations and discharge-related measures, as evaluated by signal-to-noise variance ratios. Metrics based on thalweg profile measurements were very precise, with signal-to-noise ratios ranging from 25:1 to 120:1. Design-based visual estimates of substrate characteristics, fish cover, and riparian human disturbances were also quite precise, with signal-to-noise ratios from 10:1 to 60:1. Riparian vegetation cover and presence metrics were somewhat less precise (signal-to-noise ranging 3:1 to 25:1). Flow dependent and whole-reach qualitative visual assessments tended to be imprecise, with signal-to-noise ratios less than 3:1 (Kaufmann et al., in preparation).

Scaling electrofishing effort by setting reach length proportional to wetted width reduced the variance in information return among streams of differing sizes and ecoregional settings. One-pass electrofishing (5500 shocking seconds) over reaches 40 times their baseflow width produced robust estimates of proportionate abundance of all but the rarest species, allowing accurate calculation of Indices of Biological Integrity. This rapid field technique captured >90% of fish species detected with much more intensive three-pass electrofishing (Reynolds et al., in preparation).

Macroinvertebrate total species richness increased as a function of sample size to the limits of the oversampling study (maximum of 54 surber samples over 100 channel widths). However, composite samples from 11 surber samples taken over reaches 40 times their baseflow wetted width appear adequate to stabilize the variance of diversity measures and proportional abundance metrics such as the %EPT (percent of total macroinvertebrates that are Ephemeroptera, Plecoptera, or Trichoptera orders). Furthermore, counts of at least 300 organisms were required to stabilize these proportionate abundance metrics, regardless of the number of surber subsamples making up a composite sample (Li et al., submitted).

Although relatively small in sample size, the probability sample of 35 stream reaches included a broad range of conditions, from pristine wilderness sites to locations within urban areas. By itself, the species distribution information from this study has provided valuable information. For example, it was discovered that the endangered Oregon chub [*Hybopsis crameri* (Snyder)] existed at a previously unknown location, substantially increasing its known range.

Preliminary results indicate that biotic assemblage metrics can be statistically related to habitat and human disturbances. A stepwise multiple linear regression analysis showed that five habitat structural complexity and stream chemistry variables could be used to explain more than 90% of the variance in a fish Index of Biotic Integrity (IBI) independently developed by Hughes et al. (submitted) for wadeable lowland streams in the Willamette Valley. In multivariate analysis of the same data, the first principal component was associated with riparian human disturbance and also with water quality degradation and reductions in channel habitat and cover complexity. In turn, this principal component was negatively correlated with Hughes' IBI based on fish assemblage data.

ACKNOWLEDGMENTS

The research described in this chapter has been funded by the U.S. Environmental Protection Agency. This document has been prepared at the EPA's National Health and Environmental Effects Laboratory in Corvallis, OR, through a cooperative agreement with Oregon State University (#CR818606). It has been subjected to the Agency's peer and administrative review and approved for publication. Mention of trade names or commercial products does not constitute endorsement or recommendation for use. We thank Aileen Buckley for generating the site map and Joshua Moffi, Mark Mankowski, Christian Torgersen, Edward Degner, Charley Wheeler, Margi Coggins, Jack Burgess, Sean Ollila, Kris Temple, Marci Anderson, Jenifer Lomas, and Eric Phillips for assistance with field work.

REFERENCES

Hughes, R. M., Kaufmann, P., Herlihy, A., Kincaid, T., Reynolds, L., and Larsen, D., Development and application of an index of fish assemblage integrity for wadeable streams in the Willamette Valley ecoregion, Oregon, *Can. J. Fish. Aq. Sci.*, submitted.

Kaufmann, P. R., Herlihy, A. T., Elwood, J. W., Mitch, M. E., Overton, W. S., Sale, M. J., Messer, J. J., Cougan, K. A., Peck, D. V., Reckhow, K. H., Kinney, A. J., Christie, S. J., Brown, D. D., Hagley, C. A., and Jager, H. I., Chemical Characteristics of Streams in the Mid-Atlantic and Southeastern United States, Volume I: Population Descriptions and Physico-Chemical Relationships, EPA/600/3-88/021a, U.S. Environmental Protection Agency, Washington, D.C., 1988, 397.

Kaufmann, P. R., Herlihy, A. T., Mitch, M. E., and Overton, W. S., Chemical characteristics of streams in the Eastern United States. I. Synoptic survey design, acid-base status and regional chemical patterns, *Water Resour. Res.*, 27, 611, 1991.

Kaufmann, P., Robison, G., Herlihy, A., and Beschta, R. L., Stream physical habitat: precision and biological relevance of a field survey method, *Can. J. Fish. Aquat. Sci.*, in preparation.

Li, J., Herlihy, A., Gerth, W., Kaufmann, P., Gregory, S., and Larsen, P., Quantifying stream macroinvertebrate assemblages: the relative influence of sample size and spatial distribution, *J. Natl. Am. Bent. Soc.*, submitted.

Omernik, J. M., Ecoregions of the conterminous United States, *Ann. Assoc. Am. Geographers,* 77, 118, 1987.

Paulsen, S. G. and Linthurst, R. A., Biological monitoring in the environmental monitoring and assessment program, in *Biological Monitoring of Aquatic Systems*, Loeb, S. L. and Spacie, A., Eds., Lewis Publishers, Boca Raton, FL, 1994, 297.

Reynolds, L., Gregory, S., Herlihy, A., Hughes, R., and Kaufmann, P., Spatial sampling requirements for electrofishing Willamette Valley and Cascade mountain streams in Oregon, *Trans. Am. Fish. Soc.,* in preparation.

U.S. Geological Survey (USGS), Land Use and Land Cover Digital Data from 1:250,000- and 1:100,000-scale maps, Data Users Guide 4, U.S. Geological Survey, Reston, VA, 1990.

Whittier, T. R. and Paulsen, S. G., The surface waters component of the Environmental Monitoring and Assessment Program (EMAP): an overview, *J. Aquat. Ecosys. Health,* 1, 119, 1992.

24 Assessment of Pollution Outflow from Large Agricultural Areas Based on Hydrologic Parameters and Measured Pollutant Loads

J. Rybiński, Z. Makowski, and E. Heybowicz

INTRODUCTION

It is difficult to predict pollutant loads from basins with mixed point and nonpoint sources of pollution, or to assess the impact of either source. Much data exist where measurements of pollution have been made in which both point and nonpoint sources of pollution contribute to the load. Very few measurements exist that characterize specific sources of pollution. In Poland, most river basins have a great number of point and nonpoint sources of pollution (Rybiński et al., 1993).

To define contributions from both point and nonpoint sources of pollution, a working hypothesis can be used which assumes that hydrologic conditions (precipitation and the resulting flow distribution) drive the nonpoint source and that the point source contribution can be accounted for.

HYPOTHESIS

Water outflow from drainage basins can be divided (Atlas, 1986) into two parts: the first part originating from groundwater and the second part from surface and subsurface run-off. The relation can be written:

$$Q = Q_p + Q_a \qquad (1)$$

where Q = total outflow, Q_p = underground outflow, and Q_a = surface and subsurface outflow.

Analogous to flow, the pollution load leaving the river basin for any given hydrologic state can be divided into:

$$L = L_p + L_a \qquad (2)$$

where L = total load (measured quantity), L_p = pollution load in the ground outflow (estimated quantity), and L_a = pollution load in the nonpoint outflow (estimated quantity).

In reality, only a small part of L_p originates from groundwater zone which is relatively clean water. The primary part of L_p that is polluted comes from wastes introduced into the groundwater through sewage systems. Therefore, L_p is driven by the same flux as the anthropogenic pollution effluent to the surface water.

The second component of the foregoing relation is L_a, the nonpoint pollution load. In contrast to point-source pollution, L_a is washed out of the basin by precipitation, and is therefore strongly related to stream discharge. Pollution is both washed off the basin surface and through the aeration zone in the subsurface.

The pollution load from point sources undergoes instantaneous fluctuations; however, in the longer time interval, the load changes are slow. This results in a low (small value) variation coefficient. While nonpoint pollution is a function of hydrologic conditions as well as the pollutant retention in the aeration zone, the result is a considerably higher (larger value) variation coefficient.

The foregoing discussions create two methods of estimating point and nonpoint pollution loads, (1) genetic (based on the outflow partition) and (2) statistic (assuming L_p is independent from Q).

Both methods have undergone verification based on the data from 12 significantly different river basins (Table 1). All study basins are primarily agricultural (Table 2); their soils differ in permeability (Table 1).

ANALYSIS

Data documenting the in-stream pollution were collected from 1987 to 1991, with the frequency of 50 to 100 samples per year. Data in each measurement cross section, represents several years of different hydrological conditions; however, all these years were predominantly dry. The data set used in interpretation was condensed into monthly arithmetic means of stage, discharge, and pollutant loads.

According to the assumed relative stability of L_p equal to the point pollution load, then the median loads measured at $Q < Q_p$ is accepted as a reliable quantity. It is also assumed that the difference between the total load and point load, $L - L_p$, is related to a nonpoint origin. The annual load, as well as the summary characteristic of the all the river outflow (Table 3), represents 92.7% of the area of Poland, and has been calculated based on the foregoing assumptions.

In this 12-basin data set, the definition of incidental load, consisting of the difference $L_p - L_{min}$, is attributed to the pollution discharges caused by unregulated waste management.

The average monthly diffused load of a particular pollutant in the 12 river basins studied is described in the following regression equation:

$$L_a = S_1 \cdot G \cdot P \cdot (S_2 \cdot A)^a \cdot Q_a^b \cdot H^c \tag{3}$$

and in accordance to the assumed independence of L_p from Q:

$$L_c = S_3 + S_1 \cdot G \cdot P \cdot (S_2 \cdot A)^a \cdot Q_a^b \cdot H^c \tag{4}$$

where L_a = pollution load of nonpoint origin [kg/km² per month], L_c = total load [kg/km²per month], S_1 = factor of proportionality, S_2 = correction factor of catchment area loading, $S_3 = L_p$ quantity equivalent, G = arable land ratio in the catchment area, P = soil permeability indicator, A = loading of catchment area with pollutant [t/km² per year], Q_a = surface and subsurface water outflow [mm/year], H = precipitation [mm/year], and a,b,c are exponents defined in the regression.

The foregoing relations cannot be always applied to all river basins. For example it was found that when using Equation 3, in the case of the Wieprza River basin, the equation could not predict any of the four pollution parameters within the limits of statistical significance. Unit outflow of nonpoint pollution from the Reda River drainage basin indicates low correlation for BOD₅ and phosphorus (P), and in the case of the Pasłęka River the same is true for potassium (K). In Equation 4 the situation characteristic for the Reda River catchment area for BOD₅ and phosphorus is repeated, the low correlation for nitrogen in the Łupawa River was found and in the case of the Słupia River the same is true for phosphorus.

TABLE 1
Hydrological Characteristic of River Catchment Areas

River	Total catchment area (km²)	Interpreted catchment area (km²)	Precipitation (mm)	Q (mm)	Q$_p$ (mm)	Q$_a$ (mm)	Soil percentage of the following permeability				Permeability indicator
							High	Average	Small	Very small	
Odra	116,672[a]	109,729	559.7	138.8	101.4	37.4	35.5	20.8	33.0	10.7	3.190
Ina	2189	2163	577.3	184.7	142.4	42.3	17.3	8.5	69.9	4.3	3.612
Rega	2724	2628	661.6	229.0	173.8	55.2	31.7	13.5	54.8	—	3.231
Parsęta	3151	2955	701.7	287.5	232.0	55.5	44.3	4.1	51.6	—	3.073
Grabowa	535	439	737.5	444.1	378.3	65.7	43.4	—	56.6	—	3.132
Wieprza	1634	1519	713.5	317.8	244.9	72.9	66.2	—	23.0	10.8	2.784
Słupia	1623	1599	726.0	320.7	263.8	56.9	62.5	—	37.5	—	2.750
Łupawa	924	805	720.2	311.0	266.8	44.2	70.9	—	29.1	—	2.582
Łeba	1801	1120	734.3	328.3	243.9	84.5	62.4	—	29.8	7.8	2.830
Reda	485	395	745.6	328.7	241.2	87.5	72.4	—	27.6	—	2.552
Wisła	194,424	194,376	576.9	143.3	95.9	47.4	33.5	24.2	35.5	6.8	3.156
Pasłęka	2294	2232	620.8	227.4	157.0	70.4	29.5	6.2	3.5	60.8	3.956

[a] To the Ina river outlet.

TABLE 2
Description of Riverine Catchment Areas Management

River	Total surface (km²)	(%)	Total agricultural land (km²)	(%)	Arable land (km²)	(%)	Orchards (km²)	(%)	Meadows and pastures (km²)	(%)	Woods (km²)	(%)
Odra	109,729	100	63,753	58.1	50,366	45.9	658.4	0.6	12,729	11.6	33,248	30.3
Ina	2163	100	1248	57.7	928	42.9	10.8	0.5	307	14.2	590	27.3
Rega	2628	100	1698	64.6	1319	50.2	5.3	0.2	376	14.3	628	23.9
Parsęta	2955	100	1575	53.3	1241	42.0	8.9	0.3	328	11.1	1070	36.2
Grabowa	439	100	201	45.7	149	34.0	0.4	0.1	51	11.7	198	45.0
Wieprza	1519	100	668	44.0	521	34.3	3.0	0.2	144	9.5	677	44.6
Słupia	1599	100	806	50.4	630	39.4	3.2	0.2	173	10.8	556	34.8
Łupawa	805	100	403	50.0	339	42.1	1.6	0.2	63	7.8	320	39.7
Łeba	1120	100	551	49.2	374	33.4	1.1	0.1	175	15.6	360	32.1
Reda	395	100	187	47.4	137	34.8	1.6	0.4	48	12.1	163	41.3
Wisła	194,376	100	122,846	63.2	92,912	47.8	2721.3	1.4	27,213	14.0	51,315	26.4
Pasłęka	2232	100	1341	60.1	944	42.3	2.2	0.1	395	17.7	603	27.0
Total	319,960	100	195,277	61.0	149,860	46.8	3418	1.1	42,002	13.1	89,728	28.0

TABLE 3
Distribution of Pollution Loads Discharged into Sea

Parameter	Tons/yr				
	Minimal	Incidental	Nonpoint	Underground	Total outflow
BOD_5	48,048.8	119,864.0	72,820.9	167,912.8	240,733.7
COD	422,577.4	659,845.3	462,860.2	1,082,422.7	1,545,282.9
TOC	134,360.3	176,528.3	195,501.9	310,888.6	506,390.5
Nitrogen	40,949.3	41,304.6	87,115.5	82,253.9	169,369.4
Phosphorus	4,539.9	5,633.1	3,951.3	10,173.0	14,124.3
	Percentage				
BOD_5	20.0	49.8	30.2	69.8	100
COD	27.3	42.7	30.0	70.0	100
TOC	26.5	34.9	38.6	61.4	100
Nitrogen	24.2	24.4	51.4	48.6	100
Phosphorus	32.1	39.9	28.0	72.0	100

In some cases, weak correlations, comparing mean values estimated according to the method of genetic division with the values calculated by Equation 3, still produced a high consistency (Figure 1).

Of the 12 rivers and 4 indicators studied (48 cases), in only 1 case (phosphorus in the Łupawa River) was the measured value considerably different from the calculated value. In the rest of the cases, a high correlation with a significance level above 99.9% was obtained.

A slightly larger spread of results was obtained for the complete data set over what is presented in Figure 1, the example of the Vistula River Basin. The correlation between singular results measured and calculated according to Equation 4 produces the similar effects to that of Equation 3.

Comparison of measured values and calculated values using Equation 4 is characteristic for all rivers taken together, producing a high correlation and a significance level above 99.9% for the mean values of independent variables (Figure 2). The Pasłęka River constitutes the only exception with the phosphorus outflow calculated 12% higher than the measured value. Correlations for potassium and nitrogen (N) are much stronger than for BOD_5 and for phosphorus; probably the result of different mobility and transport mechanisms between these two groups of pollutants.

The process of erosion in the transport of phosphorus transport to the receiving water (McElroy et al., 1976; Reinelt, 1990; Taylor, 1984; Bogacka et al., 1993) would suggest the presence of interrelation between phosphorus outflow and suspended solids outflow; however, such a relation has not been obtained at a statistically significant level.

If the physical meaning was addressed to the equation arguments, constant S_3 should be equal to the amount of substance ground outflow while the other part of the equation should be related to surface runoff. This is the case for potassium and nitrogen. For these pollution indicators, the discussed constants are very similar to the mentioned outflows; however BOD_5 does not display the same equality.

A relation between underground substance outflow and the constant S_3 appears in most of the river basins studied; this relation can be described by the following regression equations:

$$BOD_5 \quad y = 0.883 \quad x + 24.74 \tag{5}$$

$$N \quad y = 0.864 \quad x + 7.607 \tag{6}$$

$$P \quad y = 0.830 \quad x + 1.928 \tag{7}$$

$$K \quad y = 1.026 \quad x + 1.336 \tag{8}$$

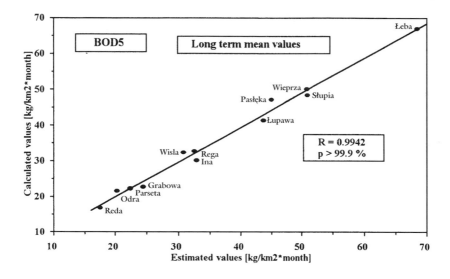

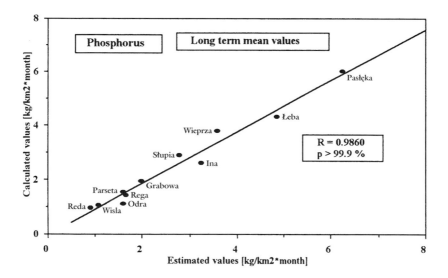

FIGURE 1.   Correlation of surface outflow loads estimated and calculated in accordance with Equation 3.

where y = underground outflow [kg/km^2 per month], and x = constant S_3 [kg/km^2 per month].

Analysis of the above equations indicates that for nitrogen, phosphorus, and potassium, S_3 is equal or slightly lower than underground substance outflow estimated by the method of genetic division. This is one method of verification. However, in the case of BOD$_5$, there is a considerable difference.

The ratio of independent variables in Equations 3 and 4 are different, and they vary from one river basin to another, and for each pollution indicator. However, using groups of variables describing:

M: catchment areas characteristic (permeability and management)
O: drainage basin loading
W: hydrological conditions (surface outflow, precipitation)

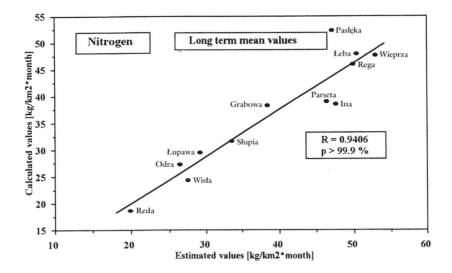

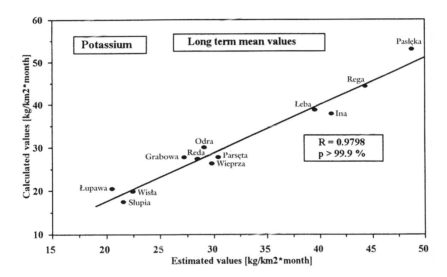

FIGURE 1.   *Continued.*

For the 12 river basins, the following order prevails:

Nitrogen outflow   W > O > M
Phosphorus outflow   M = W > O
Potassium outflow   W > O > M

In the above order, only BOD_5 has been omitted because of high variability associated with that relation. Rivers located in the eastern part of the coast are dominated by their hydrology and so are the pollutant loads. This begins with the Słupia River and ends at the Vistula and Pasłęka Rivers. In the rivers of the western coast, mainly in the Parsęta, Grabowa, and Wieprza Rivers, loading of catchment area plays the main role.

It is only in the case of phosphorus that the catchment area is the dominating factor in the relation of drainage basin loading. Numeric values characterizing the catchment area attributes are a factor in the geographical system:

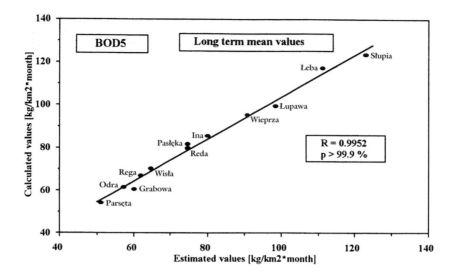

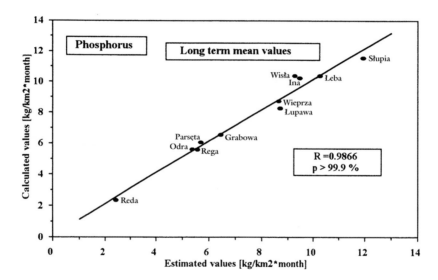

FIGURE 2.   Correlation of the riverine pollution outflow loads measured and calculated in accordance with Equation 4.

- Drainage areas of the Odra, Ina, Rega, and Parsęta Rivers have M values in the range of 1.29 to 1.62.
- Drainage areas of the Grabowa, Wieprza, Słupia, Łupawa, Łeba, and Reda Rivers display M in the range of 1.09 to 0.89.
- Drainage areas of the Vistula and the Pasłęka Rivers display M values equal to 1.47 and 1.67, respectively.

After looking at the foregoing data, one can suppose that in the drainage basins of the central coast from the Grabowa to the Reda Rivers, where the soils are highly permeable, the influence of the catchment area characteristics on the amount of nonpoint pollution outflow is small. The greatest influence is seen in the Reda and Pasłęka Rivers.

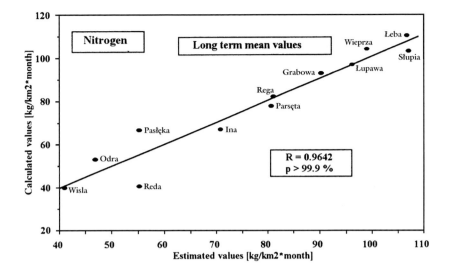

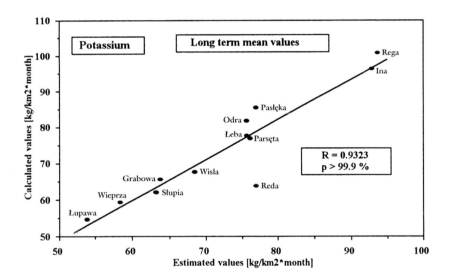

FIGURE 2. *Continued.*

Where dominance of hydrological conditions caused by surface runoff prevails, precipitation plays the smaller part. The only exception is phosphorus outflow in the Łupawa River, which is characteristic of a considerably higher ratio of precipitation in the total outflow of this element.

Taking into account the foreseen dominance of hydrological conditions in shaping the outflow of nonpoint pollutants, the following relation has also been investigated using regression techniques:

$$L_a = S_4 \cdot Q^a \cdot bH^c \qquad (9)$$

where S_4 = the factor of proportionality, a and c = exponents defined in regression and b = multiplier defined in regression.

Only a slightly lower correlation coefficients have been obtained for Equation 9 compared to Equation 4. In the case of nitrogen, the difference between the respective correlation coefficients between equations ranged from 1 to 26%, with an average value equal to 6%. This similar tendency was observed for the other indicators. It is then possible to estimate nonpoint pollution outflow based exclusively on elements of the hydrology of a particular river.

The research work presented has been done within project No. 6 6045 91 02, financially supported by Polish Committee for Science Research 1992 to 1993.

REFERENCES

Atlas Hydrologiczny Polski, Praca zbiorowa pod kier. J. Stachy, IMGW, 1986.

Bogacka, T. et al., Control of nonpoint pollution from agriculture in the Vistula Basin, WHO European Centre for Environment and Health, 1993.

McElroy, A. D., Chiu, S. Y., Nebgen, I. W., Aleti, A., and Bennett, F. W., Loading functions for assessment of water pollution from nonpoint sources, U.S. EPA 600/2-76-151, 1976.

Reinelt, L., Non-point source water pollution management (monitoring, assessment and wetland treatment), *Linköp. Studies Arts Sci.*, 57, 1990.

Rybiński, J. et al., Fluctuations of the pollutant transport in rivers to the sea and factors influencing the fluctuations. Institute of Meteorology and Water Management, Gdańsk, Committee of Scientific Research (KBN) Nr 6-6045-91-02, 1993.

Taylor, R., The Runoff of Nitrogen and Phosphorus Compounds from selected Agricultural Regions in the Vistula and Odra Drainage Basins, KBM PAN Oceanologia 18, 1984.

25 Nonpoint Sources of Pesticides in the San Joaquin River, California: Input from Winter Storms, 1992 to 1993

Joseph L. Domagalski

Organophosphate insecticides, including chlorpyrifos, diazinon, and methidathion, are applied to dormant orchards in the San Joaquin Valley, CA, during late December through January. This timeframe coincides with the period of heaviest rainfall in the valley, and rainfall mobilizes a portion of these pesticides from the orchards. The pesticides enter the San Joaquin River and have been detected along the perennial reach of the river.

A storm on the evening of February 7 and the morning of February 8, 1993, deposited more than an inch and a half of rain in the San Joaquin Valley. Two distinct peaks of organophosphate pesticide concentrations were measured at the mouth of the San Joaquin River during a single rise in discharge. Both peaks were attributed to contrasts between the soil texture and hydrology of the eastern and western valley. The fine soil texture and small size of the western tributary basins contributed to rapid runoff. Diazinon concentrations peaked within hours after rainfall ended and then decreased because of a combination of dilution with pesticide-free runoff from the nearby Coast Ranges and decreased pesticide concentrations in the agricultural runoff. Data for the Merced River, a large tributary of the eastern San Joaquin Valley, are sparse, but indicate that peak concentrations occurred at least a day after those of the western tributary streams. That delay may be due to the presence of well-drained soils, the larger size of the drainage basins, and the management of surface-water drainage networks. Runoff from a subsequent storm, on February 18 and 19, contained significantly lower concentrations of most organophosphate pesticides, indicating that runoff from the first storm had already removed most of the pesticides available for rainfall-induced transport.

INTRODUCTION

Acetyl cholinesterase-inhibiting pesticides are used extensively in the San Joaquin Valley (Figure 1), an important agricultural region of the U.S. These pesticides are used throughout the year, including the winter months when most rainfall occurs. Organophosphate pesticides affect nerve impulses in target organisms by acetyl cholinesterase inhibition (Buchel, 1983) and can be toxic to some aquatic organisms at concentrations less than one part per billion (Amato et al., 1992; Marshall and Roberts, 1978). An understanding of the transport and concentration of the pesticides that are present in surface waters during the winter months is critically important so that effective control strategies can be implemented.

The occurrence of pesticides in surface waters has been reported by numerous authors, including Moody and Goolsby (1993), Pereira and Hostettler (1993), Pereira and Rostad (1990), Richards

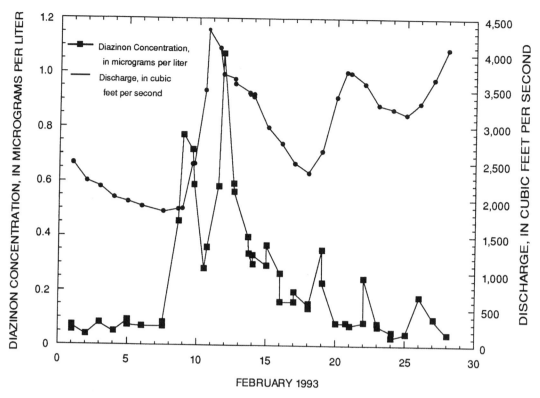

FIGURE 8. Discharge and diazinon concentrations in water samples from the San Joaquin River near Vernalis, CA, February 1993

The second peak concentration of diazinon in the San Joaquin River occurred at 1900 hours on February 11. At 1900 hours on February 9, the hydrograph for the Merced River was in the middle of the rising limb. Therefore, if the diazinon concentrations in the rising limb of the Merced River hydrograph were high, between 5 and 7 μg/L, the river could be the source of the second peak concentration of diazinon observed at the San Joaquin River near Vernalis. This hypothesis has not been tested, but is based on measured velocities in the Merced River and a tracer test conducted under a similar flow regime, which determined that the approximate travel time of a parcel of water from the Merced River sampling site to the sampling site on the San Joaquin River is 2 days. A diazinon concentration of 2.5 μg/L was measured on the falling limb of the Merced River hydrograph, and, because pesticide concentrations generally are lower on the falling limb (Richards and Baker, 1993), concentrations probably were higher on the rising limb.

Two other possible major sources of pesticides to the San Joaquin River — the Tuolumne and the Stanislaus Rivers — were sampled near peak discharge, about the same time as the peak for the Merced River. The concentrations of diazinon in these samples were negligible (Lisa Ross, California Department of Pesticide Regulation, written commun., 1993). Based on a dye-tracer study completed under a similar flow regime, it is possible that the Stanislaus and Tuolumne Rivers could have carried the diazinon measured at the San Joaquin River near Vernalis if the concentrations were low on the rising limb and high on the decreasing limb.

An examination of the data on Figure 8 for the San Joaquin River shows that there were several days when the diazinon concentrations were near or exceeded 0.35 μg/L. That concentration is significant because it is that which was reported by Amato et al. (1992) as the LC_{50} (concentration lethal to 50% of a test population) for one aquatic organism, *Ceriodaphnia dubia*, a zooplankton species. That species is typically used in bioassays of water suspected of having elevated concentrations of various contaminants (Mount and Anderson-Carnahan, 1988; 1989a,b). Therefore, the loading of diazinon from various sources within the San Joaquin Valley resulted in the presence of water that was toxic to at least one aquatic species.

Additional storm sampling was done on February 18 and 19 during the first significant rainfall following the February 7 storm (Figure 4). The Orestimba Creek site was sampled during this storm to determine if any residual diazinon was being transported to the San Joaquin River during this rainfall and to collect additional samples at peak discharge to verify relatively low concentrations attributable to dilution from Coast Ranges runoff. The highest concentration of diazinon, 0.2 µg/L, was measured at the creek site prior to the steep rise of the hydrograph and was much less than the highest concentration observed during the previous storm. This lower concentration indicates that much of the available diazinon load was transported during the February 7 storm. Following the concentration peak, diazinon concentration levels decreased by an order of magnitude and remained low throughout peak discharge. Chlorpyrifos was not detected at Orestimba Creek during this second storm.

The highest concentration of diazinon measured at the San Joaquin River site during the storm of February 18 and 19 was just under 0.4 µg/L, considerably less than the maximum concentration measured at this site during the previous storm. Basinwide response, as indicated by the diazinon concentrations measured in samples from the San Joaquin River at Vernalis, was less than during the previous storm; the reduced response indicates that, throughout the basin, most of the available diazinon had been transported previously. Chlorpyrifos was not detected at the San Joaquin River near Vernalis.

The Merced River was sampled on February 23 and 26 on the declining part of the discharge hydrograph resulting from the February 18 to 19 storm. The concentration of diazinon rose from a prestorm level of 0.009 to 0.19 µg/L. The chlorpyrifos concentration rose from less than detectable to 0.033 µg/L.

Data from this second storm indicate that much of the diazinon was transported to the San Joaquin River during the previous storm, which is supported by the relatively lower concentrations of diazinon at the Orestimba Creek, Merced River, and San Joaquin River sites. Data for Orestimba Creek support the hypothesis that low concentrations occur at peak discharge in streams of the western San Joaquin Valley. Peak discharge at Orestimba Creek that originated from the Coast Ranges resulted in the dilution of pesticide concentrations measured at the downstream Orestimba Creek sampling site.

The calculated amount of diazinon transported by the San Joaquin River, as measured near Vernalis, was 47 kg during February. This amount of diazinon represents about 0.1% of the total applied within the drainage of the perennial reach of the San Joaquin River, based on 1990 data, the most recent available. Only a relatively small amount of the applied diazinon was mobilized to the river by rainfall. Nevertheless, during the 1992 to 1993 dormant-orchard season, the amount was sufficient to result in toxic levels for at least one aquatic organism.

SUMMARY AND CONCLUSIONS

This study of pesticides for the dormant-orchard period indicates certain patterns of loading during runoff for parts of the San Joaquin Valley; however, further research is necessary. Diazinon concentrations in the western tributaries to the San Joaquin River increase during the rising part of the storm hydrograph and tend to decrease rapidly as flow from the Coast Ranges increases, relative to locally derived valley runoff. The first major storm following pesticide application mobilizes most of the diazinon that is available to be transported from the orchards. At present, specific factors, especially at the field level, that promote diazinon transport during rainfall are unknown. Further understanding of the pesticide contribution of dormant orchards to the large eastern tributaries, such as the Merced River, could increase the understanding of the San Joaquin Valley drainage system. Preliminary data indicate that pesticide input from the Merced River may be important, but this has not been verified by additional sampling.

Data indicate that toxic levels of diazinon can be present along most of the perennial reach of the San Joaquin River during the winter season. This presence occurred during the 1992 to 1993 dormant orchard period because most pesticides were applied during a narrow window of time and a heavy rainfall occurred shortly after the applications. Runoff from one major storm carried the greatest amount of pesticides to the river. The results of this study demonstrate the necessity of selecting the proper sampling frequency to assess the relative contribution of individual drainages to the loading of toxic water to the San Joaquin River. The response time of the tributaries of the western San Joaquin Valley is rapid and occurs within several hours. The only way to assess the concentrations of toxic compounds is to sample the river or streams during a storm. In contrast, assessing the response of the basin to a storm is relatively easier, because elevated pesticide concentrations usually reach the San Joaquin River at Vernalis a day after the storm. Nevertheless, twice daily sampling, separated in time, was the minimum required during the storm to document that there were at least two major sources of pesticides.

REFERENCES

Amato, J. R., Mount, D. I., Durhan, E. J., Lukasewycz, M. T., Ankley, G. T., and Robert, E. D., An example of the identification of diazinon as a primary toxicant in an effluent, *Environ. Toxicol. Chem.*, 11, 209, 1992.
Buchel, K. H., *Chemistry of Pesticides*, John Wiley & Sons, New York, 1983, 581.
California Department of Pesticide Regulation, Pesticide use data. Computer tapes available from California Department of Pesticide Regulation, 1990.
Crepeau, K. L., Kuivila, K. M., and Domagalski, J. L., Methods of Analysis and Quality-Assurance Practices of the U.S. Geological Survey Organic Laboratory, Sacramento, California — Determination of pesticides in water by solid-phase extraction and capillary-column gas chromatography/mass spectrometry, U.S. Geological Survey Open-File Report 94-362, 1994, 17.
Domagalski, J. L. and Kuivila, K. M., Distributions of pesticides and organic contaminants between water and suspended sediment, San Francisco Bay, California, *Estuaries*, 16, 416, 1993.
Edwards, T. K. and Glysson, G. D., Field methods for measurement of fluvial sediment, U.S. Geological Survey Open-File Report 86-531, 1988, 118.
Marshall, W. K. and Roberts, J. R., Ecotoxicology of Chlorpyrifos. National Research Council of Canada Associate Committee on Scientific Criteria for Environmental Quality, NRCC Publication 16079, NRCC/CNRC Publications, Ottawa, Ontario, Canada, 1978, 314.
Moody, J. A. and Goolsby, D. A., Spatial variability of triazine herbicides in the lower Mississippi River, *Environ. Sci. Technol.*, 27, 2120, 1993.
Mount, D. I. and Anderson-Carnahan, L., Methods for aquatic toxicity identification evaluations. Phase I — Toxicity characterization procedures. EPA 600/3-88-034. U.S. Environmental Protection Agency, Environmental Research Laboratory, Duluth MN, 1988.
Mount, D. I. and Anderson-Carnahan, L., Methods for aquatic toxicity identification evaluations. Phase II — Toxicity identification procedures. EPA 600/3-88-035. U.S. Environmental Protection Agency, Environmental Research Laboratory, Duluth MN, 1989a.
Mount, D. I. and Anderson-Carnahan, L., Methods for aquatic toxicity identification evaluations. Phase III — Toxicity confirmation procedures. EPA 600/3-88-036. U.S. Environmental Protection Agency, Environmental Research Laboratory, Duluth MN, 1989b.
Pereira, W. E. and Hostettler, F. D., Nonpoint source contamination of the Mississippi River and its tributaries by herbicides, *Environ. Sci. Technol.*, 27, 1542, 1993.
Pereira, W. E. and Rostad, C. E., Occurrence, distributions, and transport of herbicides and their degradation products in the lower Mississippi River and its tributaries, *Environ. Sci. Technol.*, 24, 1400, 1990.
Readman, J. W., Kwong, L. L. W., Mee, L. D., Bartocci, J., Nilve, G., Rodriguez-Solano, J. A., and Gonzalez-Farias, F., Persistent organophosphorous pesticides in tropical marine environments, *Marine Pollut. Bull.*, 24, 398, 1992.
Richards, R. P. and Baker, D. B., Pesticide concentration patterns in agricultural drainage networks in the Lake Erie Basin, *Environ. Toxicol. Chem.*, 12, 13, 1993.
Thurman, E. M., Goolsby, D. A., Meyer, M. T., and Kolpin, D. W., Herbicides in surface waters of the midwestern United States: The effect of spring flush, *Environ. Sci. Technol.*, 25, 1794, 1991.

26 Nutrient Loading in the Vistula River Course

J. R. Dojlido, E. Niemirycz, and P. Morawiec

The loadings of nutrients in the Vistula River were calculated and are discussed in this chapter. The Vistula is the largest river in Poland (length 1038.4 km, basin surface 199,813 km^2). Intensive monitoring of Vistula River water quality was carried out over a 3-year period: 1989, 1990, and 1991. Measurements were performed twice per week at three stations located in the upper, middle, and lower parts of the river. The following forms of nutrients were determined: ammonia, nitrate, organic nitrogen, total nitrogen, phosphate, and total phosphorus. The concentrations, loads, and unit loads per basin area are discussed and the point and nonpoint sources of nutrients were calculated.

INTRODUCTION

The Vistula River discharges huge loads of nitrogen and phosphorus into the Baltic Sea and may be the largest source of pollution to the Baltic Sea. Some of the calculated nutrient loadings reflect the great size of the river basin (length 1038.4 km, basin surface area 199,813 km^2), but the large population (more than 20 million), extensive industry (e.g., Silesia region), lack of tertiary sewage treatment to remove nitrogen and phosphorus, and agriculture are mostly responsible for much of the nutrients in the surface waters. The goal of this study was to estimate the loads for various nutrients in the Vistula River and to compute their discharge into the Baltic Sea.

The study was based on the data from intensive monitoring of Vistula River water quality performed by the State Environmental Monitoring Department. As mentioned, the research was performed over a 3-year period: 1989, 1990, and 1991. The measurements were performed at three stations: Tyniec 63.7 km (upper Vistula, upstream of Cracow city), Warsaw 510.0 km (middle Vistula, upstream from the discharge of sewage from Warsaw), and Kiezmark 926.0 km (lower Vistula, close to the outlet of the Vistula River to the Baltic Sea; Figure 1).

The frequency of measurement was twice a week (during a 3 year study more than 300 data measurements were collected). For nutrients, the following determinations were performed: ammonia, nitrate, Kjedahl nitrogen (sum of organic nitrogen and ammonia), phosphate (inorganic phosphorus), and total phosphorus (sum of inorganic and organic phosphorus).

The following parameters were discussed: ammonia, nitrate, organic nitrogen, total nitrogen (sum of all nitrogen forms: NH_4-N, NO_3-N and organic nitrogen; nitrites occurring in very low concentrations were omitted), phosphate (PO_4-P), and total phosphorus. The water level was measured daily. Pollutant loads were calculated from the data of pollutant concentration and water flow. Daily concentrations of pollutants and loads of pollutants changed greatly between seasons and years in ways difficult for interpretation. In this chapter the mean values for three years are discussed, i.e., mean concentration and mean load at the measuring points. In addition, the load per basin area was calculated (for three partial areas: upstream to Tyniec, between Tyniec and Warsaw, and between Warsaw and Kiezmark). The areas of the separate catchment are presented in Table 1.

○ | measuring points
————————— | rivers
— — — | basin border

FIGURE 1. Vistula River Basin.

WATER FLOW

The characteristic discharges of the Vistula River are shown in Table 2. The yearly means of river water flow from the days of water quality measurement are presented in Table 3.

AMMONIA

The mean concentration, mean loads, and a unit load are presented in Figures 2, 3, and 4. The concentration of ammonia was highest in Tyniec, much lower in Warsaw, and then a little higher

TABLE 1
Drainage Areas of Various River Catchments

From source to Tyniec	From Tyniec to Warsaw	From Warsaw to Kiezmark	From Kiezmark to Baltic Sea (+ delta)	Sum
7524 km²	77,334 km²	109,551 km²	5405 km²	199,813 km²
3.77%	38.7%	54.8%	2.7%	100%

TABLE 2
Characteristic Discharges of the Vistula River

Discharge in m³/s	Cracow (upper course)	Warsaw (middle course)	Kiezmark (lower course)
Highest	2260	5650	7840
Mean	92	592	1090
Mean low	38	198	403
Lowest	19	108	253

TABLE 3
Yearly Mean Discharge for Those Years of Water-Quality Measurement (m³/s)

Year	Tyniec	Warsaw	Kiezmark
1989	80.8	575	943
1990	59.2	455	774
1991	78.4	447	850

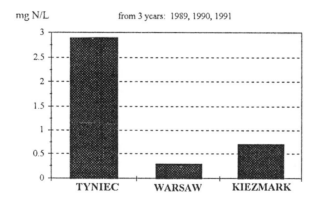

FIGURE 2. Mean concentration of NH_4-N.

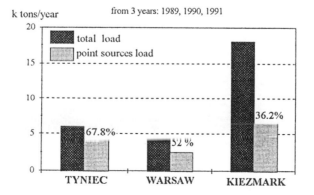

FIGURE 3. Mean load of NH_4-N in percent.

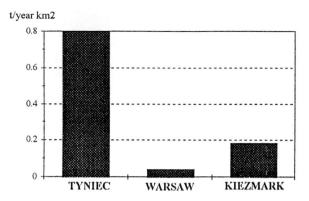

FIGURE 4.   Unit load of NH$_4$-N.

in Kiezmark. The load of ammonia was high in Tyniec (about 6000 t/yr), lower in Warsaw despite a big increase in flow, and very high in Kiezmark.

The unit load of ammonia per surface area was very high in Tyniec, small in Warsaw, and intermediate in Kiezmark. The high concentration of ammonia and large unit load in Tyniec, in the upper part of the Vistula River, reflects a very industrialized and densely populated area (Silesia region). Along the river above Warsaw, the concentration and load of ammonia decreased rapidly. This may be explained by the dilution, by less input of ammonia, and by nitrification. Simultaneously, in the reach between Tyniec and Warsaw, the concentration and load of nitrate increased. In the reach between Warsaw and Kiezmark, the concentration of ammonia increased significantly, which may be explained by the input of untreated sewage (e.g., from Warsaw where only about 25% of sewage is treated).

NITRATE

The concentration of nitrate increased between Tyniec and Warsaw, presumably by intensive nitrification in the river. Nitrate decreased in the river course to Kiezmark (Figure 5). In the reach between Warsaw and Kiezmark, the external input of ammonia and organic nitrogen was higher than that from the nitrification.

Nitrification and the direct inflow of nitrate resulted in the load of nitrate in Warsaw being 10 times higher than in Tyniec (Figure 6) (there is a nitrogen fertilizer plant situated upstream to Warsaw). The load between Warsaw and Kiezmark increased insignificantly, despite the large increase in water flow.

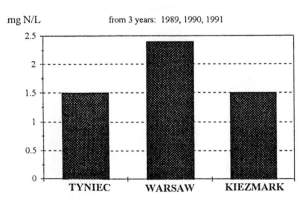

FIGURE 5.   Mean concentration of NO$_3$-N.

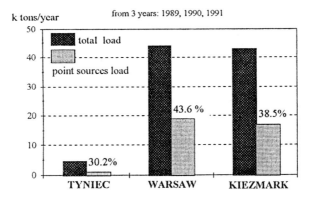

FIGURE 6.   Mean load of NO$_3$-N in percent.

ORGANIC NITROGEN

The highest and lowest concentrations of organic nitrogen were in Tyniec and in Kiezmark, respectively (Figure 7). This was the result of transformation of organic compounds into inorganic compounds along the Vistula River. The load of organic nitrogen was growing from the upper to the lower part of the river (Figure 8). The ratio of inorganic nitrogen to organic nitrogen increased from 1.7 in Tyniec to 2.4 in Warsaw and Kiezmark.

TOTAL NITROGEN

The highest and lowest concentrations of total nitrogen occurred in Tyniec and Kiezmark, respectively (Figure 9). The values of the load were reversed: the lowest in Tyniec and the largest in Kiezmark (Figure 10). The unit load was the highest above Tyniec, and decreased downstream until Warsaw (Figure 11).

PHOSPHATE

The concentration of orthophosphates were the highest in Tyniec, decreased in the middle part of the Vistula River, and increased again in Kiezmark (Figure 12). The load of phosphates increased along the Vistula River course, but the largest increase occurred in the lower part of the river (Figure 13).

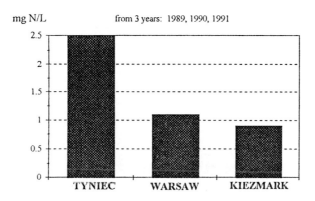

FIGURE 7.   Mean concentration of organic-N.

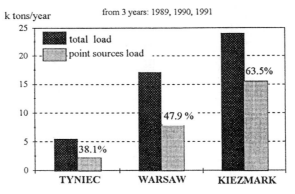

FIGURE 8.   Mean load of organic-N.

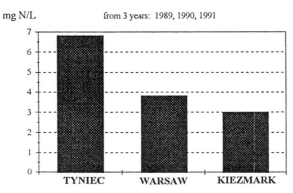

FIGURE 9.   Mean concentration of total-N.

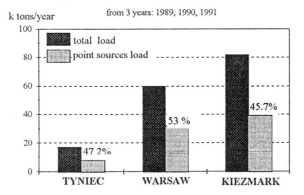

FIGURE 10.   Mean load of total-N.

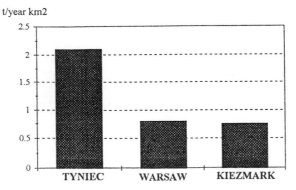

FIGURE 11.   Unit load of total-N.

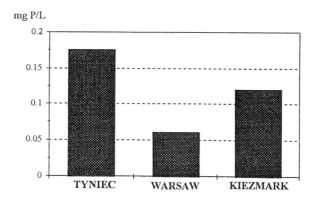

mg P/L

FIGURE 12. Mean concentration of PO$_4$-P.

The high concentration of phosphate in Tyniec could be caused by the discharge of large amounts of sewage in the upper part of Vistula River and by the small river flow. The decrease in phosphate along the river to Warsaw could be explained by processes such as dilution and sedimentation. The increase in phosphate concentration in Kiezmark may indicate a great input of phosphate in the lower part of the Vistula River. It could be also influenced by the intensive process of biodegradation of organic compounds containing phosphorus.

TOTAL PHOSPHORUS

The concentration of total phosphorus was highest in Tyniec; in Warsaw and Kiezmark the concentrations were much lower and similar (Figure 14). The ratio of phosphate to total phosphorus was Tyniec 0.28, Warsaw 0.30, and Kiezmark 0.63. The process of transformation of organic phosphorus into inorganic phosphorus (self-purification) was very intensive in the lower part of the Vistula River. The load increased along the river course (Figure 15). The unit load of total phosphorus was highest in Tyniec (discharges of raw sewage) and lower, but similar in Warsaw and Kiezmark (Figure 16).

FORMS OF NUTRIENTS

The forms (species) of nitrogen and phosphorus in the upper, middle and lower reaches of the Vistula River are shown in Figures 17 and 18. Along the Vistula River, the percentage of organic nitrogen decreased, nitrate increased in the middle part, and ammonia increased in the lower part. The proportion of inorganic phosphorus increased along the river and was especially high in the lower part.

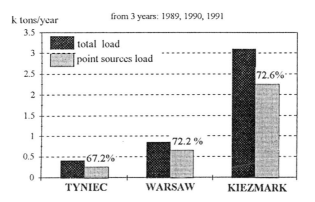

k tons/year   from 3 years: 1989, 1990, 1991

FIGURE 13. Mean load of PO$_4$-P.

TABLE 5
Mean Loads of Nutrients Transported in the Vistula River for 1989, 1990, and 1991 (Kt/yr)

Locality	NH_4-N	NO_3-N	N_{org}	N_{tot}	PO_4-P	P_{tot}
Tyniec	6.1	4.1	5.7	15.9	0.4	1.5
Warsaw	4.3	44.1	16.8	60.8	0.9	3.2
Kiezmark (to Baltic Sea)	18.5	42.8	24.0	84.0	3.1	4.9

TOTAL NITROGEN

The ratio of point sources to nonpoint source in the Vistula River is about 1:1.

PHOSPHATE

In the Vistula River, the point sources consist of about 70% of the total input.

TOTAL PHOSPHORUS

The percentage of point source is slowly growing along the Vistula River, reaching as high as 75% in the lower part.

CONCLUSION

The concentrations of all forms of nitrogen and phosphorus were the highest in Tyniec in the upper part of the Vistula River. It was caused by the large load of wastewaters discharged to the Vistula River and a low river flow. In the middle part of the Vistula River up to Warsaw, the concentrations of all forms decreased, with the exception of nitrate. In the lower part of the Vistula River, downstream to Warsaw, the concentrations of ammonia and phosphate increased, and the concentrations of other forms of nitrogen (nitrate, organic nitrogen, total nitrogen, and total phosphorus) decreased.

The load of ammonia decreased in the middle part of the Vistula River while the load of nitrate increased rapidly, despite an increase in water flow (to account for dilution) and the intensive rate of nitrification. Loads of other forms of nutrients increased gradually along the river course.

The highest values of unit load (tons/year per km²) were found in the upper part of the Vistula River. The loads of nutrients transported in the Vistula River are presented in Table 5.

ACKNOWLEDGMENTS

We thank Dr. J. Krokowski of the Institute of Meteorology and Water Management in Cracow for permission to use the data from the monitoring station in Tyniec.

27 Nutrient Nonpoint Pollution in Experimental Watersheds in the Vistula Basin

Teresa Bogacka and Regina Taylor

Nonpoint pollution generates a very serious problem. The average river outflow of nitrogen (N) and phosphorus (P) from Polish rivers into the Baltic Sea (over 200,000 t of N and 20,000 t of P) constitutes, about 30% and 40% respectively, of total load transported by all rivers flowing into the Baltic. According to estimated calculations, about 60% N and 30% P of the total outflow from Poland comes from nonpoint sources.

A study was performed from 1977 to 1990 in small watersheds of the Vistula drainage basin selected in different regions of Poland (Figure 1). The watersheds represent areas of typical physiographic features, with agricultural use varying from 26 to 100% of the individual small watershed. None of the watersheds had any significant point pollution sources. Data from these investigations were used to estimate nutrient losses from agriculture into surface waters, and also constituted the basis for a surface-water protection program against nonpoint pollution. The results of experiments carried out in the watersheds are presented in Table 1. Agricultural land use, coefficients of the unit nutrient runoff from watershed, topography, and soil permeability were characterized for each watershed. The last two features strongly affect nutrient leaching from watershed into surface waters in the hydrologic cycle.

The experimental stream watersheds selected in the Raba (Stachowicz, 1990) and Rudawa (Żeglin, 1990) drainage basins are typical of the piedmont areas and middle hills of great slopes, which are covered with the poorly permeable soils. As shown in Table 1, the unit nutrient runoff was high and ranged from 6.8 to 14.7 kg N/ha per year and from 0.5 to 2.20 kg P/ha per year. This value is dependent on contribution of the arable land to the total area of watershed and on the atmospheric conditions (Table 1).

The Brynica River Watershed (Pudo et al., 1979) in the Czarna Przemsza drainage basin represents typical agricultural area of the piedmont plateau covered with highly permeable soils. The nutrient runoff was low (2.2 to 3.0 kg N/ha per year and 0.12 to 0.19 kg P/ha per year; Table 1).

The watersheds of Wilga (Guberski and Szperliński, 1990), Sucha (Jakubowska, 1979), and Skierniewka Rivers (Dojlido et al., 1990) represent the Mazowiecka Lowland District which includes a large part of the upper and lower Vistula drainage basin (Table 1). These lowland agricultural watersheds (75 to 100% of arable land) are characterized by different lithology. The nutrient runoff from these watersheds ranged from 1.06 to 16.68 kg N/ha per year and 0.025 to 1.69 kg P/ha per year and it was mainly dependent on the soil permeability.

The watersheds of Wda and Radunia Rivers — tributaries of the Vistula (Taylor et al., 1979) — are typical for the Middle Pomeranian Lake Districts. They are characterized by diversified surface sculpture created during ice age as well as by unconsolidated and highly permeable soils consisting of postglacial gravels, sands, and clays. These formations of various granulation occur in different proportions in areas of the watersheds. Because of this, the soils in the catchment areas are characterized by differential infiltration properties. Therefore, the soil permeability of the Wda

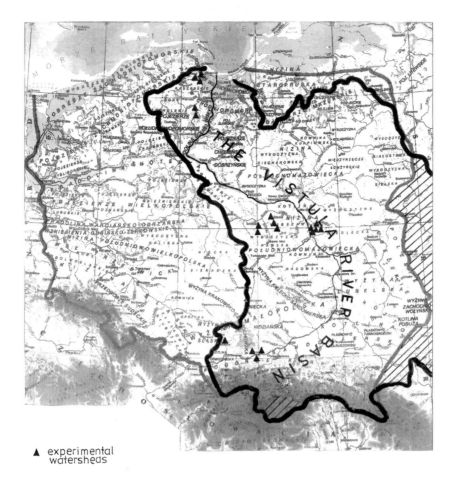

▲ experimental
   watersheds

FIGURE 1. Location of experimental watersheds in the Vistula Basin.

River watershed is high, while in the Radunia River watershed it is low (Table 1). In this region, a large resource of groundwater together with water stored in lakes causes low-flow augmentation and the lowering of peak-flow amplitudes compared to other parts of Poland. Loads of nitrogen and phosphorus transported by the Wda and Radunia Rivers ranged from 0.44 to 2.08 kg N/ha per year and 0.16 to 0.31 kg P/ha per year, respectively.

In watersheds devoid of point pollution sources, the quality of river water depends on agricultural intensity. Of all the investigated rivers, the waters of typical forest rivers had the best quality. Mean concentrations of pollutants and nutrient concentrations in these forest waters indicated high water quality. Watersheds that were totally or mainly utilized for agricultural purposes had the most polluted river waters. This is particularly true for watersheds with fertile and poorly permeable soils, as well as those situated in the piedmont regions of steep slopes and covered with soils having poor infiltration. It was found that ammonia nitrogen (NH_3-N) is not present in high enough concentrations to be a point pollution problem. The NH_3-N concentrations in the water runoff from all investigated watersheds were lower than 3 mg/L. The NO_3-N concentrations in the river waters vary more than concentrations of other types of nitrogen. This is closely related to the transport dynamics and to the physicochemical properties of nitrates. During the investigation period, the concentrations of total nitrogen in the river waters ranged from 0.05 to 18.27 mg N/L. The amplitude of the observed fluctuations not only depended on the watershed nature, but also on the water relations prevailing in watershed. Moreover, the maximum concentrations of nitrates and total nitrogen depended on soil fertility, as well as the amount of fertilizer used. The lowest nitrate and total nitrogen contents in the river waters were found in the watersheds of highly permeable soils

TABLE 1
Outflow of Nitrogen and Phosphorus from the Experimental Watersheds in the Vistula Basin

River, basin, description	Watershed area (km²)	No. of subwatershed	Soil permeability	Agricultural land (%)	Year	Water runoff (m³/ha/yr)	Load (kg/ha per year)	
							N	P
Brynica, Czarna Przemsza basin, piedmont plain, dominated slopes 0 to 3%	100	I	High (sands, loamy sands)	40.9	1977–1978	1835	2.12	0.190
					1978–1979	2197	2.98	0.120
Brezówka stream, Raba basin, middle mountains, slopes <20%	5.3	I	Low (sandstones, slates, clays)	43.5	1979	3217	9.30	0.600
					1980	2586	6.80	0.400
					1981	2807	9.60	0.300
Wolnica stream, Raba basin, middle mountains, slopes 10 to 20%	5.9	I	Low (sandstones, slates, clays)	83.2	1979	3847	9.10	1.100
					1980	4257	10.50	2.200
					1981	3154	11.90	0.500
Rudawa river, Vistula tributary, piedmont, dominated slopes 5 to 15%	290	I	Low (loess, loams, limestones)	67.5	1987–1988	2468	14.60	1.000
					1989–1990	1912	14.50	1.270
Wilga, Vistula tributary, lowland, average slope 2.7%	13.9	I	High	78.9	1977		1.06	0.045
					1978		1.20	0.045
	20.2	II	High	83.8	1977		1.44	0.066
					1978		1.55	0.089
	43.6	III	High	75.4	1977		2.09	0.025
					1978		2.37	0.045
	22.0	IV	High	78.7	1977		1.86	0.050
					1978		1.88	0.078
	31.9	V	Middle	78.4	1977		1.26	0.246
					1978		1.54	0.268
Sucha River, Bzura Basin, lowland, average slope 0.5%	22.7	I	Middle	81.0	1977–1978	2937	10.19	0.900
					1978–1979	2988	10.46	0.410
	13.6	II	Low	100.0	1977–1978	3187	16.68	1.690
					1978–1979	3000	16.51	0.720
	59.2	III	High	79.0	1977–1978	2364	9.70	1.100
					1978–1979	2320	9.56	0.520

TABLE 1 *(Continued)*
Outflow of Nitrogen and Phosphorus from the Experimental Watersheds in the Vistula Basin

River, basin, description	Watershed area (km²)	No. of subwatershed	Soil permeability	Agricultural land (%)	Year	Water runoff (m³/ha/yr)	Load (kg/ha per year) N	Load (kg/ha per year) P
Skierniewka River, Bzura Basin,	100.8	I	Low	84.1	1981–1983	1041	2.20	0.100
lowland, average slope 0.5%	14.9	II		83.5	1981–1983	657	1.50	0.100
Wda River, Vistula tributary,	940	I	Very high (gravels, sands)	30.0	1977–1978	1777	0.94	0.171
outwash area, domin. slopes					1978–1979	1615	0.91	0.162
0 to 3%	1386	II		26.0	1977–1978	1825	0.89	0.178
					1978–1979	1670	0.82	0.162
Radunia River, Motława Basin,	210	I	High	71.0	1977–1978	3877	2.08	0.260
diversified sculpture of the earth					1978–1979	4092	1.85	0.250
surface, slopes 5 to 7%	328	II		64.7	1977–1978	3274	1.93	0.250
					1978–1979	3527	1.92	0.270
	31	III		36.0	1977–1978	2205	1.68	0.230
					1978–1979	3022	2.04	0.210

(e.g., the rivers of the Middle Pomeranian Lake District); the highest contents occurred primarily in some rivers of the Warsaw Valley.

Phosphorous compounds are a greater threat to the agricultural pollution than nitrogen compounds. It is only in waters of Mazowiecka Lowland District watershed, with its highly permeable soils, that all measured concentrations of phosphorous compounds were lower than 0.100 mg P/L. In some few cases (mainly related to the watersheds of small arable area), the average concentrations of phosphorous compounds were as high as 0.400 mg P/L, while the maximum concentration was 2.6 mg P/L. Such high concentrations were found mainly in agricultural watersheds covered with loamy soils. This is particularly true of the piedmont watersheds which are characterized by high precipitation and intensive soil erosion. A great part of the phosphorus was transported by suspended matter. The proportion of suspended phosphorus to the total phosphorus amount in the river waters differs in values for the watersheds according to the soil type. This proportion varied from 26 to 35% for watersheds of boulder clays, and from 6.3 to 10.6% for watersheds of sands and loamy sands.

Normally, average nutrient concentrations vary in response to varying meteorological conditions. Figure 2 illustrates the annual variability of nitrate and phosphate concentrations in the water

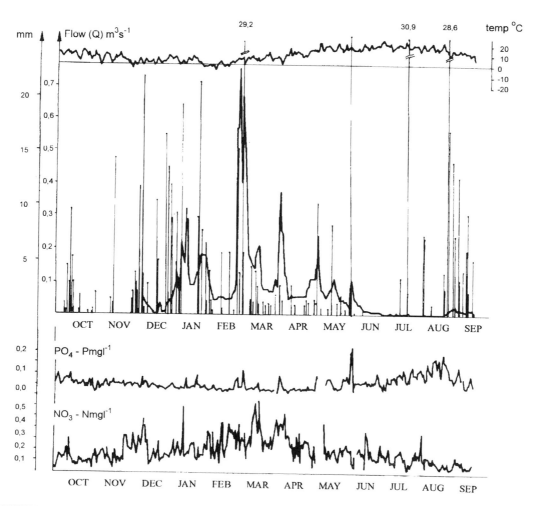

FIGURE 2. Influence of the meteorological and hydrological conditions on the nitrate and phosphate concentration forming in the Krezeszowska Stream from 1982 to 1983.

TABLE 4

Heavy Metals Lixiviation from Wastes

Element	Zn	Cu	Pb	Cd	Cr	Ni	Hg
Extraction %	0	25	39	0	0	74	45

Heavy metals leached from the wastes were also analyzed. The amount of leaching was determined based on differences between heavy metal content in ash before and after filtration (Table 4). The results of experiments show that the intensity of metal leaching depends on the element; the intensity ranged from 0% for Zn, Cd, and Cr to 74% for Ni.

The amount of soluble ash components that can penetrate to the groundwater is very high and, therefore, it could cause a serious threat to the environment. For example, the soluble portion that can be leached from the ash stored on the storage yard during 1 year is equal to about 1500 m^3.

In spite of imperfections in the storage yard simulation, the investigations of the filtration column prove that the amount at heavy metals passing to the oversediment water is extremely low in comparison with the amount leached out during the filtration process. The annual dose of heavy metals that can penetrate to the groundwater, as estimated based on the results shown in Table 4, amounts to 190 m^3. In this regard, the existence of a hydraulic connection between the groundwater under the storage yard and that of the Żuławy region is a concern.

THREAT TO SURFACE WATER FROM THE OVERSEDIMENT WATER

The discharge of oversediment water from the wet storage yard of grate waste to the receiving water may cause varying effects. Oversediment water of high alkalinity may cause coagulation of mineral and organic compounds. As a result, large quantities of residue may be precipitated and, at the same time, purification of the receiving water may occur. In the case of low alkalinity, the oversediment water discharge should be considered a source of pollution. Laboratory experiments performed to evaluate effects of oversediment water (Tables 5 and 6) show that dilute oversediment waters change the composition of the Martwa Vistula water insignificantly. The results also suggest

---

TABLE 5

Lixiviation of Grate Wastes by Water from the Martwa Vistula

Parameter		Martwa Vistula	Wastes:water					
			1:250	1:100	1:50	1:25	1:10	1:1
pH	—	8.0	8.3	8.55	8.9	9.0	9.55	9.3
Total alkalinity	mEq/dm^3	3.1	3.3	3.3	3.2	3.0	2.1	1.2
Total hardness	mEq/dm^3	16.5	7.9	6.3	6.3	7.3	6.3	12.8
Total dissolved matter	mg/dm^3	984	998	1009	1044	1083	1268	2038
SO$_4^{2-}$	mg/dm^3	78.0	77	84	104	116	187	381
Cl$^-$	mg/dm^3	425	411	418	425	425	482	695
Fe	mg/dm^3	1.4	0.0	0.0	0.7	0.7	0.7	5.6
Zn	μg/dm^3	85	21	—	24	61	162	167
Cu	μg/dm^3	17.3	10.2	—	8.8	10.9	13.8	16.2
Pb	μg/dm^3	68	40	—	48	100	46	62
Cd	μg/dm^3	4.5	4.5	—	3.7	4.2	6.0	3.3
Cr	μg/dm^3	110	102	—	114	110	122	144
Ni	μg/dm^3	10	10	—	20	10	10	10
Hg	μg/dm^3	0.4	0.7	—	0.2	0.15	1.7	1.1

TABLE 6
Chemical Composition of the Oversediment
Water in Laboratory Model

Parameter		Step of filling	
		I (1:10)	II (1:0.5)
pH	—	8.45	9.0
Total alkalinity	mEq/dm³	1.6	0.8
Total hardness	mEq/dm³	7.5	0.4
Total dissolved matters	mg/dm³	3600	4985
OH^-	mg/dm³	0.0	5.7
HCO_3^-	mg/dm³	74	0.0
CO_3^{2-}	mg/dm³	7	0.0
SO_4^{2-}	mg/dm³	392	639
Fe	mg/dm³	11	6
Zn	µg/dm³	36	181
Cu	µg/dm³	11.5	20
Pb	µg/dm³	10	74
Cd	µg/dm³	2.9	3.0
Cr	µg/dm³	172	206
Ni	µg/dm³	10	10
Hg	µg/dm³	1.7	3.8

that only large quantities of ash deposited in the water of the storage yard cause significant changes in the chemistry of this water. Ash extraction with large quantities of water (1:250) actually decrease concentrations of heavy metals below the values found in the Martwa Vistula water. Increases in ash quantity per unit of water volume resulted in increasing concentrations of the various elements.

Model investigations that simulated the initial and final stages of the storage yard filling (Table 6) provided more detailed information on composition of produced oversediment water. The initial phase of storage was characterized by decarbonization of water and an insignificant increase in the pH of the water. An increase in the sum of highly soluble constituents was also found at the same time that the concentrations of metals in the oversediment water were lower than in the input water (Martwa Vistula). This fact, as well as the decarbonization process, indicates that the reaction between the Martwa Vistula water and the grate waste results in precipitation of components of both reactants.

The results of experiments simulating the final phase of the storage yard filling show that, at this stage, the concentrations of soluble ash components (mainly sulfates) are very high and heavy metals concentrations are higher than for the initial phase. The quantities of metals are within the range of permissible concentration values for surface waters.

On the basis of the presented experimental results, the degree of contamination of the receiving water can be estimated. In the case of discharge into the Vistula (the Vistula mouth), increased heavy metal concentrations should be expected. Nevertheless, this increase will be insignificant, taking the quantitative ratios of mixed waters into consideration. However, absolute doses are also important. Despite the predicted insignificant changes in concentrations, input of high doses of toxic substances into the Vistula water and then to the Bay of Gdańsk may result in effects that are not readily anticipated.

CONCLUSIONS

Experimental results suggest the possibility of groundwater pollution by water filtration through the unsealed bottom of the storage yard, as well as by the influence of the oversediment water on receiving water.

1. Analysis of leaching of soluble ash components (including heavy metals), indicates that, at most, one third of the total amount leached substances passes to the oversediment water. The remaining two thirds can penetrate to the groundwater. It can be assumed that during 1 year about 190 m^3 of heavy metals can penetrate through the storage yard bottom into the groundwaters as a result of the filtration process. This fact poses a serious threat to the environment. This problem may be potentially even more important when the hydraulic connection between the storage yard and groundwater of Żuławy is taken into consideration.

2. When analyzing the effects of oversediment water on receiving water, it is important to consider two stages of storage: (a) the initial stage of storage, characterized by low concentrations of leached ash components in the oversediment water, which results in insignificant effects on chemistry of the receiving water, and (b) the final stage, which is characterized by high concentrations of leached ash components in the oversediment water (about 1 kg of heavy metals and 10 m^3 of other chemical compounds per day).

These are small values when compared to the amount of compounds transported by the Vistula River. Despite this fact, one cannot conclude that the oversediment water will not adversely affect the receiving water. The results of shortened experiments performed on some organisms typical of the Gdańsk Bay and the Vistula River indicate that the oversediment water can stimulate or inhibit different physiological processes.

REFERENCES

Elder, L., Recommendations of methods for marine biological studies in the Baltic Sea, *Baltic Mar. Biologist Publ.*, 5, 1, 1979.

Kulik-Kuziemska, I., Chemical and biological aspects of fly ashes and the phosphogypsum in the coastal region of Gdańsk, Proc. Sem. Management of industrial wastes in the coastal region. PAN, NOT, IHPG Gdańsk, 89-115, 1981.

29 Metal Loadings to the Baltic Sea from the Rivers of Poland

Elżbieta Heybowicz and Romuald Ceglarski

The measurement of metal outflow loads from the mouths of Poland's rivers is of great importance for two reasons: (1) it provides the balance of loadings of Polish inland surface waters with metals, and (2) it helps to determine metal loads from Polish territory into the Baltic Sea.

It is possible to determine the metal loads of Polish rivers to the Baltic Sea because the drainage areas of the studied rivers include 89.7% of the territory of Poland and 90.3% of the drainage areas of the rivers within the borders of Poland flow to the Baltic Sea. Polish river basins constitute 20.2% of the total area of the Baltic Basin, and 12% of the investigated drainage areas of the Vistula and Odra Rivers are located outside the borders of Poland.

The following interpretations are based on the results of river monitoring for the Vistula, Odra, and 10 smaller Pomeranian Rivers, carried out by the Department of Water Pollution Control of the Institute of Meteorology and Water Management in Gdańsk. Data are from 1987 to 1992 (Rybiński et al., 1993) for the Vistula and Odra Rivers and 1988 to 1992 for the remaining rivers. The data set consists of about 100 samples per year for the Vistula and Odra Rivers, and about 50 samples per year for each of the remaining rivers. Sodium, potassium, manganese, iron, zinc, copper, lead, chromium, nickel, and cadmium were analyzed; however, data for chromium and nickel are based on the results of measurements collected from 1990. All metals were analyzed by atomic absorption spectrometry (flame and electrothermal method) directly in water samples or after mineralization.

Concentration ranges of metals found in the river water extend over as many as five orders of magnitude (Figures 1, 2, and 3). The largest ranges of concentrations occur in samples from the Pomeranian Rivers, while the smallest ranges occur in samples from the Odra River. The lowest variability of results was found for potassium, and the highest for zinc. A comparison of mean metal concentrations in the studied rivers with Polish standards for surface waters (Table 1) (Dz. U. No. 35, 1990; Dz. U. No. 116, 1991) shows that the most exceedances are for manganese. Only mean concentrations of manganese exceed the permissible limits for the class I water quality. Surface waters utilized as a source of drinking water and for food industry should meet the requirements of class I. This fact suggests that the accepted standard for manganese may be too strict in comparison with the natural pollution of our rivers caused by this element.

Figures 1, 2, and 3 also show that the concentrations of iron, zinc, copper, and lead in the river waters only occasionally exceed the limits for class III water quality. This fact can be explained by acceptance of the same standards (except for iron) for all three classes of water quality, besides the incidental pollution of river waters, caused by the above-mentioned elements. Water resources in Poland generally are in poor condition, so it is important to protect river waters from metal pollution because they may be used as a source of drinking water in the near future. Taking all of this into consideration, it was decided to introduce the same standards for all classes of water quality. The standards of permissible metal concentrations for these wastewaters are given in Table 1.

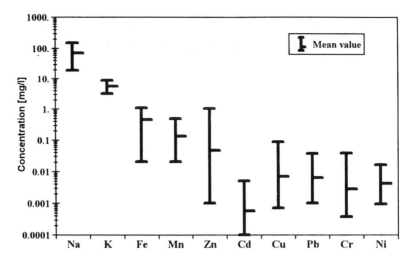

FIGURE 1. Range of metal concentrations in the Vistula River.

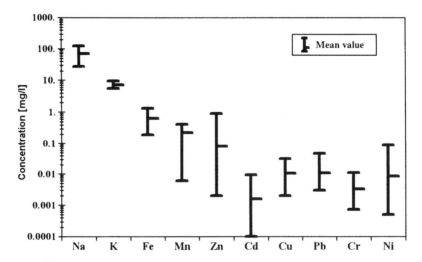

FIGURE 2. Range of metal concentrations in the Odra River.

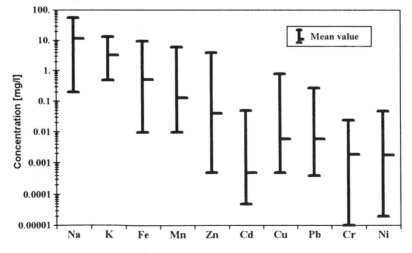

FIGURE 3. Range of metal concentrations in 10 Pomeranian Rivers.

TABLE 1
National Permissible Limits of Metals in Different Types of Waters (mg/L)

Metal	Waste water	Surface waters			Drinking water
		Class of water quality			
		I	II	III	
Sodium	800.0	100.0	120.0	150.0	200.0
Potassium	80.0	10.0	12.0	15.0	—
Iron	10.0	1.0	1.5	2.0	0.5
Zinc	2.0	0.2	0.2	0.2	5.0
Chromium^{3+}	0.5	0.05	0.1	0.1	—
Chromium^{6+}	0.2	0.05	0.05	0.05	0.01
Cadmium	0.1	0.005	0.03	0.1	0.005
Manganese	—	0.1	0.3	0.8	0.1
Copper	0.5	0.05	0.05	0.05	0.05
Nickel	2.0	1.0	1.0	1.0	0.03
Lead	0.5	0.05	0.05	0.05	0.05

Our investigations have also shown that chloride concentrations in the Vistula and Odra Rivers occasionally exceed permissible limits for surface waters. Chloride is contributed by discharges of industrial brines and saline waters from mines to these rivers in the upper parts of the Vistula and Odra Rivers.

The use of frequently measured metal concentrations and river flows results in greater accuracy in the determination of the metal loading. By calculating instantaneous loads as the product of water flow and metal concentration in a random sample, and assuming linear variability between adjacent measurements, the mean or total loads for selected time intervals were derived from the following equation:

$$L = \frac{m}{m} \sum_{i=1}^{n} C_i q_i$$

where L = load, C = measured concentration, q = measured flow, n = number of measurements, and m = conversion factor of units.

Variability of the loads was defined by coefficient of variations, i.e., the standard deviation expressed as percentage of the mean value.

Potassium represents the lowest variability of the instantaneous loads among metals (Table 2), while the Odra River has the lowest variability among rivers. The highest variabilities were observed for cadmium and for the Pomeranian Rivers, where incidental discharges of wastes containing metals cause considerable increases in sample standard deviations at low mean loads.

Analysis of seasonal distributions of the riverine metal outflows shows, in all cases, that the highest values occur during spring months (March to May) and that outflows are reduced in September and October. Hydrological conditions are responsible, to a large extent, for these differences in metal outflows. Loads increase during spring freshets and are limited during fall low-water levels.

The contribution of each river to the total metal load varies considerably. It is obvious that the Vistula and Odra Rivers dominate (Table 3), carrying 90.6 to 98.2% of the metal loads.

However, outflows are related to catchment area, and consideration of unit loads change the above-mentioned picture considerably (Figure 4). Large catchment basins indicate significantly

TABLE 2
Variability Coefficients of Monthly Metal Loads

Metal	Vistula	Odra River	Pomeranian Rivers
Sodium	35.4	26.1	53.7 ÷ 102.4
Potassium	38.8	38.1	35.8 ÷ 74.4
Iron	100.0	42.6	48.0 ÷ 141.0
Manganese	74.5	32.4	41.7 ÷ 196.1
Zinc	113.5	97.6	89.7 ÷ 248.7
Cadmium	142.3	79.8	109.7 ÷ 390.8
Copper	85.7	55.8	75.1 ÷ 166.7
Lead	80.2	66.9	80.5 ÷ 272.7
Chromium	158.8	70.9	81.8 ÷ 245.2
Nickel	75.5	88.4	85.7 ÷ 223.2

lower unit outflows than the small ones (except for unit load of cadmium and nickel in the Odra River). This probably results from metal accumulation in the bottom sediment of reservoirs and stagnation areas.

Discussion on the variability of metal loads as a function of water flow enables, to a certain extent, an estimation of the load sources. A more detailed discussion of this problem was covered in Chapter 24. Here, the definitions of the applied descriptions are given. Total mean annual load (outflow) can be divided into surface (nonpoint) and underground (point) origin. The latter can be further differentiated into minimal outflow, and subtracting minimal outflow from the underground outflow yields incidental outflow.

Pollution load in underground outflow is found to be caused most commonly by the flux of pollutants from point sources, i.e., wastes. Incidental load originates from poor waste management, resulting in periodic discharges of untreated wastes.

If the above divisions are applied to the total loads in the riverine outflow (Table 4), the highest contribution of point sources are from sodium, manganese, and potassium. The rest of the metals, except for chromium, also come primarily from point sources (52 to 63%), and 43 to 54% from

TABLE 3
Contribution of Each River to Total Metal Load

River	Na	K	Fe	Mn	Zn	Cd	Cu	Pb	Cr	Ni
					% of total load					
Odra	31.0	35.6	33.9	40.3	42.6	51.8	38.0	40.5	32.4	45.8
Ina	0.30	0.96	1.22	0.84	0.76	0.55	0.74	0.75	0.36	0.36
Rega	0.33	1.18	1.62	1.24	0.68	0.41	0.79	0.63	0.77	0.44
Parsęta	0.41	1.07	1.90	1.30	0.88	0.70	1.20	1.01	1.14	0.63
Grabowa	0.06	0.13	0.30	0.24	0.18	0.12	0.22	0.17	0.20	0.10
Wieprza	0.14	0.39	1.21	0.85	0.49	0.61	0.58	0.78	0.54	0.26
Słupia	0.18	0.46	0.79	0.74	0.55	0.48	1.19	0.95	0.61	0.30
Łupawa	0.06	0.19	0.32	0.31	0.19	0.22	0.25	0.38	0.30	0.09
Łeba	0.11	0.39	0.82	0.65	0.40	0.63	0.88	0.52	0.46	0.22
Reda	0.04	0.14	0.21	0.15	0.15	0.12	0.18	0.23	0.10	0.10
Wisła	67.2	58.6	56.7	51.9	50.3	43.3	54.6	52.0	61.3	51.0
Pasłęka	0.20	0.80	0.95	1.42	2.80	1.11	1.39	2.09	1.93	0.67

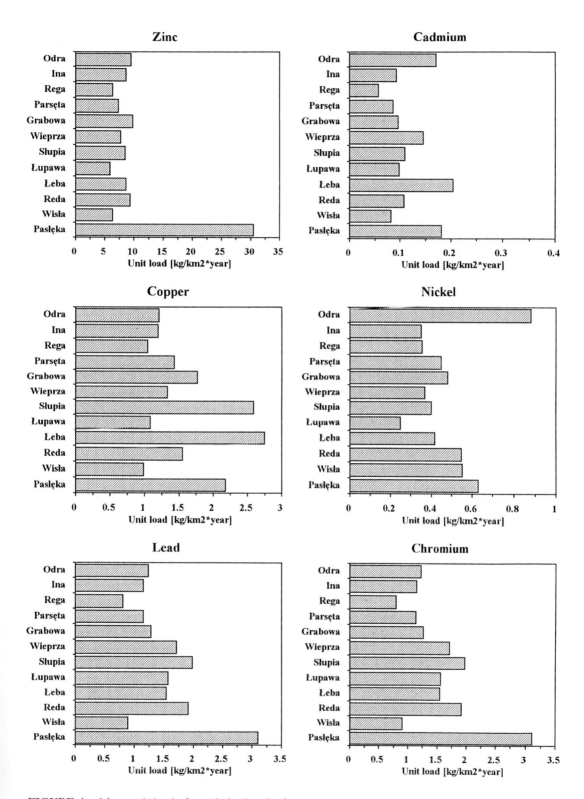

FIGURE 4. Mean unit load of metals in river basins.

TABLE 4
Load of Metals from Polish Rivers Entering Baltic Sea

Parameter	Unit	Years 1988 to 1991									
		Na	K	Fe	Mn	Zn	Cd	Cu	Pb	Cr	Ni
Mean annual outflow	t/year	2,820,966	275,675	27,771	7570.1	2757.8	36.2	374.0	366.7	121.4	219.0
Mean annual unit outflow	kg/km² per year	8696.9	849.9	85.6	23.3	8.50	0.111	1.153	1.131	0.374	0.675
	%	100.0	100.0	100.0	100.0	100.0	100.0	100.0	100.0	100.0	100.0
Minimal outflow	t/year	1,411,360	138,542	3877.5	1414.0	217.7	2.43	41.7	55.6	17.2	27.9
Minimal unit outflow	kg/km² per year	4351.1	427.1	12.0	4.4	0.67	0.008	0.128	0.171	0.053	0.086
	%	50.0	50.3	14.0	18.7	7.9	6.7	11.1	15.2	14.1	12.7
Underground	t/year	2,242,463	183,015	14,354.6	5240.4	1729.8	20.0	201.7	225.1	54.7	135.7
Underground unit outflow	kg/km² per year	6913.4	564.2	44.3	16.2	5.33	0.062	0.622	0.694	0.169	0.418
	%	79.5	66.4	51.7	69.2	62.7	55.2	53.9	61.4	45.1	62.0
Incidental outflow	t/year	831,103	44,473	10,477.1	3826.4	1512.1	17.5	160.0	169.5	37.6	107.8
Incidental unit outflow	kg/km² per year	2562.2	137.1	32.3	11.8	4.66	0.054	0.493	0.523	0.116	0.332
	%	29.5	16.1	37.7	50.5	54.8	48.5	42.8	46.2	30.9	49.2
Nonpoint outflow	t/year	578,503	92,660	13,416.1	2329.8	1028.0	16.2	172.3	141.6	66.7	83.3
Nonpoint unit outflow	kg/km² per year	1783.5	285.7	41.4	7.18	3.17	0.050	0.531	0.437	0.206	0.257
	%	20.5	33.6	48.3	30.8	37.3	44.8	46.1	38.6	54.9	38.0

incidental outflows. The very low incidental outflow of sodium and, particularly, potassium is characteristic. A comparison of these values shows that 50% of riverine metal loads could be removed by improved waste water disposal or metallurgical plants.

To describe the magnitudes and directions of changes in riverine metal outflows, the data sets were reduced to 90% span in order to eliminate extreme values, and the results were used in linear regressions of loads against time. With few exceptions, the time trends seem to be very weak. Setting the significance level of no correlation at $\alpha = 0.05$ eliminated 46.5% of the cases, which indicate no time trends. Of the remaining 77 cases, 54 indicate negative trends and 23 indicate positive trends. The Odra outflow showed significant correlations with negative trend slopes for all tested metals, except cadmium. The Vistula River, showed six cases of significant correlation, of which four trend slopes were negative (Rybiński et al., 1993).

The lack of negative trends appears to be not only a national problem but also a Baltic problem. This is very clearly indicated by the Second Baltic Sea Pollution Load Compilation on the examples of zinc, cadmium, copper, and lead (Second Baltic Sea Pollution Load Compilation, 1993). The contribution from our country to the total load of metals discharged into the Baltic Sea in 1990 is as follows: Zn ~48%, Cd ~61%, Cu ~25.5%, and Pb ~23%.

ACKNOWLEDGMENTS

This work has been done within the project No. 6-6045-91-02 financially supported from 1992 to 1993 by the Polish Committee for Science Research.

We thank Msc. Eng. Rybiński for his essential help, Dr. Makowski for his help in data processing, and Mrs. Megger for typing the English text.

REFERENCES

Dz. U. No. 35, pos. 205, Regulation of the Ministry of Health and Social Care, dated May 4, 1990.

Dz. U. No. 116, pos. 502 and 503, Regulation of the Ministry of Environmental Protection, Natural Resources, and Forestry, dated November 5, 1991.

Rybiński, J., et al., Fluctuations of the pollutant transport in rivers to the sea and factors influencing the fluctuations, KBN Report No. 6-6045-91-02, Gdańsk, 1993.

Rybiński, J., Heybowicz, E., and Ceglarski, R., Metals, environmental conditions in the Polish zone of the southern Baltic Sea during 1992, Maritime Branch Materials, IMWM, Gdynia, 1993.

Second Baltic Sea Pollution Load Compilation, Baltic Sea Environment Proceedings, No. 45, Helsinki Commission, 1993.

30 Upper Clear Creek/Standley Lake, Colorado Water-Quality Assessment

Timothy D. Steele and Russell N. Clayshulte

During 1993, detailed deliberations and negotiations regarding changes in water-quality standards were conducted among entities of the Upper Clear Creek Watershed, middle and lower Standley Lake tributary areas, and several northwestern Denver (CO) suburban cities (Westminster, Northglenn, and Thornton). Deliberations, negotiations, and associated technical studies were the result of a perceived need to protect water quality in Standley Lake, a water supply for approximately 250,000 inhabitants living in these suburban cities. Flows of Clear Creek are conveyed through canal diversions near Golden, CO, to Standley Lake. Consultant studies developed for both the upper-watershed entities as well as the Standley Lake cities provided technical data and information used during these deliberations. The upper-watershed entities are represented by the Upper Clear Creek Watershed Association, which is responsible for watershed management. Areas of the upper watershed are highly mineralized and have been designated as a Federal Comprehensive Environmental Response and Recovery Act (CERCLA) Super-fund site as a result of past mining activities. In addition, two upper-basin wastewater treatment facilities had wastewater-discharge compliance problems, and several stream segments do not comply with specified water-quality standards as adopted by the State of Colorado Department of Public Health and Environment. A cooperative monitoring program executed by interested entities has provided watershed data for general information use as well as input into a QUAL2E modeling application. A narrative water-quality standard was adopted by the Colorado Water Quality Control Commission (Upper Clear Creek Watershed Association, 1994; Appendix B); the agreement has the needs of the state for certain conceptual water-quality goals, for watershed and in-lake management plans, and a directed continuing watershed and in-lake monitoring program. The agreement uses a voluntary approach to water-quality management rather than imposing a mandatory standard and control regulation. This procedure of addressing and resolving conflicts in goals and information serves as an exemplary case study for technical and institutional interactions of a similar nature.

INTRODUCTION

BACKGROUND

Standley Lake, a water-supply source for three municipalities with a population in excess of 250,000 inhabitants in the northwestern Denver metropolitan area, obtains a major percentage of its water from the Upper Clear Creek Watershed (Figure 1). Abandoned mining activities have caused parts of this watershed to be designated as a priority Superfund area with about 20 specific cleanup sites proposed for remediation. The watershed is also a priority nonpoint source watershed in the Denver metropolitan area. In addition, point-source wastewater discharges in the watershed are being evaluated for potential advanced treatment to remove nutrients. Abandoned mining activities are the primary cause of nonpoint-source trace metals. Point- and nonpoint-source sediment, metal, and nutrient contributions can be attributed to urban and construction activities, stormwater runoff along roadways, and wastewater effluent discharges regulated under the Colorado Discharge Permit System (CDPS).

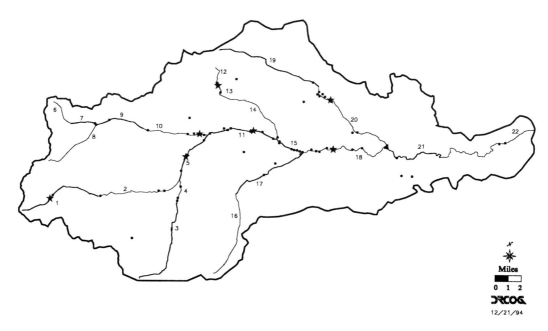

FIGURE 2. Clear Creek QUAL2E model stream reaches; •, monitoring data points; ★, treatment facility outfall locations.

adjusted so model predictions matched field data at selected points. The model calibration focused on total nitrogen and total phosphorus as indicator nutrient species for two time periods. The model calibrations for these two species were judged by the modeler to be adequate, particularly when considering implications of data constraints and the complex watershed hydrology.

The calibrated fall (low-flow) and spring (high-flow) QUAL2E model results can serve as a useful resource-management tool. The model was used to evaluate future conditions under various loading options (DRCOG, 1994). Model runs were made with all wastewater treatment plants at design flows with and without nutrient controls. The facilities currently discharge on an annual-average approximately 19,300 lb of total phosphorus. Based on predicted growth in the Upper Clear Creek Watershed by the year 2015 (DRCOG, 1993 in Table 9), these facilities would produce about 37,100 lb of total phosphorus without controls.

There are about 14,000 lb of total phosphorus and 242,000 lb of total nitrogen in Clear Creek at Golden on an average annual basis. If the wastewater treatment plants discharge at design capacity without any nutrient control, then the total-phosphorus loadings could increase by 46%; whereas, the total-nitrogen loadings would increase by about 16%.

Table 1 shows the predicted monthly total phosphorus and total nitrogen point-source loadings for existing flow and nutrient conditions and for design flows with nutrient controls. The model estimates that nutrient controls would reduce total phosphorus loading by about 37%, while no predicted decrease in total nitrogen is shown in the model-application results for Clear Creek at Golden. A substantial nonpoint-source nitrogen loading appears to occur in the Lower Clear Creek stream reaches in order to fit field-survey data for calibration of the QUAL2E model.

The model estimates that most of the total phosphorus loading in Clear Creek at Golden, occurring during the months of October through April, is derived from point sources. In the model, most of both total phosphorus and total nitrogen loadings was presumed to be originating from nonpoint sources in the spring/summer higher-runoff months of May through September. The model results estimated that the total nitrogen loadings in Clear Creek at Golden during October through April were divided about equally between point and nonpoint sources. Continuing water-quality monitoring will serve, in part, to confirm or refute these and other model assumptions.

TABLE 1
Estimated Monthly Point-Source Nutrient-Species Loadings in Clear Creek, Golden, Colorado

Month	Existing flow and nutrient conditions		Design flows with nutrient controls	
	Total phosphorus (lb)	Total nitrogen (lb)	Total phosphorus (lb)	Total nitrogen (lb)
January	212	1534	175	3579
February	186	1347	154	3144
March	206	1495	171	3489
April	341	2473	283	5770
May	495	5442	247	5937
June	1207	13,273	603	14,479
July	757	8328	379	9085
August	336	3695	168	4031
September	201	2206	100	2407
October	415	3004	343	7010
November	292	2116	242	4938
December	240	1741	199	4062
Annual	4888	46,654	3064	67,931

The uncertainties in the model prediction and the estimated nutrient loadings, along with the general lack of other relevant information, indicated to the Standley Lake cities and the Upper Clear Creek Watershed Association that determining a watershed control regulation was not appropriate at this time. As a result, the 24 parties to the control regulation negotiated an agreement (see below), which calls for a 3-year voluntary and cooperative program to better establish watershed-management requirements. This includes a comprehensive water-quality monitoring program, watershed-model refinement, identification of appropriate best management practices (BMPs), and delineation of source-problem areas and restoration projects.

STANDLEY LAKE EVALUATIONS

Various water-quality evaluations of Standley Lake have been conducted, including those by Mueller and Ruddy (1993), Horne (1993), and as summarized by Tetra Tech, Inc. (1994). The USGS (Mueller and Ruddy, 1993) study included comparative analyses with earlier in-lake monitoring and investigations conducted by the consultant of the cities (Richard P. Arber Associates) during the 1980s. The USGS study estimated nutrient loadings, including bottom-sediment internal loadings of nitrogen and phosphorus. During the USGS study, no taste and odor problems occurred. The Horne (1993) evaluation of Standley Lake concluded that relatively high winter-period nutrient loadings were a major cause of in-lake eutrophication. This conclusion was supported by the year-by-year variations of hypolemnetic anoxia that have occurred, where oxygen-depletion durations of between 8 and 12 weeks have occurred in 1982 and from 1986 through 1991; whereas, oxygen-depletion durations averaged less than 4 weeks from 1983 to 1985, when winter inflows from the Croke Canal were curtailed (Tetra Tech, Inc., 1993 in Figure 12). Horne's (1993) conclusion of Standley Lake being characterized as moderately eutrophic is mainly due to dense and long-lived spring phytoplankton blooms dominated by *Asterionella* diatoms, which resulted in quite shallow Secchi depths during April to June 1990 of the USGS study. However, Tetra Tech questioned the trophic-state-index (TSI) analysis and associated inferences by Horne (1993) for Standley Lake.

Control-Regulation Proposals/Deliberations/Negotiations

Between about September 1993 and January 1994, agreement proposals and counterproposals were made regarding water-quality standards and associated rulemaking for control of conditions of Standley Lake. Representatives of the cities, tributary areas, and Upper Clear Creek Watershed exhibited considerable willingness to resolve conflicting viewpoints and to resolve their differences through mutually agreed on language of the final Agreement. The terms of this Agreement allow for more study, evaluation, and in essence understanding in the complex blend of processes and conditions affecting the water quality of Standley Lake. However, specific milestones and indications of progress are to be demonstrated during the tenure of this agreement, as is summarized below.

Settlement Agreement

By February 1994, a Settlement Agreement was successfully negotiated among 24 entities with interests in the Upper Clear Creek Watershed, tributary areas, and Standley Lake. This Agreement was adopted as in the settlement agreement presented to the Colorado Department of Public Health and Environment's Water Quality Control Commission. Terms of the Agreement are being implemented, which include BMP manuals, a continuing water-quality and biological monitoring program (data collection and reporting), a semi-annual report to parties, and an annual report to the Commission.

Ongoing and Future Water-Quality Monitoring

Between October 1992 and July 1993, a water-quality monitoring program was implemented in the watershed. These data were used to model fall (low-flow) and spring (high-flow) runoff conditions. During this period, a voluntary monitoring program was conducted at 17 surface-water sites that were monitored throughout the watershed and at 8 wastewater-treatment facilities. During 1994, this monitoring program was modified to conduct 8 upper-watershed surveys, along with diversion-canal and in-lake samples (Clear Creek Watershed/Standley Lake Monitoring Committee, 1994).

SUMMARY AND CONCLUSIONS

The decision-support process implemented among affected parties, coupled with technical studies summarized herein, resulted in a consensus agreement and a future plan of action. This plan will continue to assess water quality conditions in the Upper Clear Creek Watershed as well as changes and impacts of Clear Creek flows diverted into three canals which meander through a rapidly developing tributary-basin area and provide water to Standley Lake, which is used as a municipal water supply by the cities of Northglenn, Westminster, and Thornton. Salient points of the resultant agreement, settled in February 1994, include the following terms:

1. For the interim period (at least until the next triennial review), a narrative standard for Standley Lake requires maintenance of this impoundment in a mesotrophic state.
2. A jointly designed and operated water-quality monitoring program would be implemented over the next 3 years (1994 to 1996); basic data summary reports would be prepared annually.
3. In the Upper Clear Creek Watershed, an evaluation of nutrient-reduction options for both point and nonpoint sources will be conducted (considering technologies and cost effectiveness).
4. Management plans would be developed for Standley Lake and the tributary areas (where Clear Creek diversions flow through three canals to Standley Lake).

5. BMPs would be developed and adopted by the three major entities: Upper Clear Creek Watershed Association (for nonpoint sources), tributary-basin entities, and the Standley Lake cities.

6. In 1997, results of the above-summarized activities and associated progress on water quality enhancements will be evaluated and appropriate action taken regarding adopting a control regulation for Standley Lake. This might include setting a total-phosphorus effluent limitation (1.0 mg/L, 30-day average) and implementing in-Lake treatment to reduce internal nutrient loadings to the reservoir.

ACKNOWLEDGMENTS

The work discussed herein was funded through grants from the Upper Clear Creek Watershed Association. Mr. R. L. Jones served as Technical Advisor to projects contracted to ASI (T.D. Steele, water-quality assessment) and to DRCOG (R.N. Clayshulte, water-quality modeling). Dr. Steele currently is affiliated with HSI GeoTrans and continues to provide technical services to the Association. Results and opinions provided herein reflect the perspectives of the authors and do not necessarily represent consensus by any of the entities with vested interests in water-quality conditions of the Upper Clear Creek Watershed, tributary areas, and Standley Lake.

REFERENCES

Advanced Sciences, Inc., Upper Clear Creek Basin/Standley Lake water-quality assessment, Final Report prepared for the Upper Clear Creek Basin Association, September 22, 1993, 11.

Brown, L. C. and Barnwell, T. O., The enhanced stream water quality models QUAL2E and QUAL2E-UNCAS: documentation and user manual, U.S. Environmental Protection Agency, Environmental Research Laboratory, Office of Research and Development, Athens, GA, 1987, 189.

Camp Dresser & McKee Inc. (CDM)/RBD, Inc. Engineering Consultants (RBD), Final Report — Clear Creek/Standley Lake water management study, Prepared for the Cities of Northglenn, Thornton, and Westminster, April (June) 1994.

Clear Creek Watershed/Standley Lake Monitoring Committee, Clear Creek watershed management sampling program, upper Clear Creek basin, Standley Lake supply canals, and Standley Lake, June 10, 1994, 6.

The Colorado Water Quality Forum, Colorado watershed protection approach, Working Paper, July 1, 1994, 55.

Denver Regional Council of Governments (DRCOG), Clean Water Plan, Vol. II, Assessments and management plans, Updated December 1993, 197.

Denver Regional Council of Governments (DRCOG), Upper Clear Creek Watershed QUAL2E Model, Prepared for the Upper Clear Creek Watershed Association, 1994, 42.

Horne, A. J., An evaluation of algae blooms in Standley Lake, Colorado, Completion Report for the Cities of Westminster, Thornton, and Northglenn, Alex J. Horne Associates, El Cerrito, CA, with assistance from Commins Consulting, Lafayette, CA, 1993, 45.

Mueller, D. K. and Ruddy, B. C., Limnological characteristics, nutrient loadings and limitation, and potential sources of taste and odor problems in Standley Lake, Colorado, U.S. Geological Survey Water-Resources Investigations Report 92-4053, Denver, CO, 1993, 55.

Tetra Tech, Inc., TMDL SWAT team review, nutrient loading and eutrophication, Standley Lake, Colorado, Prepared for Denver Regional Council of Governments, Denver, Colorado, under contract to U.S. Environmental Protection Agency, Office of Wetlands, Oceans and Watersheds, Washington, D.C. 20460, 1994, 38.

Upper Clear Creek Watershed Association, Clear Creek Watershed agreement semi-annual report, August 1994, 7.

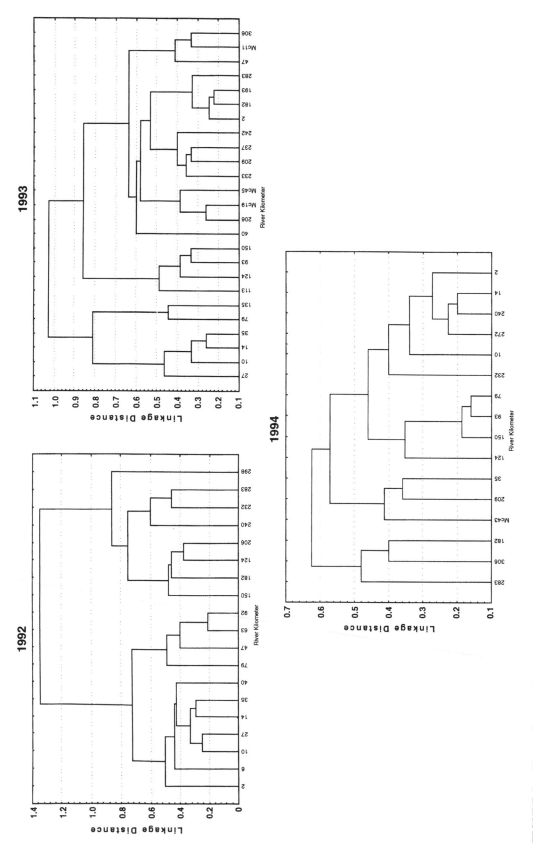

FIGURE 1. Cluster diagrams for Willamette fish abundance data for 1992 to 1994.

IBI	Integrity Class	Characteristics
58-60	Excellent	Comparable to pristine conditions, exceptional assemblage of species
48-52	Good	Decreased species richness, intolerant species in particular; sensitive species present
40-44	Fair	Intolerant and sensitive species absent; skewed trophic structure
28-34	Poor	Top carnivores and many expected species absent or rare; omnivores and tolerant species dominant
12-22	Very Poor	Few species and individuals present; tolerant species dominant; diseased fish frequent

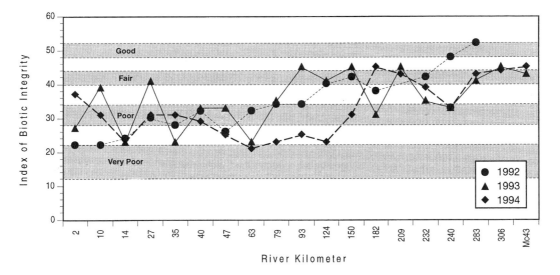

FIGURE 2.   Comparison of index of biotic integrity values in Willamette River from 1992 to 1994.

low scores for two similar metrics: number of native species and percent introduced species. The highest IBI scores were found at RKs 182 and 209 and from McKenzie RK 43 to RK 306. The higher IBI scores from the latter stations may be attributed to the number of salmonids collected from these stations. The range of IBI values for all 3 years overlaps for many of the stations, but the mean IBI scores for the stations downstream of Newberg are 8 to 15 points lower than the mean IBI scores for the stations upstream of Corvallis. The range of IBI scores are the most variable (i.e., largest range) for stations between Newberg and Corvallis (e.g., RK 93 and RK 124).

Multiple regression analyses were conducted on IBI individual metric scores from all 3 years (1992 to 1994) to determine the relative contribution of each metric to the overall score. These analyses indicate that a small number of metrics can explain most of the variability in the observed IBI scores. For 1992 and 1994, the top three metrics accounted for 95% or more of the total IBI score, while in 1993 the top four metrics accounted for 90% of the overall IBI score. The identity of the top metrics, however, is not the same from year to year. In 1992 and 1993, the number of native species and the percent of introduced species were the most important metrics. In 1994, neither of these two metrics were in the top two; instead, the percent catchable salmonids and the percent insectivores were the two most important metrics. The latter two metrics explained virtually none of the variability in 1992 and 1993 IBI scores. Conversely, of the top two metrics in 1992 and 1993 (the number of native species and the percent of introduced species), only the percent introduced species explained any of the IBI variability in 1994 and it was relatively minor (6%).

Many of the IBI metrics used for the Willamette River are replacements or modifications of those developed by Karr et al. (1987) for midwestern streams and rivers. This modification was necessary if the metrics were to be a meaningful measure of the health of local fish communities and is commonly undertaken for different regions of the country [Leonard and Orth, 1986 (West Virginia); Bramblett and Fausch, 1991 (Colorado); Lyons, 1992 (Wisconsin); Gatz and Harig, 1993 (Ohio); Minns et al., 1994 (Great Lakes)]. Although 13 metrics were used for the Willamette River, only three or four of those metrics appeared to be sensitive indicators of fish community health (as measured by the IBI) for a particular year. The most important metrics for the Willamette River appear to be those measuring species richness (e.g., native species and salmonids) and trophic composition (e.g., percent insectivores).

The fact that these three or four metrics were not the same from year to year, however, argues for the continued inclusion of metrics which have the potential to yield useful information, but may not have done so for a particular dataset. Although the temptation exists to eliminate metrics which have not yet proved to be useful, the dynamic nature of the ecosystem suggests that different metrics can be important during different seasons or years. The theoretical underpinnings of the IBI have undergone considerable scrutiny in recent years, so the elimination of metrics should be done only after careful consideration. The scores for each of the metrics are thought to be functions of the underlying biotic integrity of the system, rather than the reverse (Karr et al., 1986). Thus, it is possible to imagine that there are other ways (i.e., metrics) in which the biotic integrity is manifested in a quantifiable manner.

The relative contribution of individual metrics to the IBI for other fish community studies in different regions might be expected to differ from the Willamette River, but some generalizations can be observed. In two other studies in which a similar analysis of metric contribution was performed, measures of species richness were also very important. The two most important metrics in southern Ontario and three midwestern streams were presence/absence of brook trout and the number of native species (Steedman, 1988) and total fish species and number of sucker species (Karr et al., 1987), respectively.

Several authors have used the tentative integrity classes proposed by Karr et al. (1986) and Plafkin et al. (1988) to assign a qualitative label (e.g., excellent, good, fair, poor) to each IBI score (Crumby et al., 1990; Bramblett and Fausch, 1991). Virtually all of the fish community research utilizing the IBI approach has been conducted on smaller river systems. This approach has merit if the metrics used to create the IBI accurately reflect the biotic integrity of the system; if they do not, the labels may be misleading. Despite the uncertainty in the predictive power of the metrics, the IBI scores calculated in this study are useful when viewed in the context of spatial and temporal variation within the Willamette River. Therefore, these qualitative labels have been used in this study for comparative purposes. Additional monitoring is needed to develop a more extensive database to refine these qualitative classifications for a river as large as the Willamette.

No single metric could account for the variation observed in the IBI scores obtained for a particular station between years. When 1993 and 1994 IBI values are compared, similar scores (F13) were obtained for seven of the common stations. When 1992 and 1994 values are compared, similar scores were obtained for 6 of the 16 stations in common. For all 3 years, similar values were obtained for only 2 of the 16 common stations.

The lack of a consistent temporal trend in IBI scores exists in spite of seasonal and/or hydrologic variability in the river between the three sampling years. The 1992 sampling occurred in August, while the 1993 and 1994 sampling occurred in October. The river flow in 1992 was approximately 50% of the flow in 1993, which in turn was approximately 20% greater than 1994. The water temperature was also several degrees lower in 1993 and 1994 compared to 1992. These differences in river conditions could help explain the overall lower abundance in the latter 2 years. However, the determining factors of adult fish abundance at a particular point in time are typically more complex than river flow and water temperature.

The mean IBI scores in the Hughes and Gammon (1987) study were significantly higher than from 1992 to 1994 at the upriver stations. This may be attributed, in part, to the significantly higher numbers of native, cottid, and catostomid species and numbers of individuals observed by Hughes and Gammon (1987). No consistent relationship was noted between the mean IBI scores from 1983 to 1992 to 1994 in the middle portions of the river. However, the mean IBI scores for the 1992 to 1994 data were significantly higher than the 1983 scores for the lower river. This may be attributed to the significantly higher percentages of carp and omnivores (e.g., largescale suckers, carp) observed in recent surveys.

The spatial trend in IBI scores is more pronounced than the temporal trend. Both the current study and Hughes and Gammon (1987) noted consistently (and significantly) higher IBI scores at the upriver stations. Several explanations for these differences can be made. The characteristics of the fish habitat at the upriver stations were notably different from the lower river stations, particularly those in the Portland metropolitan area. The upriver stations tended to have cooler, faster-moving water, a more narrow and shallow channel, and a larger sediment grain size than the lower river stations. The degree to which these differences can explain the spatial variation in the IBI scores was explored in detail by Tetra Tech (1995).

Another hypothesis that has not been tested is that the water quality at the upriver stations supports a higher level of biotic integrity than the lower river stations. Many industrial facilities are located within the Portland area and other large point sources of pollution are located throughout the river below Corvallis. On the other hand, given the densely populated nature of the Willamette River Basin, nonpoint sources of pollution are likely to be present throughout the main stem of the river. At present, it is not possible to make a definitive statement about the water quality at upriver vs. downriver stations.

Skeletal Abnormalities

The incidence of skeletal deformities in juvenile fish in the Willamette River was initiated as a pilot study in 1992 to assess the potential utility of this biological indicator. Based on the results obtained in 1992, the use of this indicator was expanded in 1993. Additional sampling was conducted in 1994 to confirm previously identified problem areas and to provide additional coverage of the river. Because each year's sampling design was based on the previous year's sampling, the results are presented chronologically.

In 1992, juvenile northern squawfish were collected from four locations along the main stem of the Willamette River: RKs 298, 202, 80 (west bank: away from the City of Newberg), and 5. The total length of the fish analyzed from these locations ranged from 21 to 60 mm; mean fish length ranged from 27.5 to 39.5 mm. The incidence of skeletal deformities was low (<3.2%) for the samples collected downstream of the cities of Eugene (RK 298), Corvallis (RK 202), and Portland (RK 5). A significantly higher (2×2 frequency table, p <0.001) incidence of skeletal deformities (25.8%) was measured for fish collected along the west bank of the Newberg Pool at RK 49.7 (Figure 3).

The 1993 fish collections were designed to: (1) verify the results obtained during 1992; (2) intensify sampling efforts in the Newberg Pool area where elevated incidence of skeletal deformities had been measured in 1992; (3) measure the incidence of skeletal deformities upstream and downstream from a major industrial discharge from a bleached kraft pulp and paper mill at RK 237, and (4) measure the incidence of skeletal deformities in a "field reference" location to assess background levels of skeletal deformities.

Fish collections during 1993 occurred from August 17 through September 4, approximately 1 to 3 weeks later in the summer than the 1992 collections. The total length of the fish ranged from 13 to 65 mm; mean fish length at each collection site ranged from 18.5 to 37.3 mm. Fish collected at the same four sites analyzed during 1992 showed a similar pattern to that observed during 1992 (Figure 3). The incidence of skeletal deformities was low (<3.0%) for fish collected downstream of Eugene, Corvallis, and Portland. A significantly higher (2×2 frequency table, p <0.001) incidence of skeletal deformities (30.8%) was measured for fish collected along the west bank of the Newberg Pool at RK 80. The incidence of skeletal deformities showed good agreement between 1992 and 1993 (2×2 frequency table, p >0.05), with values differing by 0.5 to 5.0%.

Nine locations were sampled from RK 41 (upstream of the mouth of the Clackamas River) to RK 82 (near the upstream end of the Newberg Pool) to provide a more intensive evaluation of the incidence of skeletal deformities occurring in the Newberg Pool. The incidence of skeletal deformities in this stretch of river ranged from 22.6 to 52.0%. The highest values occurred in a 4-km

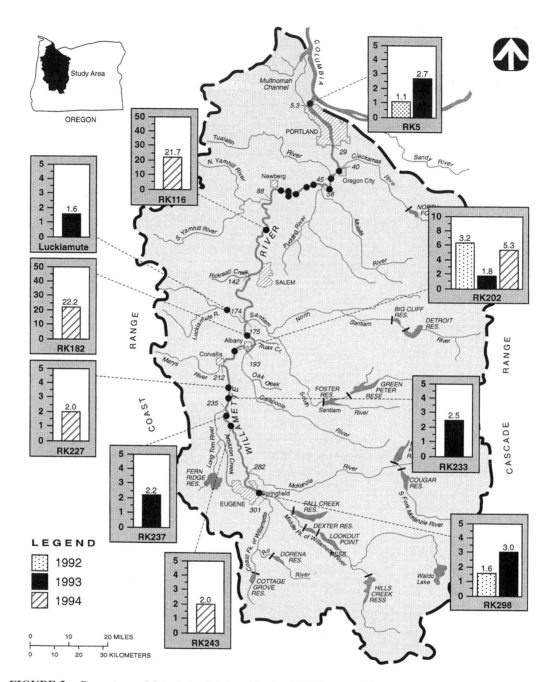

FIGURE 3. Percentage of fish skeletal deformities in the Willamette River for 1992 to 1994.

stretch of river in the upstream end of the pool (RK 78 to RK 82) with values declining gradually in the downstream direction (Figure 4).

During 1993, two samples were collected on opposite banks at RK 80. The incidence of skeletal deformities measured for fish collected on the west bank (across from the city of Newberg) was 30.8%, while a significantly higher incidence of deformities (52.0%) was measured for fish collected on the east bank of the river (side nearest the city of Newberg).

Skeletal deformities were measured upstream (RK 237.2) and downstream (RK 233.0) from a bleached kraft pulp and paper mill discharge at RK 237.0. This major industrial discharge was targeted for evaluation because there are numerous scientific reports that have linked increases in

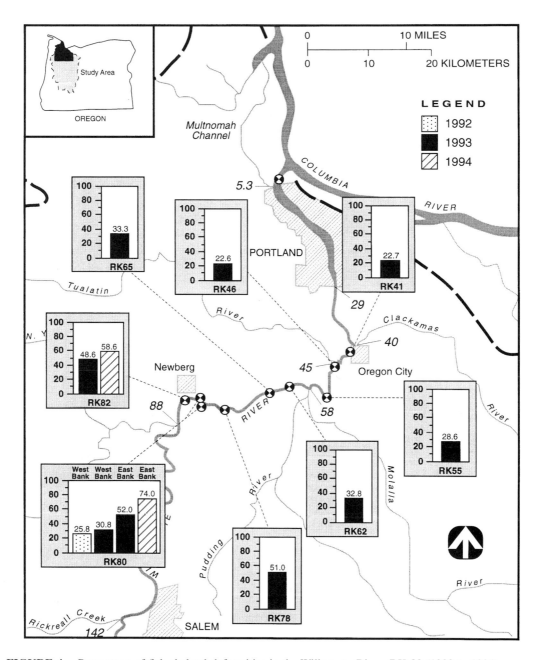

FIGURE 4.   Percentage of fish skeletal deformities in the Willamette River, RK 80 (1992 to 1994).

the incidence of fish skeletal deformities to exposure to effluent from Swedish bleached kraft pulp and paper mills (Bengtsson, 1988; Härdig et al., 1988; Lindesöö and Thulin, 1992; Mayer et al., 1988; Thulin et al., 1988). The incidence of skeletal deformities upstream and downstream of the discharge was 2.2% and 2.5%, respectively. These values were not significantly different from each other (2×2 frequency table, $p > 0.05$).

A site located near Helmick State Park in the Luckiamute River was selected as a likely "field reference" due to the absence of any nearby point sources and the proximity of relatively undisturbed forested land in the riparian zone, which may reduce local contributions from nonpoint sources. The incidence of skeletal deformities at this site was 1.6%. No significant difference (2×2 frequency table, $p > 0.05$) was seen between the incidence of skeletal deformities measured near Helmick

State Park and the values measured within the stretch of river from Eugene to downstream of Corvallis, and downstream of Portland.

The relationship between fish size and the incidence of skeletal deformities was evaluated by pooling the data collected at five sites [RKs 41, 46, 55, 62, and 80 (west bank)] and examining the incidence of skeletal deformities as a function of fish total length. Over a range of total length from 15 to 33 mm there was no significant relationship ($R^2 = 0.035$) between the incidence of skeletal deformities and total fish length.

The 1994 survey was designed to: (1) provide additional data for sections of the river that were not sampled during 1992 or 1993; (2) repeat sampling at two sites in the upstream end of the Newberg Pool where the highest incidence of skeletal deformities were observed in 1993; and (3) collect additional data upstream and downstream from the bleached kraft mill discharge at RK 237.0.

Fish collections during 1994 occurred from July 21 through August 1, approximately 2 to 3 weeks earlier in the summer than the 1992 or 1993 collections, respectively. The total length of the fish ranged from 15 to 50 mm; mean fish length ranged from 23.3 to 45.5 mm. The pattern of the incidence of skeletal abnormalities was generally consistent with the results obtained during 1992 and 1993. The incidence of skeletal deformities in the upper river above RK 202 was low (<5.3%) with the highest incidence of deformities occurring in the upstream end of the Newberg Pool: 58.8% at RK 82 and 74% at RK 80 (east bank: side nearest Newberg). The incidence of skeletal deformities upstream (RK 243) and downstream (RK 227) from the bleached kraft mill discharge at RK 237 were low and not significantly different (2×2 frequency table, $p > 0.05$). Elevated percentages of skeletal deformities were measured in the previously unsampled stretch of river from RK 201 (downstream of Corvallis) to the Newberg Pool (RK 82). The incidence of skeletal deformities was 22.2% approximately 10 km downstream of Albany at RK 182 and 21.7% near Wheatland Ferry at RK 116 (Figure 3).

Evaluating all 3 years of data, the incidence of skeletal deformities observed in the upper river (Eugene to Corvallis) and in Portland is within the range of 2 to 5% reported for unstressed natural fish populations and laboratory stocks (Gill and Fisk, 1966; Wells and Cowan, 1982). Between Corvallis and Willamette Falls (RK 43), the incidence of skeletal abnormalities was much higher (>20%) than for the upper river (Figure 3). The highest incidence of skeletal deformities was measured in a 4-km stretch of river (RK 78 to 82) located in the upstream portion of the Newberg Pool (mean = 56.0%, SD = 14.1%, $n = 8$).

The high frequency of skeletal defects in the upper Newberg Pool and the gradual decline in values in the downstream direction strongly suggests that a local source is the cause of skeletal defects in this region of the Willamette River. This hypothesis is also supported by the observation that the incidence of deformities at RK 80 is significantly higher for fish collected along the east bank (side nearest Newberg) of the river than fish collected along the west bank. Skeletal defects have been attributed to a wide range of causes including genetic factors (Gill and Fisk, 1966), nutritional deficiencies (Mayer et al., 1978; Roberts and Shepherd, 1974), parasitism (Bucke and Andrews, 1985), elevated water temperatures (Brungs, 1971; Gabriel, 1944; Hubbs, 1959), low oxygen concentrations (Blaxter, 1969; Turner and Farley, 1971), heavy metals (Bengtsson et al., 1975, 1988; Bengtsson, 1974; Holcombe et al., 1976; Muramoto, 1981; Pickering and Gast, 1972), herbicides (Couch et al., 1979; Wells and Cowan, 1982), pesticides (Couch et al., 1977; Hansen et al., 1977; McCann and Jasper, 1972; Mehrle and Mayer, 1975; Meyer, 1966; Weis and Weis, 1976), PCBs (Mauck et al., 1978; Mehrle et al., 1982), bleached kraft pulp and paper mill effluent (Bengtsson, 1988; Härdig et al., 1988; Lindesjöö and Thulin, 1992; Mayer et al., 1988; Thulin et al., 1988), and ore smelter effluent (Bengtsson and Larsson, 1986; Mayer et al., 1988).

The sampling reported here was not designed specifically to identify the source of the observed abnormalities. Possible explanations for the results are given elsewhere (Ellis, in preparation). In general, no definitive cause could be identified. The dramatic differences between "background" and Newberg Pool levels of skeletal abnormalities warrants additional sampling in the future.

Fish Health

Northern squawfish were collected at Willamette RKs 10 and 79 and Multnomah Channel RK 6. Largescale suckers were collected at Willamette RKs 35, 40, 79, 206, and 298. All parameters except for length, weight, Ktl (condition factor), fat, bile, and the blood parameters are reported as percentage of the sample population that was abnormal. Abnormality is uniquely defined for each parameter.

For northern squawfish, the length and weight of fish obtained at RK 10 and Multnomah RK 6 were considerably lower than those for fish from RK 79, indicating that at least two different year classes were incorporated in this study (Wydoski and Whitney, 1979). It was likely that fish from the Portland region were 1 to 2 years younger than fish captured at RK 79. Percent abnormality was less than 20% for all external features (Figure 5). For internal features, percent abnormality was less than 20% for spleen, kidney, and hindgut at all stations, except for hindgut at RK 10, at which 40% were considered abnormal. No significant fat deposits were noted for any of the squawfish, although this index is probably not meaningful for this species (see discussion below). The bile was yellow or straw color and the gall bladder was full for almost all specimens. The mean hematocrit and plasma protein values were quite similar at all three stations and ranged from 40.5 to 49.8% and 5.3 to 5.9%, respectively. Since typical values for these parameters are unavailable for this species, it is difficult to assess the implication of these data. It should be noted, however, that the coefficient of variation (CV) for hematocrit was relatively high at RK 10 (30%) and Multnomah RK 6 (43%). It has been observed that CVs over 15% for hematocrit values could indicate that some of the fish in the population may be unhealthy (Goede and Barton, 1990). Leukocrit was not detectable for most of the blood samples.

For largescale sucker, the mean length and weight ranged from 222 to 470 mm and 136 to 1145 g. The largest fish were found at the two most upstream sites (RKs 206 and 298). The smallest fish were found at the downstream-most site at which suckers were caught (RK 40). It was likely that the largest suckers were at least 3 to 4 years older than the smallest suckers (Wydoski and Whitney, 1979). The percent abnormality for all external indices was less than 25% of each sample population, with the exception of gills at RKs 206 and 298 (80 and 40% abnormal, respectively) and thymus at Stations RKs 206 and 298 (35 and 40% abnormal, respectively) (Figure 5). In addition, four individuals at RK 206 were completely missing one eye (20% abnormal). This condition appeared to be congenital and not the result of disease or injury. The percent abnormality for all internal indices was less than 30% of each sample population. At RKs 206 and 298, none of the internal indices had percent abnormalities less than 15%, while at the other three stations, one or two of the parameters showed 0% abnormality. Fat deposits were more prominent for suckers than they were for squawfish. Fat deposits surrounding all internal organs were noted in some specimens. The bile was almost always a yellow/straw color and the gall bladder was usually empty. No major differences between stations were noted for either of these two indices. Hematocrit values ranged from 28.8% at RK 40 to 43.9% at RK 206. A high hematocrit value can result from acute stress, while a low value may indicate disease (Goede and Barton, 1990). The CV for the hematocrit values was usually relatively high. Only RK 40 had a hematocrit CV (14%) of less than 27%. Leukocrit and plasma protein ranged from 1 to 2.5% and 3.9 to 5.4%, respectively. The normal values for these variables in this species are not known.

The fish health assessment protocols used in this study were designed for salmonid fishes. Interspecies comparison of fish health data is probably not appropriate given the physiological differences both between these two species and between these species and salmonid species. The degree to which a given index reflects a meaningful physiological condition for nonsalmonid fishes is unknown. For example, the fat index developed for salmonids is a measure of the fat surrounding the pyloric cecae. Because squawfish do not have pyloric cecae and do not store significant amounts of fat in their body cavity, the low value for this parameter (mean = 1; i.e., little or no fat) could

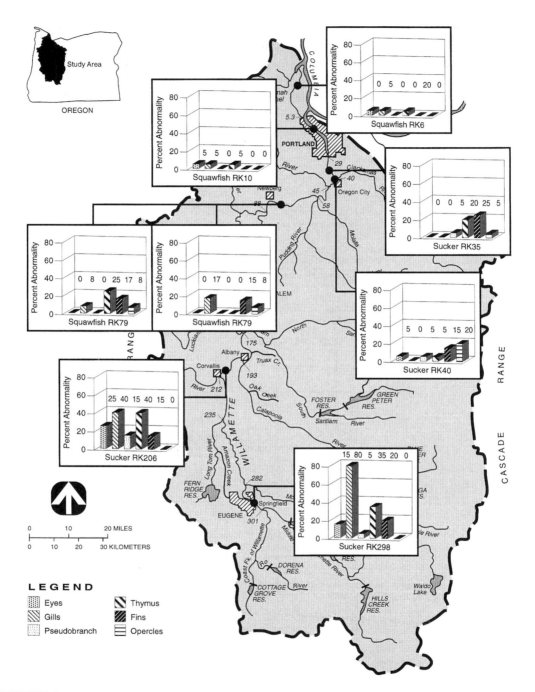

FIGURE 5. Percent abnormalities of fish health/condition indices for northern squawfish and largescale sucker collected in August to September 1992.

be misinterpreted as an indication of food availability and/or assimilative capacity, when in actuality, the index is probably meaningless for this species.

In the Willamette River, fish health assessment has only been performed once before, using squawfish (ODEQ, 1994b). The fish health assessment protocols have not been utilized for suckers prior to this study. Without any historical data on suckers, it is difficult to know whether the parameters measured are within the "normal" variation for this species in this region. The data

MONITORING-SOURCES OF WATER POLLUTION

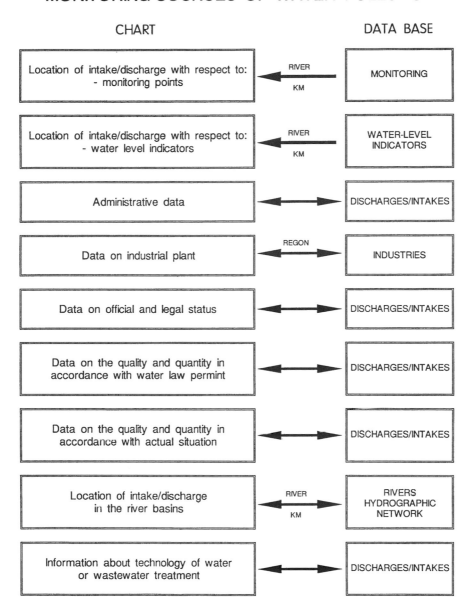

FIGURE 4. Monitoring sources of water pollution.

- classification of the object,
- data on the firm,
- location in the river basin (river side, category of the river basin),
- the type and the destination of the object,
- antifiltration devices,
- control-measurement devices,
- data on the design documentation,
- data on maintenance of the object, and
- evaluation of the technical condition of the object.

HYDROTECHNICAL OBJECTS

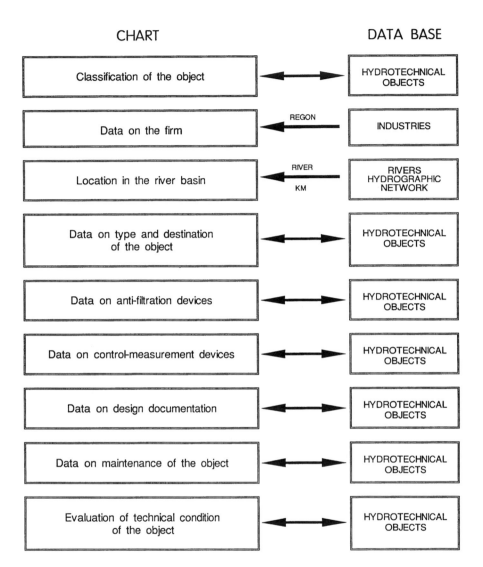

FIGURE 5. Hydrographical objects.

Like in the database "Intakes/discharges," the data concerning the units responsible for exploitation of hydrotechnical objects and their location on the area of river basin are automatically taken from the additional databases of "rivers," "industries," "monitoring," and "water-levels indicators" (Table 3).

IMPLEMENTATION OF MIS

A water-resource information system is consistent with the program activities and guidelines of State Environmental Monitoring. As soon as MIS is implemented throughout Poland, it will enable rational management of all water resources and facilitate the following:

• systemic management of water resources with respect to river basin system,

---

TABLE 3
Sample Output of MIS

Institution name	Rzeszowske Przedsiebiorstwo "KRUSZGEO"
Facility name	Wydzi. Produkcji Kruszywa — Maniowy
River kilometer	178.60
Withdrawal	Intake
Permitted by	Wojew.nowosadeckiego
Permit date	08-08-1990
Permit number	OS.III 7211/20/90
Permitted discharge (m^3/day)	3091.20
Permitted BOD_5 (kg/day)	4.00
Permitted COD (kg/day)	0.00
Permitted SS (kg/day)	20.0
Permitted chloride (kg/day)	0.00
Permitted sulfate (kg/day)	0.00
Actual discharge (m^3/day)	2060.80
Actual BOD_5 (kg/day)	4.90
Actual COD (kg/day)	0.00
Actual SS (kg/day)	11.0
Actual chloride (kg/day)	0.00
Actual sulfate (kg/day)	0.00

Note: Example of complete indentification of objects of data collected at IMGW Katowice, March 7, 1994. BOD_5, Biochemical Oxygen Demand for 5 days; COD, Chemical Oxygen Demand; SS, Suspended Solids; m^3/day, cubic meters per day; kg/day, kilograms per day.

- rational decision making on investments and water exploitation,
- optimization of decision making in external conditions such as drought, floods, and ecological calamities,
- determination of fees and fines for trespassing standards, and
- analysis of quality changes in the river (basin) in case of beyond limit wastewater discharges or in case of opening new wastewater treatment plant.

Thus the system can be widely used by:

- the Ministry of Environmental Protection, Natural Resources and Forestry, and other central units active in the field of water management for statistical purposes,
- specialized agendas of State Administration dealing with water management,
- Regional Boards of Water Management,
- Voivodship Inspections of Environmental Protection,
- Main Statistical Office,
- State Environmental Monitoring,
- design offices, and
- universities and research institutes.

EVALUATION OF MIS

In 1992, MIS was evaluated by a specialist who analyzed the water management information programs already existing and under evaluation in the field. The outcome of this analysis follows:

1. The system rationally selected an appropriate range of information without overloading the system with unuseful data.
2. The system was relatively simple to work within the programs offered and the types of words a new user would use.
3. The system delivered clear text presentations that met the actual expectations of the user.
4. The system allowed for the use of computer graphics showing spatial relationships as simplified schematics.
5. The system was verified for practical use.

34 Control of Point and Nonpoint Sources of Pollution by Use of Airborne Imagery of Wastewater Effluent into Polish Rivers and Reservoirs

A. Dobrowolski, B. Głowacka, and M. Smoręda

Imagery taken from aircraft has been used to monitor wastewater effluent into Polish streams and reservoirs since the 1970s. In the 1970s and 1980s, the imagery of heated water discharged from thermal power plants was done with thermovision cameras using the infrared spectrum. Since 1990, a technique using video cameras and natural colors in four channels (red, green, blue, and noninfrared) was initiated in the Vistula Basin. The use of imagery for locating wastewater effluents, ecotonal zone destruction, and the effects of agricultural nonpoint return flow to rivers and reservoirs is demonstrated in this chapter. Examples of video images showing effluent plumes contrasted to the receiving water are presented. In addition to color differences between the effluent and the receiving water from chemistry differences or differences in suspended matter, the effects of the plumes can also be seen by increases in algae growth or the absence of ice cover. Video images are processed with specialized computer hardware and software that allows for the analysis of spreading and dilution characteristics of the effluent as the images record its dispersion in the receiving water.

DISCUSSION

Airborne imagery can be useful for the location and control of point and nonpoint sources of surface-water pollution. It is possible to recognize the points of discharge, observe the spreading of the pollutant, and the extent to which it effects the receiving water body. This methodology operates on the premise that polluted water differs from receiving water in color, concentration of suspended matter, and temperature, and that the different water can cause physical and biological changes in the receiving water.

In Poland since the 1970s, the imagery of thermal plumes created from the discharge from thermal power plants has been taken from aircraft (Ciołkosz and Dobrowolski, 1974; Dobrowolski et al., 1985). Imagery was done using thermovision cameras and scanners. In Figure 1, the thermal image is shown for a stretch of the Vistula River downstream from the outlet of Kozienice thermal power plant. The image makes it possible to determine configuration of the plume and to quantitatively define the transversal distribution of surface temperature. Accuracy of the temperature determination was about ±0.2°C, and receiving water temperature is about 9°C.

Since 1992, the Department of Hydrology of the Institute of Meteorology and Water Management (IMWM), Warsaw, has used video imagery taken from aircraft to document the state of Polish rivers and reservoirs in both summer and winter periods. Simultaneous to these video images, hydrometric, actinometric, and additional photometric measurements were made to obtain better

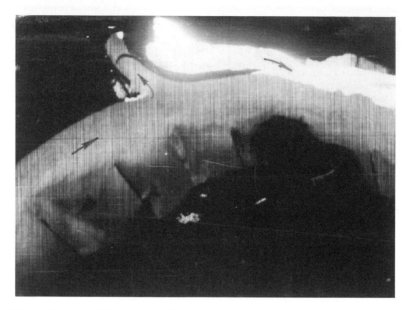

FIGURE 1. Thermal image of the stretch of the Vistula River downstream from the outlet of a thermal power plant.

interpretation of the aerial videos (Dobrowolski, 1992). The following four examples illustrate the utility of using video imagery to monitor waste water effluent:

1. The municipal waste water discharge into Vistula River several kilometers downstream from Warsaw during a time of drought is presented on Figure 2.
2. Figure 3 shows the incidental rise of bed sediment to the surface during pipe-line construction across Bug River.

FIGURE 2. Municipal wastewater discharge into Vistula River downstream Warsaw.

FIGURE 3.   Incidental rise of the bed sediment to the surface of Bug River.

3. Agricultural and domestic wastewater discharge to Zegrze Reservoir from a pumping station (Figure 4). The Zegrze Reservoir, situated about 40 km north of Warsaw. It is a source of drinking water for the capital of Poland. Reservoir embankments block the direct runoff from adjacent farmland, but small streams, still connected for drainage purposes, carry large quantities of nutrients into the reservoir.

4. Figure 5 shows the nonpoint affects of agriculture and domestic water effluent (0.26 m³/s) into the Sulejów Reservoir. An algal bloom along the shore is an indicator of the high concentrations of nutrients carried by the wastewater.

The video airborne technique is also helpful for identifying discontinuities in the ecotone zone, where intensive nonpoint leakage occurs from nutrient-laden groundwater in the agricultural catchment (Zalewski and Dobrowolski, 1994). The Sulejów Reservoir is the source of drinking water for the large towns of Łódź, Piotrków, and Trybunalski. The ecotonal zone is a buffer between the pollutant and the rest of the reservoir. A regular observation and quantification of this discontinuity is important in the management of the reservoir. The area neighboring the Sulejów Reservoir, with its buffering zone of forests, meadows, and sandy coast, was recorded using the airborne video technique and color photography during a low water period (about 3 m below the maximal level). Figure 5 is an exemplary frame of the image data set. Algal blooms (*Microcistis* sp.) occur along the shoreline of the reservoir indicating the location of the ecotonal zone which is a result of the leakage of nutrients from cultivated farmland.

Thermal infrared is usually used to identify heated water discharges; however, when ice cover exists in a river or reservoir, video techniques also seem to be helpful. Airborne imagery of the Vistula River during a winter period shows ice on the river surface. An absence of ice cover near heated water discharges allows for the determination of the part the reach affected by heated water. The ice free stretch of the Vistula River shown in Figure 6 is in the neighborhood of the thermal plant in Warsaw.

After registration, video images can be processed and analyzed. A special computer system for image processing and analysis is used by the Hydrology Department of the IMWM. This system is equipped with multimedia hardware which allow the input of video film in real time and in true colors. After processing, the hardware allows output to be written on video tape or floppy disks.

FIGURE 4. A plume from agriculture wastewater and mats of algae indicating upstream agricultural influences into the Zegrze Reservoir.

Images in the format of frozen video frames are saved as a disk files for later analysis and quantification.

The most important processing operations are radiometric corrections, image enhancements, and classification. The horizontal and vertical scale of image is calculated on the basis of fixed datum points. For a flight ceiling of about 1000 m, width of imagery was about 900 m, and a pixel size was 1 m × 1 m. After scaling, measurements of the longitudinal and transverse sections of wastewater plumes can be made. Computer software allows for the presentation of luminance along any straight line on the image as a histogram (Figure 7A,B). The analyses of histograms within the wastewater plume allows for determination of the dilution of wastewater in the receiving stream.

Four separate images result from splitting the true color image into basic channels of Red, Green, Blue, and NIR, and can enrich video information about wastewater. The choice of color image depends on the kind of pollutants and their spectral characteristics. In the case of municipal

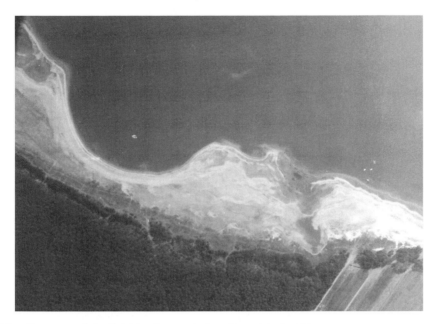

FIGURE 5.   Fragment of the Sulejów Reservoir shoreline with algal bloom.

FIGURE 6.   Disappearance of ice cover near heated water discharge.

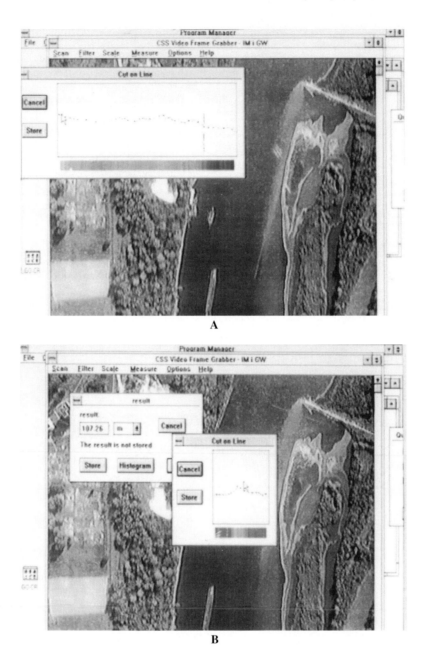

FIGURE 7.   Computer analysis. Luminance histogram. (A) Along wastewater stream; (B) in cross section of the stream.

waste water discharges from Warsaw, the best differentiation from river water was obtained by using the NIR and Red wavebands. The best results for nonorganic suspension can be obtained by using the Red and Green wavebands, and sometimes the NIR waveband. For identification of algal blooms, the best result uses the combination of NIR and Red wavebands.

The level and range of tonal scale for natural color video images depend on hydrological and meteorological conditions as well as the resolution of the video camera. Therefore, control measurements at known points of the reservoir are necessary for obtaining quantitative interpretation of video images.

REFERENCES

Ciołkosz, A. and Dobrowolski, A., Using of the airborne thermal imagery for the investigation of temperature distribution in the Vistula River downstream Kozienice Thermal Power Plant, 1974.

Dobrowolski, A., Halemba, B., and Rudowski, G., Application of the airborne remote sensing to measurements of water surface temperature KGB PAN, Studia i materiały oceanologiczne, No. 47, Gdańsk, 221, 1985 (in Polish).

Dobrowolski, A., Airborne remote sensing of the Vistula River with use of video technique. Wiadomości IMGW, 4, 1992 (in Polish).

Engman, E. T. and Gurney, R. J., *Remote Sensing in Hydrology,* Chapman and Hall, New York, 1991.

Zalewski, M. and Dobrowolski, A., The manipulation and monitoring of water level in reservoir as a tool for improvement of water quality by control of biotic mechanisms. Proc. of XVIII Congress on Large Dams, Durban, South Africa, Q.69, R.11, 1994.

35 Bull Run River Quality: A Natural Urban Ecosystem

Faith E. Ruffing

Water-quality data of the Bull Run system collected by the author and the city of Portland Bureau of Water Works have been studied and analyzed. Several areas have been identified where water quality is being affected. These include increased bacteria concentrations below a logging operation, logging related turbidity changes, and increased chlorine demand within the distribution system. Different technical approaches taken regarding water quality are discussed to provide an environmental characterization of water quality for this natural/urban ecosystem. Water-quality questions developed for this ecosystem are posed against the data available to illustrate these impacts. Suggestions for future areas of water-quality data requirements and study are presented.

INTRODUCTION

HISTORY

For the last 100 years, the city of Portland has received its water from the Bull Run River located 35 miles east of the city in the Mt. Hood National Forest. Access to the watershed, reservoirs, and headworks has been related to timber harvest and construction and maintenance of water bureau facilities. The water quality of the Bull Run system is exceptional and the city is able to purvey some of the highest quality water in the country to its customers.

SCHOOLS OF THOUGHT ON WATER QUALITY

Timber harvest activities have occurred in the Bull Run over the last century. Recently, considerable discussion has taken place regarding the impacts of timber harvest on the water quality of the Bull Run River. Several schools of thought on water-quality impacts from timber harvest in the Bull Run Watershed have been discussed.

HYPOTHESES

1. Timber harvest and other activities associated with timber harvest had no impact on water quality.
2. Timber harvest and other activities associated with timber harvest impacts water quality in several ways.

APPROACHES

1. Monitoring close to the timber harvest site is required in order to determine water-quality impacts. This is project monitoring.
2. Monitoring at key stations is important to determine quality of the water entering the reservoirs. Changes at key stations such as increases in bacteria or turbidity would require

0-56670-138-4/97/$0.00+$.50

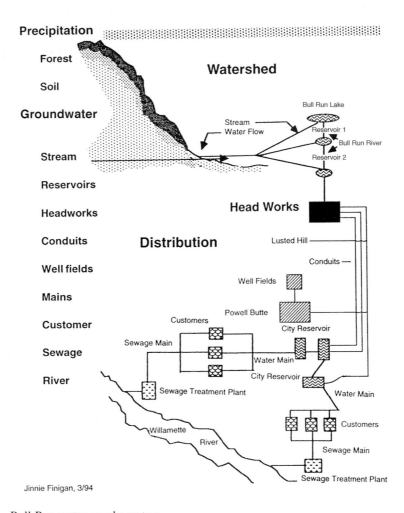

Jinnie Finigan, 3/94

FIGURE 1. Bull Run water supply system.

increased sampling at stations sited upstream at the confluence of tributaries in order to locate the source of these increases.

3. The quality of the water entering the distribution system at the headworks needs to meet drinking water standards established by the EPA. Monitoring of the intake and distribution system is necessary.

ENVIRONMENTAL SCIENCE APPROACH

The environmental science approach looked at the basin in terms of the entire water system. Figure 1 is a schematic representation of the Bull Run water supply system depicting location of the water in the *natural ecosystem* and the *urban distribution system* as it flows from Bull Run Watershed to the Willamette River.

The source of the water is precipitation and groundwater. As part of the hydrological cycle, the water comes inland from the coast in the clouds, falls to the forest, and is stored in the soil and the groundwater. The rain travels through the forest, the soil, and some going into the groundwater but most eventually collects in the rivers and streams.

The water from the streams is stored in Bull Run Lake, and two reservoirs in the Bull Run Watershed which flow in sequence. The water is withdrawn from the lower reservoir through inlet towers and goes to the headworks where chlorine is introduced into the supply. Ammonia is introduced at the Lusted Hill facility after sufficient chlorine contact time has occurred. Three conduits carry it to reservoirs in the city. The majority of the water goes through these reservoirs to the distribution system and the customers. During periods of supply shortages, the Bull Run source is augmented with the backup supply from Wellfields along the Columbia River. After use, the wastewater goes from customers, to the sewage collection system and treatment plants, and from there to the Willamette River.

The data presented are a preliminary analysis of select observations which raise many questions for further analysis. It is not intended to be a rigorous analysis but rather to point the way for further investigation. It will focus on the events in the watershed and the distribution systems.

WATER-QUALITY MONITORING PROGRAM

The city of Portland and the U.S. Forest Service monitor the watershed through a joint effort with an extensive monitoring program dating back to the 1960s. Figure 2a is a map of the Bull Run Watershed monitoring sites. Station 2 is at the headworks. Stations 15, 18, 35, and 44 are the Key Stations. The Source Search Stations 36 and 37 and the Project Stations AO1 and 69 in Cedar Creek for WY 1986 have been added (U.S. Forest Service, 1986). The monitoring program is updated periodically to meet changing monitoring needs.

Additionally, Portland monitors the reservoirs, the headworks, the conduits, the town reservoirs, and the distribution system. Considerable monitoring of the watershed has occurred since 1960 to address water quality and timber harvest issues. However, the majority of the data analysis has focused on standards development and compliance. This report is a preliminary analysis of selected water quality data and changes related to timber harvest in the Bull Run Watershed.

WATER QUALITY IN THE BULL RUN SYSTEM

The Water Bureau chose this location for Portland's water supply 100 years ago. It was an excellent choice as the precipitation is abundant. An annual average of 70 in. is recorded at the headworks. This is twice that which falls in the city. In other parts of the watershed the precipitation is about 160 inches/year.

The quality of the rainwater is quite high. It is very dilute and carries very little contaminants because of the very clean air in the Pacific Northwest. Rain quality data has been collected as part of the NADP program and analyzed for chemical constituent input into the Bull Run system (Ruffing, 1990).

The quality of water that is held in the forest, soil, groundwater, and stream part of this system is also very high quality with low biological activity and minimal mineral and nutrient concentrations (Aumen et al., 1989).

The water is collected in Bull Run Lake and two reservoirs with a total storage of 13.5 billion gallons. The water stratifies in the summer months with the development of a thermocline which separates into the warmer productive epilimnion and the hypolimnion with lower temperatures and minimal growth making it high in quality. Suspended materials carried down the river settles to bottom of the reservoirs. An anaerobic layer of low quality water develops on the bottom leaving the highest quality water in the middle.

The Water Bureau is able to withdraw this middle layer high quality water for delivery to its customers with only a minimal amount of disinfection required to meet drinking water standards.

TIMBER HARVEST ACTIVITY IMPACTS

Possible water-quality changes that may occur if the forest were removed through timber harvest are described in Tables 1 and 2.

TABLE 1
Possible Water Quality or Quantity Changes in Watershed

Precipitation

No change

Forest

Loss of forest canopy
Loss of canopy storage capacity
Loss of nutrient exchange in the fog drip and rain through fall
Loss of protection of the forest floor from sun, wind, and rain
Increased forest floor temperature in deforested acres
Increased rate of flow through remaining system from road ditches and culverts
Different ratio of nutrients leaving the forest
Increased flow during storms
Decreased flow during dry periods due to loss of forest canopy storage

Soil

Increased temperature and drying of the soil layer
Increased channel and road erosion bringing soil, soil water, soil nutrients, and soil bacteria
  into the water system
Change in nutrient ratios reflecting forest floor filtered soil water to one reflecting increased
  or complete soil water dominance

Groundwater

Decrease in recharge from slow percolation through the forest and soil layers for compacted
  surfaces near flowing streams

Streams

Increased temperatures from loss of shade
Increased runoff during storms
Decreased runoff during dry periods because of loss of canopy storage
Increased runoff until vegetation uses soil moisture
Increased sediment, turbidity, nutrient concentration, and organic matter from eroded soil from
  disturbed areas near flowing streams
Increased proportion of soil bacteria to total bacteria reflecting changes in soil water dominance

SELECTED OBSERVATIONS

The observations selected for this chapter were those that demonstrated changes that could be attributed to timber harvest activities, those for which data was available, and those from the different locations in the system.

Watershed

Bacteriological changes represent the sum total of variations in physical parameters and chemical constituents. Increases in total bacterial concentrations or changes in the proportion of soil bacterial to total bacteria in the water below timber harvest activities could reflect water-quality changes.

Bacteriological data collected throughout the watershed as part of the monitoring program conducted by the U.S. Forest Service and the city of Portland included samples collected at Project Stations above and below timber harvest activities. Standard plate count (SPC) were selected for the general bacteriological population and total coliform tests were chosen because many of the atypical (ATY) colonies growing on the plates are thought to be soil bacteria.

TABLE 2
**Possible Changes in Water Quality of the Supply
and Distribution System**

Reservoirs

Increased productivity in the epilimnion due to increased stream temperatures
  and nutrient concentration
Lowering of thermocline from increased stream temperatures increasing volume
  of the epilimnion (poor quality water) decreasing volume of the hypolimnion
  (high quality water)
Increased turbidity

Headworks

Increased temperature
Increased turbidity
Increased bacteria
Increased chlorine demand

Distribution system

Increased temperature
Increased turbidity
Increased bacteria
Increased chlorine demand
Increased turbidity accumulation
Increased bacterial regrowth

Reservoirs

Increases in productivity in the epilimnion of the watershed reservoirs could indicate water-quality changes in the watershed that could result from timber harvest activities. Changes in the total productivity of the epilimnion of the reservoirs could possibly be determined by relating the bacteriological data to the depth of the thermocline and therefore the volume of the epilimnion. This analysis was deemed too complex and beyond the scope of this chapter.

Headworks

Increases in turbidity at the headworks could indicate increased erosion in the watershed due to timber harvest which carried through the natural river and reservoir system and withdrawn at the intake towers. Turbidity data has been collected at the headworks since 1965. Turbidity is an important standard for the distribution system and one which theoretically would change as a result of timber harvest.

Distribution System

In order to comply with drinking water standards, the water supply is chloraminated. Increases in chlorine demand to maintain disinfectant requirements in the distribution system could be a result of increases in temperature, nutrients, bacteria, and turbidity. The chlorine dosage at headworks is adjusted to maintain chlorine residuals throughout the distribution system. Both bacteria and turbidity impact the effectiveness of the treatment. Increases in turbidity and bacteriological accumulation in the system would require increased dosage.

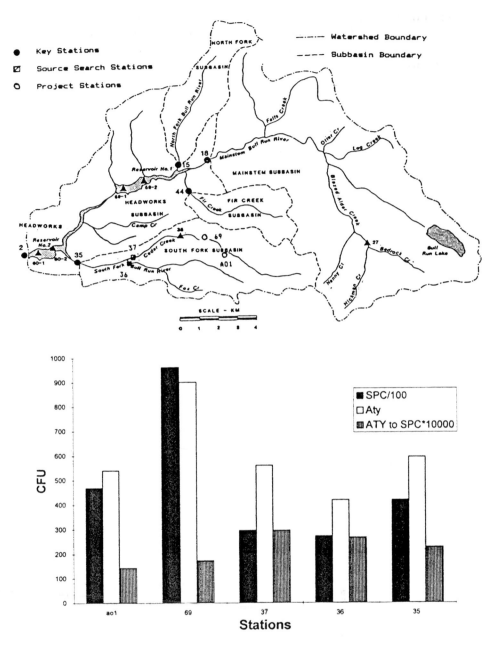

FIGURE 2. Monitoring stations in the Bull Run Watershed (top); bacteriological data June to November 1986 (bottom).

RESULTS

WATERSHED

Sampling stations in the watershed from June to November 1986 are shown in Figure 2 (top). Of interest are Project Stations AO1, above a timber harvesting operation on Cedar Creek; Station 69, below the operation; Station 37, a Source Search Station downstream; Station 36, a Source Search Station; and Station 35, the Key Station for the South Fork of the Bull Run River (U.S. Forest Service, 1986).

The bottom of Figure 2 is a graph of the data collected from June to November 1986 as the average SPC (times 100), ATY, and average percentage of ATY to SPC from stations above and below and downstream from the timber harvest operation. The difference between the SPC for the upstream and downstream stations is not significant; that for the ATY is statistically significant (City of Portland Bureau of Water Works, 1987).

TIMBER HARVEST

Records of the total annual acres of timber harvested, reforested and reestablished in the Bull Run Watershed are published by the U.S. Forest Service. The top part of Figure 3 shows the historical record of accumulated acres of clear cut, the accumulated acres of clear cut minus the accumulated acres of reforested land, and the accumulated acres of clear-cut minus the accumulated acres of reestablished forest (U.S. Forest Service, 1992).

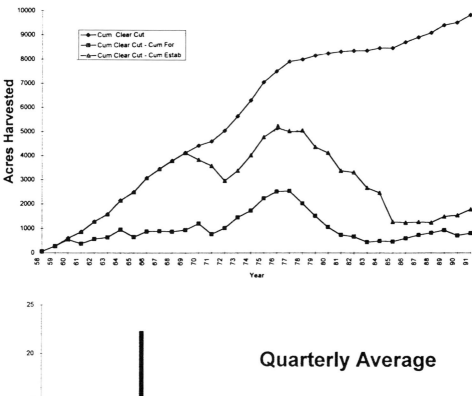

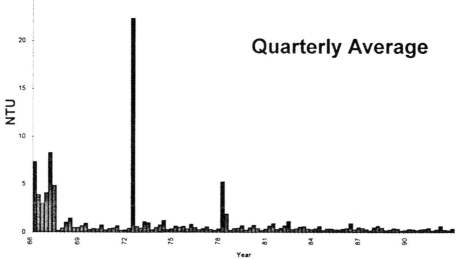

FIGURE 3.   Bull Run timber harvest history 1951 to 1991 (top); mean turbidity at the headworks 1965 to 1991 (bottom).

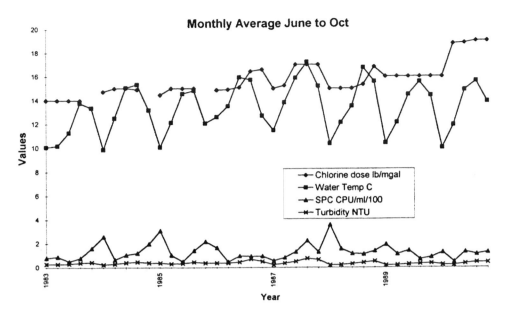

FIGURE 4. Water quality and chlorine dosage at the headworks 1983 to 1990.

HEADWORKS

Figure 3 (bottom) graphs the mean quarterly turbidity data from samples taken at the headworks for the entire record 1965 to 1992 (City of Portland Bureau of Water Works, 1992). Note the higher turbidity reported for 1965 to 1967. After 1968, the data gives a relatively straight line except for 1972 and 1978 and seasonal variation. The drop in turbidity at the headworks may reflect the reestablishment of the forests, or it may reflect better management practices by the Forest Service and the City of Portland.

DISTRIBUTION SYSTEM

The raw water is disinfected with chlorine at the headworks. Since 1983, the dosage is adjusted to maintain chlorine residuals in the distribution system. The chlorine dosage from 1983 to 1990 from June to October is shown in Figure 4. There is a steady increase in the dosage except for 1988 and 1989. This increase in chlorine demand may reflect increased nutrients, turbidity, and bacteria accumulating in the distribution system which could increase the demand (City of Portland Bureau of Water Works, 1991).

DISCUSSION

WATERSHED

The data for the Project Stations are of short duration covering the period of timber harvesting. No follow-up data are available. Data from two other Project Stations do not show statistically significant differences between the above and below sites for SPC or ATY. It is interesting that the high concentrations observed at Station 69 are diluted out as the water moves downstream and are no longer observable at the Key Station.

Differences between the AO1 and 69 may be due to geomorphological or ecosystem changes between these sites. An investigation into the source of these bacteria is needed. The identity of the atypical bacteria needs to be established in more detail.

HEADWORKS

Station 2 turbidities show rather stable measurements from 1968 to 1993. The quarterly averages are within the old Standard of 1 NTU for the most part and within the new Standard of 5 NTU except for 1972.

High turbidities in 1972 are believed to be due to the failure of an earthen dam at Boody Lake on the North Fork of the Bull Run River above Reservoir 1. The increase in 1978 is not yet explained.

High turbidities prior to 1968 may be due to lag time between clear cutting and reestablishment of the forest in 1967 or to the 1964 flood — a 500-year event. It may also be due to other watershed activities such as road building. Increased turbidities during the peak logging from 1973 to 1985 are not reflected in the data, raising additional doubt regarding the relationship between timber harvest and increased turbidities.

Headworks data reflects the sum total of the watershed forest management practices, conditions in the reservoirs, and the collection system. Changes in water quality as determined by turbidity do not seem to be above background except prior to 1968. From the bacteriological examination on the South Fork changes occurring close to the operation may not show up at the headworks.

Examination of turbidities at the Key Stations and integration with other parameters such as flow and precipitation needs to be performed to further elucidate the relationship. A review of the historical record for incidents needs to be reviewed to identify other events related to the peaks in the turbidity data.

The goal of managing the watershed and operating the distribution system to reliably meet the drinking water standards is being met as attested to by the low turbidities observed at the headworks.

DISTRIBUTION SYSTEM

The increased dosage may be due to increased demand but it may also be due to increased diligence by the Water Bureau to protect the supply from standards violation by increasing the chlorine residual throughout the system. This would provide a larger buffer between the actual chlorine residual and that required by the standards.

CONCLUSIONS

General conclusions from the selected preliminary data analysis are

1. Indications that changes in water quality may be occurring below a logging operation as indicated by shift in the ratio of soil to total bacteria.
2. Quarterly averages of turbidity at Station 2 show a drastic decline from 1966 to 1992.
3. The turbidities seem to reflect the acreage of clear cut minus the acreage established after reforestation.
4. Increases in chlorine demand in the distribution system result in increased dosage.
5. These changes may be the result of timber harvest activities, including road building and reforestation in the watershed.
6. None of the changes observed indicate a long-term impact on water quality due to timber harvest if clear cuts are reforested and reestablished quickly.
7. Bacterial changes are limited to increases in ATY coliform bacteria. There is no indication of increases in the total bacterial concentrations or of pathogenic bacteria which can be separated from the natural variation due to meteorological conditions.
8. The turbidity levels at Station 2 are well below the present standard of 5 NTU and below the previous standard of 1 NTU. Occasions where the standard has been exceeded are more indicative of nonlogging incidences, such as mud slides.
9. The Bull Run Watershed is the most thoroughly monitored watershed in the country. Thirty years of data collection has failed to produce strong evidence of permanent water

quality degradation related to logging that cannot be mitigated by careful forestry management practices that require early reestablishment of healthy stands of young trees.

10. A rigorous analysis of all of the environmental factors including meteorological, geomorphological, and ecological conditions of the Bull Run water-quality data should be performed to further elucidate impacts of logging on water quality in this system.

REFERENCES

Aumen, N. G., Grizzard, T. J., and Hawkins, R. H., Water Quality Monitoring in Bull Run Watershed, OR, Task Force Final Report submitted to the City of Portland, Oregon Bureau of Water Works, 1989.

City of Portland Bureau of Water Works, Personal Correspondence: Analysis of Bacteriological Data Collected at Cedar Creek Project Station June to November 1986, 1987.

City of Portland Bureau of Water Works, Personal Communication: Transmittal of Turbidity Data Collected at Headworks from 1965 to 1991, 1992.

City of Portland Bureau of Water Works, Personal Communication: Transmittal of Raw Water Quality Data and Disinfection Dose at Headworks 1983–1990 from June to October, 1991.

Ruffing, F. E., *Precipitation in the Bull Run Watershed*, Prepared for the City of Portland, Oregon by Sun Mountain Reflections, 1990.

U.S. Forest Service, *Bull Run Watershed Management Unit Annual Activities Schedule Water Year 1987*, 1986.

U.S. Forest Service, *Bull Run Watershed Management Unit Annual Activities Schedule Water Year 1992*, 1992.

36 South African Approaches to River Water-Quality Protection

P. J. Ashton and H. R. Van Vliet

Increased concern over the deterioration in the quality of surface-water resources in South Africa during the past few decades has led to fundamental changes in the approaches to national water-quality management. The new approaches promote sustainable development and focus on procedures to ensure that surface waters remain fit for use by all recognized water users. In addition, strict precautions limit the entry of toxic substances into the aquatic environment. Domestic and industrial effluent is recognized as an important water resource and current management practices promote the concept of effluent recycling and reuse within strict safety limits.

Water-quality management is conducted on a catchment-wide basis and is the responsibility of a single central authority. The management process consists of a series of stages where all water users within the area of concern assist with the development of use-specific water-quality objectives for each variable of concern. Water-quality management objectives for each river are reached and these are combined into an overall catchment water-quality management plan. A detailed water-quality monitoring program is used to evaluate compliance with management objectives.

INTRODUCTION

South Africa is an arid country with limited freshwater resources; rainfall is strongly seasonal and unevenly distributed across the country. There are few perennial rivers and natural lakes; river flows are characterized by large seasonal variability and are also subject to prolonged periods of low or below-average flows. The locations of major centers of population and industrial development were determined by the presence of mineral deposits or natural harbors and these centers are seldom located near readily available water supplies. The water demands of the country's major metropolitan areas have been met by a complex system of water resources projects, which include extensive interbasin water transfers and the reuse of treated effluent flows (DWAF, 1991). Inevitably, this has been accompanied by a progressive deterioration in water quality in virtually every river system.

The vital importance of adequate freshwater supplies for sustainable economic and social development has been frequently emphasized (DWAF, 1986, 1991). Recent estimates indicate that the population of South Africa has grown at a rate of approximately 2.85% per year, from 22 million in 1970 to approximately 40 million in 1994. The implications of this population growth for the available water resources are highlighted in Figure 1.

It is predicted that the proportion of the available surface water resources that can be exploited with existing technology will be fully utilized by the year 2000 at the highest rates of population growth and water use. At the lowest rates of population growth and water use, these resources will be fully exploited by the year 2020 (Figure 1). Similarly, if additional technology becomes available, all of South Africa's surface and groundwater resources will be fully utilized by 2015 and 2040, for the lowest and highest rates of population growth and water use, respectively.

The problems caused by scarce-water resources and increasing rates of population growth have been compounded by marked changes in demographic patterns and living conditions which have

0-56670-138-4/97/$0.00+$.50
© 1997 by CRC Press, Inc.

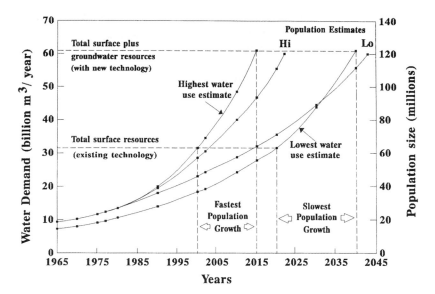

FIGURE 1. Comparison of predicted growth in water use with predicted population growth in South Africa, using two estimates of population growth rate and two estimates of water use rate.

led to increased per capita demands for water. In several regions, the anticipated demands for water are expected to exceed conventional economically exploitable sources within two to three decades (DWAF, 1986). The quantity and quality of water supplies will play a critical role in the future socioeconomic development of South Africa. In view of this situation, very careful management will be required to ensure that the maximum benefits can be derived from the scarce resources available.

The limited nature of the available water resources has also given rise to intense competition between the different water use sectors, each of which has its own needs and expectations. Where each water use group has attempted independently to follow a course of water resources development, the result has been inefficient and ineffective use of a scarce resource. Local and overseas experience has shown that a carefully coordinated and holistic approach, which considers all aspects of water resource management within a region, is the only way in which maximum efficiency of resource utilization can be achieved (Ashton et al., 1995).

WATER-QUALITY MANAGEMENT IN SOUTH AFRICA

Water resource management is a responsibility of central government and the Department of Water Affairs and Forestry (DWAF) is regarded as the custodian of this resource. The Department's earlier water-quality management strategies were based on the requirement that effluent discharges conform to a single uniform effluent standard (DWAF, 1986). Recently, the Department revised its approaches to water-quality management (DWAF, 1991). A receiving water-quality (RWQ) based approach was implemented for nonhazardous substances and a pollution prevention approach was adopted for hazardous substances. In addition, the Department also committed itself to applying a precautionary principle as an integral part of its water-quality management strategies (Van Der Merwe, 1991).

The Department's water-quality management philosophy has changed from "end-of-pipe" pollution source control to a system based on maintaining the fitness for agreed or specified uses of particular water bodies. This enables agreed limits to be placed on the levels of pollution that can be allowed or tolerated by recognized categories of water users in each aquatic system. The ultimate quality objective for the receiving water depends on agreement between users who may be potentially disadvantaged by effluent discharge and those who may not. The approach relies on knowledge of the quality requirements for each water user and on our understanding of the ability of aquatic systems to tolerate waste discharges (Ashton et al., 1995). This policy shift represents a significant change from a relatively simple and standardized system, to a more complex and information-intensive process, with a greater degree of integrated resource-based management.

An important milestone in the RWQ approach to water-quality management has been the development of water-quality guidelines for domestic, industrial, agricultural, and recreational water uses (DWAF, 1993). These guidelines define the effects of changes in the concentration of different water-quality variables on the water's fitness for each use and support the need for knowledge of users' quality requirements. A fifth set of water-quality guidelines for aquatic ecosystems is currently in preparation.

A further feature of the RWQ approach is the acceptance of the concept of environmental capacity. This implies that receiving waters have an inherent capacity to assimilate an effluent and thus provides natural limits for the quantity of an effluent that can be discharged safely. In addition, the concept also implies that environmental capacity can be predicted readily in anticipation of proposed effluent discharges (Stebbing, 1992). The environmental capacity of a water body is therefore an integral and inseparable part of the natural resource and must be managed accordingly (Ashton et al., 1995).

The RWQ approach requires specific water-quality management objectives to be set for each water user in each river section, at levels which ensure sustained fitness for recognized uses. Appropriate strategies are then developed to manage the quantity and quality of water in each river section and to ensure that the set objectives are maintained. Patterns of land use, water abstraction, and effluent return flows all interact with natural catchment processes to exert a range of complex effects on water quality; all of these must be taken into account (DWAF, 1991). Ideally, the entire river system is managed as a unit so that the spatial and temporal availability of water of a suitable quality is matched with the water-quality requirements of each water user. The component steps of the RWQ process are shown in Figure 2.

Effective development and execution of catchment wide water-quality management strategies in terms of the RWQ approach requires continual collation and interpretation of large quantities of information from diverse sources. Recent developments in information technology now facilitate the rapid analysis of spatially oriented information on personal computer systems. These systems provide an efficient and cost-effective platform for the development and implementation of integrated water-quality management strategies. Currently, geographic information systems (GIS) pro-

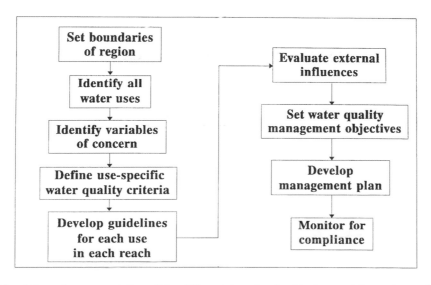

FIGURE 2. Schematic representation of the different steps involved in the receiving water-quality (RWQ) approach to water-quality management.

vide a cost-effective way to access, display, and interrogate such information in an easily understood format.

The sweeping socioeconomic changes that are occurring in South Africa have led to greater emphasis on issues around equitable sharing of resources and public participation in resource management decisions. There is now broader acceptance of the need for open and effective communication between water users and resource managers (Ashton and Van Zyl, 1994). In common with trends in other parts of the world, the general South African public has shown increased concern for the environment and its effective management. The RWQ approach enables all water users to enter into negotiations regarding their water-quality requirements (DWAF, 1991) and management of resources in which they may have a stake.

RECENT DEVELOPMENTS

The ability of a water resource to meet the demands made on it by users is closely linked to the "health" of the component aquatic ecosystems (Chapman, 1992; Roux et al., 1993). This implies that aquatic ecosystems have to be managed in such a way that they remain diverse, productive, and resilient in the face of natural and artificially induced changes in their major driving forces (MacKay and Ashton, 1994). Hence, a water resource can no longer be defined simply as the water flowing in a river channel, but rather as the entire aquatic environment and all its biotic components. In essence, therefore, the water resource management goal can be summarized as follows:

> ... to maintain natural processes and ecological functioning within aquatic ecosystems at a level of health sufficient to meet the water quantity, water quality, aesthetic and conservation demands of the people who rely on these resources, without unreasonable costs, and without disadvantage to future generations.

This presents a daunting task for water resource managers and requires yet another change in the development and application of water-quality management strategies and policies. The Department of Water Affairs and Forestry now manages the aquatic environment as a resource, where the biota, the physicochemical instream habitats, and the processes that link biota and habitats, are inseparably part of the water resource. This shift in emphasis is supported by a revision of all water-related legislation and the implementation of a carefully coordinated national water-quality monitoring and data-management system. Biological monitoring and ecotoxicological studies provide additional support (Roux et al., 1993).

PROBLEM AREAS ENCOUNTERED

DEFINING USER REQUIREMENTS

In terms of the South African Water Act, five categories of water use were recognized by the Department of Water Affairs and Forestry. These recognized uses are domestic, industrial, agricultural, recreational, and environmental conservation (DWAF, 1991, 1993; Van der Merwe, 1991). The different water-quality requirements of each user group imply that water, which would be ideally fit for one specific user group, may not be ideally suited for another (DWAF, 1993). In addition, water very seldom becomes totally unfit for use when its quality deteriorates. Quality is not an intrinsic property of water but is linked to the specific use made of the water. A definition of precisely what constitutes fitness for use for each water user is thus a key issue in the evaluation and management of the quality of water resources (DWAF, 1993). This process is complicated by the wide variety of possible water users and their often conflicting demands. Furthermore, where issues of the aquatic environment are concerned, there is often very little local information available regarding the water-quality requirements of different environmental components.

The general international trend has been to report guidelines or criteria as single values where there would be no impairment on any use (Train, 1979; Kempster et al., 1980). However, effective water-quality management should be based on more comprehensive information. Accordingly, water quality guidelines for South Africa have been expressed in terms of the effect of changes in water quality on specific water uses (DWAF, 1993). Thus, the South African water-quality guidelines for domestic, industrial, agricultural and recreational uses have three important characteristics:

1. They are expressed as a range of values.
2. Descriptions of the fitness for use of each range are associated with the guideline ranges.
3. The descriptions indicate effects on the water user from the ideal water quality to the point of unacceptability.

It is important to remember that water quality is affected in different ways by local variations in physicochemical and biological conditions. Therefore, general water-quality guidelines must be adapted to account for site-specific conditions whenever necessary. This can best be achieved by involving all water users within a specific region in the decision-making process, allowing maximum benefit to be gained from their knowledge of local conditions.

USER PARTICIPATION

The general public's interest and active involvement in the management of the country's scarce water resources has been welcomed and the Department will involve the public as far as is practically possible in decisions regarding water-quality management. As public perceptions concerning the safety of water for drinking and recreational purposes demand increasing attention, public involvement will become more common. Public participation has become the overriding reality and all those involved in water-quality management must accept this and learn to deal with it (DWAF, 1991, 1993).

At present, there is no single structure or formal process that determines how each water user should be involved in the decision-making process. In most cases, water users and the general public are invited to attend one or more public meetings at which the Department of Water Affairs and Forestry and external consultants present an overview of the appropriate water-quality information. This is usually followed by the presentation to representatives of each water user group of formal documentation that outlines the water-quality issues at stake and the management actions or plans which are envisaged. However, this type of information sharing process falls far short of true public participation.

These current practices appear to be somewhat ad hoc in nature, with greatest attention being paid to assessments of what is deemed to be "acceptable" and "unacceptable" water quality for different user groups. The situation becomes particularly complicated when the audience is comprised of individuals who may not have a clear understanding of water-quality issues. While the new water-quality guidelines provide an effective means of communicating with the general public on the impact of development on water quality, many people still do not understand much of the scientific terminology or the ways in which the information will be used (MacKay, 1994).

The existing processes of involving water users and the general public in decisions around water-quality management need to be examined and the Minister for Water Affairs intends to initiate a more structured formal process for public participation. This is intended to devolve decision making and responsibility to local authority and community level, followed by eventual ownership and operation of the resource management process. At this stage, the issues around the precise levels of responsibility that will be given to provincial authorities, local authorities, and communities have not been resolved.

Particular attention will need to be paid to providing all water users with sufficient information on which to base an informed opinion. The process of public involvement will become easier once

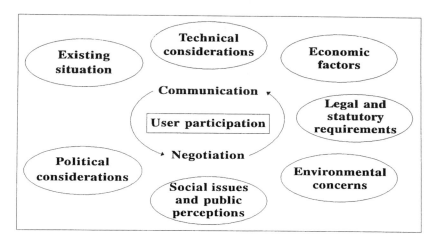

FIGURE 3.   Diagramatic representation of the component issues involved in public participation processes around water-quality management.

the users have a greater understanding of the issues at stake. The component issues which must be evaluated are highlighted in Figure 3. Recent developments in the use of three-dimensional visualization technologies have facilitated the process of communication with nontechnical audiences (MacKay, 1994).

Public participation in decision making is an effective instrument for conflict resolution because it places a greater emphasis on resolving issues of collective interest before individual concerns. However, the participants should include groups and individuals who are likely to be affected or influenced by the specific development or management actions. However, no single method or approach toward public involvement is able to satisfy all the requirements for information exchange, consensus building, and consultation. Nevertheless, the approaches followed should seek to enhance consensus and understanding, encourage dialogue and feedback, while displaying flexibility and ensuring honesty (MacKay, 1994).

ACHIEVING A BALANCE

Achieving an equitable balance between the often conflicting demands of different water users presents enormous challenges to our water resource managers. Public opinion, social norms and values, and modern principles of sound business practice are the major driving forces which balance the perceived costs and benefits of any development (Figure 4). Considerable debate has been made regarding the issues that affect the ability and willingness of water users to pay for specific assured levels of water quality. Similarly, while it is widely accepted that "the polluter must pay" principle should be strictly enforced by the regulatory authority, there is less consensus as to the desired level of water quality that should be attained. This debate will continue in the future.

The interactions and driving forces illustrated in Figure 3 focus attention on the development and implementation of appropriate legislation and policies for the rational deployment of available technology to satisfy water user requirements and safeguard the aquatic environment. Ultimately, the choice of which technological solution to use for a particular application will be driven by a society's system of values and its ability to afford a particular level of benefit. The great disparities in economic status shown by South Africa's unique mix of underdeveloped, developing, and highly developed components of society have made it very difficult for the regulatory authority to resolve this issue.

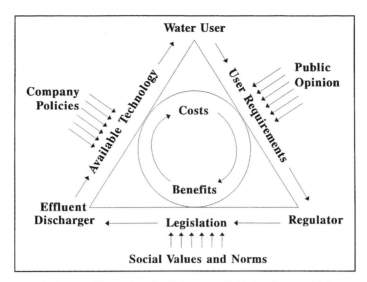

FIGURE 4. Conceptual diagram illustrating the linkages and driving forces which control the interactions between water users, regulatory authorities, and effluent dischargers.

FUTURE DEVELOPMENTS NEEDED

It is never easy to project what will happen in the future when the present is in a state of change. Nevertheless, clues can be gleaned both from current intentions and past experiences. The RWQ approach to water-quality management provides a scientific basis for management decisions and, even though it is technically complex and requires large amounts of information, it can provide a foundation for future actions. The utilization of environmental capacity concepts in the RWQ approach is not an ideal situation. There has to be a progressive shift toward greater emphasis on pollution prevention approaches. There is also an urgent need to focus attention on managing aquatic systems in such a way as to ensure sustainable utilization of the water resources. The development of appropriate management tools is therefore a high priority.

In view of the vulnerability of our water supplies, the management of water quantity and water quality should remain within the jurisdiction of a single authority. However, it is important to remember that the water quality of rivers and lakes reflects the results of man's activities in their catchments; ideally, water management should be conducted on a catchment-wide scale. Therefore, it is logical to predict the development of greater coordination between the Department of Water Affairs and Forestry and those authorities responsible for other environmental issues. This collaboration should also extend to joint water resource management with neighboring countries where common river systems are shared.

In the future, we can also look forward to more frequent and effective participation of all water users and the general public in decision-making processes which affect the quality of their water supplies. This will require the development of structured participative processes that will ensure fair and just representation. In turn, this process will have to be driven by a greater acceptance of the need to share knowledge and information. Regular public disclosure of environmental information is an accepted and widely acclaimed process in the U.S. and the U.K. where it provides an enormously powerful incentive to ensure that water-quality management remains efficient and effective.

There is also a need to develop and implement policies and procedures that can reconcile financial and environmental goals. Here, more emphasis will need to be placed on improving our understanding, use, and analysis of resource economics. For too long now we have grossly under-estimated the value of our water resources; this imbalance needs to be corrected as soon as possible.

CONCLUSIONS AND RECOMMENDATIONS

The challenge to maintain the fitness for use of water, South Africa's most limiting natural resource, depends on the successful implementation of an effective and dynamic water-quality management policy. This policy must be sufficiently flexible to meet the increasingly complex and changing demands of South African society which, in turn, are driven by rapid social, political, and economic developments. Cooperation among all concerned and joint acceptance of responsibility are prerequisites for success.

The RWQ approach to water-quality management provided a sound scientific basis for water-quality management policies and strategies in South Africa. Now, greater attention needs to be paid to continued development and effective implementation of pollution prevention strategies. This should include an effective legal framework and, ideally, the implementation of suitable economic incentives. This will place even greater responsibility on the Department of Water Affairs and Forestry, as custodian of our water resources. In addition, this will require the development of effective processes of communication and negotiation between all water user groups.

As an integral and important part of the water-quality management process, water users and their representatives must participate actively in the communication and information-sharing process. This will provide water resource managers with information regarding the characteristics and processes which affect the fitness for use of water in their areas of responsibility. In turn, improved processes of communication and negotiation will give water users a better appreciation of their roles and responsibilities in water resource management. Future generations of South Africans rely on us to start the process and to ensure that our water resources continue to be managed on a sustainable basis.

REFERENCES

Ashton, P. J., Van Zyl, F. C., Heath, R. G., and King, N. D., Catchment-based water quality management: a case study of the Crocodile River, Eastern Transvaal, *Proc. Third Biannual Conf. Water Inst. South. Africa*, 1, 14, 1993.
Ashton, P. J. and Van Zyl, F. C., Water quality management in South Africa: protection of a natural resource, in *New World Water 1994*, Sterling Publications, London, 1994, 43.
Ashton, P. J., Van Vliet, H. R., and MacKay, H. M., The use of environmental capacity concepts in water quality management: a South African perspective, in *New World Water 1995*, Sterling Publications, London, in press.
Chapman, P. M., Ecosystem health synthesis: can we get there from here?, *J. Aquat. Ecosyst. Health*, 1(1), 69, 1992.
CSS, *Population Census 1991. Geographical Distribution of the Population with a Review for 1970 to 1991*, Central Statistical Service Report No. 03-01-02, 1992, 284.
DBSA, *South Africa: An Inter-Regional profile*, Centre for Information Analysis, Development Bank of South Africa, Midrand, 1991, 60.
DWAF, *Management of the Water Resources of the Republic of South Africa*, Department of Water Affairs, Pretoria, 1986, 459.
DWAF, *Water Quality Management Policies and Strategies in the RSA*, Department of Water Affairs and Forestry, Pretoria, 1991, 38.
DWAF, *South African Water Quality Guidelines, 1st ed., Vol. 1 to 4: Domestic Use, Recreational Use, Industrial Use and Agricultural Use*, Department of Water Affairs and Forestry, Pretoria, 1993, 447.
Kempster, P. L., Hattingh, W. A. J., and Van Vliet, H. R., *Summarized Water Quality Criteria*, Department of Water Affairs and Forestry, Pretoria, Technical Report TR 108, 1980, 45.

MacKay, H. M., Community-based planning for utilisation and management of natural resources: lessons from the Chatty River floodplain project, Port Elizabeth, South Africa, *J. Environ. Manage.*, submitted, 1994.

MacKay, H. M. and Ashton, P. J., The concept of resource capability, and its application in the development, management and allocation of water resources, *J. Environ. Manage.*, submitted, 1994.

Roux, D. J., Van Vliet, H. R., and Van Veelen, M., Towards integrated water quality monitoring: Assessment of ecosystem health, *Water S.A.*, 19 (4), 275, 1993.

Stebbing, A. R. D., Environmental capacity and the precautionary principle, *Marine Pollut. Bull.*, 24, 287, 1992.

Van Der Merwe, W., A review of water quality management in the Republic of South Africa, *Water Sewage Effluent*, 11 (2), 11, 1991.

37 The Oregon Watershed Health Program: Local Empowerment to Restore Watersheds

Mary Lou Soscia*

INTRODUCTION

In 1993, the state of Oregon created the Oregon Watershed Health Program as a part of new natural resource strategy acknowledging the critical importance of watersheds to Oregon's livability and economic health. This program was a commitment to encourage government and citizens to work together in developing voluntary plans for improving watershed health. This grew from a recognition that many Oregon watersheds no longer have the capacity to satisfy all demands placed on them by a growing population and economy. New listings of endangered species, widespread shortages for irrigation needs, growing disputes over water rights, and degraded water quality were signs that a new cooperative approach to managing watersheds was seriously needed.

To provide funding support, the 1993 Legislature allocated $10 million in lottery funding to two areas: the Grande Ronde in Northeastern Oregon and the South Coast and Rogue Basins in Southwestern Oregon (Senate Bill 81). The Grande Ronde Basin already had Snake River Chinook listed under the Endangered Species Act, and the South Coast and Rogue Basins had potential listings looming around the corner with coastal coho and winter and summer steelhead. This pilot program, the Oregon Watershed Health Program, was intended to bring better cooperation among state agencies and with local citizens in managing resources using a watershed framework and promoting voluntary actions to improve watersheds. The Strategic Water Management Group (SWMG), 13 state and 4 federal agencies and the Governor's office, was given oversight authority for watershed councils and the targeted basins.

The Oregon Watershed Health Program completed its work of allocating money for restoration, enhancement, and protection by June 30, 1995. As a result of this pilot initiative, nearly 100 miles of stream banks have been planted with native vegetation to shade stream banks and hold runoff back. More than 300 in-stream enhancement structures have been placed in streams to slow currents and create fish habitat. More than 100 miles of fence have been erected to protect stream banks from grazing livestock. Seventeen fish screens have been installed on major Rogue River Basin water diversions to prevent fish from being stranded in ditches and fields. This pilot initiative has also set up a process that deeply involves citizens in voluntary actions that improve water conditions and repair the damages of historic activities.

LOCAL WATERSHED COUNCILS

The cornerstone of this new initiative was the creation of local voluntary watershed councils to provide a forum for citizens to work in conjunction with local, state, tribal and federal agencies to

---

* From 1993 to 1995, the author served as the Program Manager of the Oregon Watershed Health Program.

help solve watershed problems in their homes — the watershed in which they live. Legislation enacted in 1993 established state policy to support a watershed based approach by encouraging the formation of local watershed councils. Under the process described in the law, House Bill 2215, local watershed councils are appointed by local governments to work in partnership with natural resource agencies to develop and implement watershed action plans.

House Bill 2215 provided guidance for council membership. Local watershed councils may include: representatives of local government, nongovernmental organizations, and private citizens. This may include representatives of local boards and commissions, and agencies, Indian tribes, public interest group representatives, private landowners, industry representatives, academic, scientific, and professional community, and representatives of state and federal agencies. House Bill 2215 did not include any funding but specifically referred to Senate Bill 81 which funded the Oregon Watershed Health Program. As a result of this policy direction, approximately 50 local watershed councils have been formed throughout Oregon which are currently engaged in the management of their watersheds, and funding and completion of protection and restoration projects. Hundreds of people throughout Oregon are now participating in local watershed councils.

A LOCAL/STATE PROCESS

The Oregon Watershed Health Program was based on the development of locally prepared watershed assessment and action plans which would be the blueprint for watershed restoration and protection efforts. With the program inception in January 1994, recognizing the need for time for local partnerships to form, the first activity was to fund early action projects. These early action projects included fish screening, tree planting, fencing, and other restoration projects. Local watershed partnerships soon started forming and 13 local watershed councils were officially recognized by the state in the two areas. Multiagency and multidisciplinary field teams were established in each basin to provide an intensive technical boost to local efforts. Local watershed councils worked with these field teams, existing state agency field and central staff, interest groups, and other government agencies to characterize their watersheds, assess problems, and find ways to solve them.

Working in partnership with the diverse groups in each basin, the watershed councils developed watershed action plans that identified watershed problems and needs and provided a blueprint for solutions. These plans characterized watershed conditions, identified priority areas of restoration and protection, set out public involvement strategies, and identified funding sources to implement the plan. Action plans were drafted for all basins in the Rogue and South Coast. The Grande Ronde Model Watershed Board of Directors has developed a basinwide operations plan and is in the process of developing action plans on a subbasin level.

WATERSHED RESTORATION AND PROTECTION

The Legislature allotted $6.5 million in lottery funds for projects in the two targeted basins and directed state agencies to get projects underway as soon as possible. Funding was matched by a wide amount of federal, state, local, and private sources, at approximately $.30 to every $1.00 in lottery funding. More than 150 projects have been completed or are underway in the two areas. Here are a few examples of the projects begun with the aid of this funding:

- Fifteen thousand tree seedlings were planted along an eroded 10 mile stretch of the Illinois River where land practices (logging, draining, diking, and land clearing) had created a degraded riparian area.
- In the Upper Grande Ronde River in Union County, approximately 300 miles of roads are proposed to be closed on federal lands to reduce impacts to sensitive salmon habitat.
- Logs and boulders were placed in stream along four miles of the South Fork of Little Butte Creek in Jackson County to improve fish habitat. Four off-channel alcoves were constructed to aid in the rearing of salmonids.

- More than 100 students participated in a study to improve habitat for salmon along a six mile stretch of Five Points Creek in the Upper Grande Ronde River.

NEXT STEPS

The important role of watershed councils was reaffirmed by the 1995 Oregon Legislative Assembly in House Bill 3441 which supplants the 1993 House Bill 2215. The 1995 Legislature unanimously passed House Bill 3441 to support the continuation of support to watershed councils. House Bill 3441 emphasizes the implementation of watershed-wide conservation, restoration and enhancement, using a cooperative approach, including assessing the condition of the watershed, developing a priority based action plan and executing the plan, using a broad range of financial resources. The Legislature also directed the Governors Watershed Enhancement Board* (a citizens group, in existence for 7 years, providing statewide grants for watershed enhancement) to grant funds to support watershed councils in assessing watershed conditions. develop action plans, implement projects and monitor results.

LESSONS LEARNED

The Oregon Watershed Health Program is a bold, new successful experiment that provides an important model for working on a watershed-wide basis through local voluntary watershed councils. With the tremendous enthusiasm generated at the local level and the difficult natural resource controversies existing in communities around the country, it is widely recognized that this process is the future of natural resource management. As with any experiment, there are many lessons to be learned. These lessons may help states and communities develop successful local/state partnerships for watershed management.

- Trust takes time and is necessary for watershed partnerships to be successful.
- Watershed assessment and action plan development are essential to understand the natural and present conditions of the watershed and to select projects which protect high quality areas or are directed to areas with the most chance of recovery. Monitoring is then necessary to gauge the impact of actions that have been taken.
- Watershed recovery requires a long term political commitment.
- Communication and education are essential: the public must be involved in community-based watershed protection and restoration.
- Pilot efforts can suffer if not integrated into ongoing efforts.

---

* More information on the Governor's Watershed Enhancement Board can be obtained by contacting the Board at (503) 378-3589, ext. 823.

38 Appropriate Livestock Management Facilities Protect Water Quality in Suburban Watersheds

Derek C. Godwin and J. Ronald Miner

Recreational animals are an important feature of the Tualatin River Basin in Washington County, OR. In addition, this suburban Portland region has experienced recent population growth. Many of the new residents are seeking a home with the ability to raise animals and have the possibility of employment in the urban area. Many of the animal facilities have limited facilities for the care of the animals or for the protection of water quality. This project involved the construction of facilities on cooperating landowner's property to determine the acceptability of alternate practices and to demonstrate the availability of practices that would protect water quality while also facilitating animal rearing. Covered manure storage is an important addition to small animal enterprises in areas of high winter rainfall. Off-stream watering provisions can protect riparian areas from physical damage and can prevent the direct deposit of manure into the stream. A water tank or an animal operated, commercially available diaphragm pump was demonstrated to reduce animal impact either with or without the presence of fencing along the stream to physically restrain the animals.

INTRODUCTION

The Tualatin Watershed, particularly the suburban regions of Washington County, OR serve both as agricultural production areas and as residential areas for persons preferring to live in a rural environment. These areas are particularly attractive to people employed in the urban area who want to own or maintain a small animal enterprise as part of their lifestyle. Some of these landowners maintain one or more horses, a few cattle, pigs, or sheep. In attempting to understand the relationship between land use and water quality, it was deemed important to understand the extent of animal raising and the techniques used to store, process, and dispose of animal manure. Related to this interest was a perceived need to launch an educational effort that would lead to improved animal management and housing practices in order to reduce the frequency and amount of manure escaping the landowners' premises and entering the stream- or groundwater of the area.

ASSESSING THE NUMBER OF ANIMALS

In attempting to better understand the relationship between land use and water quality, it was deemed important to understand the extent of animal raising and the techniques used to store, process, and dispose of animal manure. The overall goal was to describe the water quality impacts of small animal enterprises on water quality of the Tualatin River Basin streams. In order to do this, it was decided to conduct a telephone survey of the area. The design of that survey was to

0-56670-138-4/97/$0.00+$.50

TABLE 1
Results of the Presurvey

Response	With letter	Without letter
Have livestock	1	1
Do not have livestock	3	4
Refused to answer	1	1
Unable to contact	2	0

select appropriate areas to survey and then to contact a representative sampling of the residents of the selected area.

A telephone survey was planned. It was decided that a pretest or presurvey would be run to determine if a mailing should precede the telephone call. The presurvey consisted of four questions: (1) Does the respondent own livestock? (2) Is there a creek on his/her property? (3) What is the source of feed for the animals? and (4) How is manure collected and handled? Two groups of six property owners were arbitrarily selected using property size as the criteria. In each group there were two each of small (less than 5 acres), medium (5 to 20 acres), and large (more than 30 acres) owners. One group of six was sent a letter with information about the survey. Both groups were called within ten days of the mailing. The results of the presurvey are shown in Table 1.

The sample selection strategy was to identify three one mile wide strips through the watershed; one each from the east, middle, and west portion. Strips were sought that had sections which included a creek, a variety of suburban, farm and forestry activities, and did not include an extensive commercial area. Because there were multiple strips that were acceptable, the final selection was by coin toss. Once the strips were selected, a list of all land parcels in the strips was obtained from the Tax Assessor's Office. The number of parcels in each section was counted. A representative sample size was selected to be 100 parcels per strip. On that basis, the fraction of land parcels to be sampled from each section was a function of the number of parcels per section as outlined in Table 2.

The three strips, consisting of 51 sections were sampled with a total of 301 responses. Of the responses, 78 (26%) of the participants owned animals. The number of animals ranged from 1 to 257. Fifty-nine (20%) of the landowners were willing to provide information on their manure handling methods. Forty-nine of the landowners (16%) of these either stored manure in uncovered

TABLE 2
**Number of Parcels Sampled
in Each Section**

No. of parcels per section	Proportion sampled
0–5	all
6–15	0.5
16–35	0.33
36–55	0.25
56–75	0.20
76–105	0.17
106–145	0.14
145–195	0.12

piles or did not handle manure at all. Sixteen of the landowners (5%) maintained animals that had direct access to a creek.

ASSURING AN ANIMAL-OPERATED PUMP WOULD SUPPLY SUFFICIENT WATER

As the earlier survey indicated, many small commercial and noncommercial farm animal enterprises allow animal access to the streams for drinking. Because manure is a source of bacteria, nitrogen, and phosphorus that adversely impacts water quality, a reduction of the amount of animal manure deposited in the stream and riparian area is desirable. This can be done by reducing the amount of time an animal spends in the stream. One way to accomplish this is to provide an off-stream source of water. Unfortunately, this is not always easy. In some locations, it is possible to install one or more watering tanks which require pumping water from the creek or diverting water from the household supply. In some locations this may prove an expensive procedure.

An alternative watering option involves the use of an animal-operated diaphragm pump. This type of pump, also referred to as a pasture pump, has a basin of water (1–2 pints) that animals can see. The basin is partially covered by a lever. Thus for the animal to fully access the water it must push the lever with its nose or muzzle. When the animal releases the lever, the pump pulls water from the source and refills the basin. The pump remains primed by the use of a check valve at the end of the supply pipe which is submerged in the water source (stream, pond, or well). This pump requires no electrical connection and the animals pump the amount of water they need.

Among the questions raised by animal owners were: (1) How long does it take an animal to learn to use the pump? (2) Does having the pump limit water consumption? and (3) Since only one animal can use the pump at a time, will the animals drink less water than they would from an open tank? To answer these questions, a trial was conducted in which the amount of water consumed per day by a group of 27 Holstein dairy heifers using one open water tank was compared to the amount of water consumed when they had a single pasture pump available. In addition, we wanted to determine how long it took the heifers to learn to use the pump. The pasture pump used in this trial was a Utina M Pasture Pump distributed by Farm Trol Equipment Co. of Therese, WI.

The 27 heifers averaged 850 lb. They were alternately allowed to water directly from an open 100-gallon water tank and from the pasture pump. On days they had access to the pump, the pump plywood platform was placed on the water tank making the tank inaccessible to the heifers. The heifers were allowed approximately 2 weeks of alternating water tank-pasture pump before comparative data were collected. Animal behavior was monitored during this period. Once acclimated, the daily water consumption was recorded for a 12-day period. Temperatures were relatively stable throughout this period with highs ranging from 65 to 75°F and lows from 55 to 60°F. The daily water consumption values are presented in Table 3.

The learning period for the heifers to operate the pump seemed to range from 0.5 to 1 day. Within 2 days it appeared that even the less dominant heifers had established a routine of when to get water. The water consumption data did not indicate any difference between the pasture pump and the open tank.

EVALUATING THE EFFECTIVENESS OF OFF-STREAM WATERING

GRAZING BEEF CATTLE

A study was designed to monitor the time a group of four beef cows, 625 pounds each, spent in or in the vicinity of a stream with and without an alternate water source available to them. A water trough located approximately 75 ft from the stream was the alternate water source. The cattle were grazing the area between the water trough and the stream and were generally closer to the stream than to the trough.

TABLE 3
Daily Water Consumption
of 27 Holstein Dairy Heifers
(July 1993)

Day	Water consumption (gallons)	Method
1	360	Pump
2	809	Pump
3	494	Pump
4	ND	—
5	628	Tank
6	424	Tank
7	398	Tank
8	525	Tank
9	632	Tank
10	605	Pump
11	418	Pump
12	473	Pump

The time the animals spent at the stream was measured continuously by using a data logger and two light beam counters. A passage narrow enough to allow only one animal to enter at a time was constructed. Within the stream access area there was sufficient space, 100 ft², for all the cows to be there at once. The pasture in the vicinity of the stream was thoroughly grazed prior to the study.

The study was begun by monitoring the behavior of the cows for 17 days while they had an off-stream watering facility and were required to go to the creek to water. After these 17 days, a water trough was made available and a 6-day period allowed for the animals to adjust to this situation. Cattle behavior was then monitored for the next 11 days while the cattle had both the stream and the water trough available to them. During the period when the cows had only the stream available to them as a water source, the average amount of time spent at the stream was 15 min per cow per day. When the cows had both a water trough and the stream, they spent an average of 4 min per cow per day in the vicinity of the stream. This represents a 75% reduction and was demonstrated to be statistically significant at the 99% confidence level (Godwin, 1994). This monitoring program also confirmed the earlier assertion by Miner et al. (1992) that cattle do not typically drink during the night.

GRAZING HORSES

This set of observations is based on the behavior of two mature horses having access to a pasture pump (Utina M, as described earlier) and part of the time both the pump and a stream adjacent to the pasture. The pump was located 175 ft from the area of creek access. Provisions were made to measure the amount of water the horses were pumping. Water consumption was monitored under three conditions. For the control, the horses had access to the pasture pump and two pasture areas, one pasture was decidedly wetter than the other. We will refer to them as the wet and the dry pastures. Water consumption from the pump was monitored for 30 days of the control condition. A second condition was used for 7 days. The horses had access to the creek, the pasture pump, and the wetter of the two pastures. Water consumption from the pump was monitored for 8 days under a third condition. The third condition was one in which the horses had access to the drier

---

TABLE 4
Average Amount of Water Pumped Through a Pasture Pump by Two Mature Horses under Various Pasture and Alternate Water Supply Situations

Pastures and water sources available	Water pumped (L/day)
Both the wet and dry pastures were available. The only source of water was the pasture pump.	24.4
The wet pasture was available to the horses. Both the pasture pump and the creek were available.	11.6
The dry pasture was available to the horses. Both the pasture pump and the creek were available.	20.3

---

pasture, the pasture pump, and the creek. The amount of water pumped through the pasture pump by the two horses is recorded in Table 4 for the alternatives evaluated.

Due to the variability in daily water consumption and the relatively short data collection period, it is not possible to conclude that there are differences in the extent to which the pasture pump is used depending on the availability of the creek as an alternate source or the difference in water consumption based on the dryness of the pasture. What is evident from these measurements is that with or without access to a stream, the horses drank a significant portion of their water from the pasture pump. It is safe to conclude that this reduced the dry weather impact of the animals on the stream quality.

EVALUATING THE EFFECTIVENESS OF COVERED MANURE STORAGES

Another controllable source of water pollutants on the NCAEs is the runoff of rainfall from manure storage piles. The rural landowners involved as cooperators in this project expressed considerable interest in constructing roofed composting facilities as an alternative to the uncovered manure storage. Although more expensive, each of the landowners expressed the thought that they were getting something of value from the manure by having composted it. They were of the opinion that during previous years, their manure had been of no value to them because of the extensive extraction that had occurred. Smaller operators constructed manually stirred composted facilities. The larger operators wanted facilities into which tractors and front-end loaders could be driven to facilitate compost handling.

CONCLUSIONS

This study and work by Miner et al. (1992) confirm that off-stream watering sources reduce the amount of time that animals spend in and adjacent to the stream. Under dry weather conditions, no overland runoff, the off-stream watering is almost as effective as fencing and avoids the construction and maintenance costs of fencing. The pasture pump proved to be an effective alternative to a water trough and may be more cost-effective in locations where power is not readily available.

The construction of roofed manure storage and composting facilities provides an absolute benefit in avoiding the transport of nutrients, bacteria, and organic material into surface waters. For the suburban residents participating in this project, the covered manure storage and compost units also provided the landowner access to a useful product.

REFERENCES

Godwin, D. C., Implementing best management practices in small commercial and non-commercial animal enterprises, Master's thesis, Department of Bioresource Engineering, Oregon State University, Corvallis, 1994, 185.

Miner, J. R., Buckhouse, J. C., and Moore, J. A., Will a water trough reduce the amount of time hay fed livestock spend in the stream (and therefore improve water quality)?, *Rangelands,* 14 (1), 35, 1992.

39 The Restoration of the Clyde River

Desmond Hammerton

There is well-documented evidence to show that the Clyde River, which was in a clean and relatively unpolluted condition at the beginning of the nineteenth century, deteriorated in a dramatic fashion over the next 50 years to become one of the most polluted rivers in Britain. A Royal Commission Report published in 1872 showed that the river and its principal tributaries around Glasgow were grossly polluted by sewage and effluents from chemical, mining, and manufacturing industries. Salmon, formerly abundant, had virtually disappeared by 1860; indeed almost all fish species had disappeared from the lower reaches around Glasgow. Despite many efforts to control pollution, the river remained severely polluted for over 100 years. This chapter describes the work of the Clyde River Purification Board, set up in 1956, which culminated in the return of the salmon in 1983. The author describes the obstacles which had to be overcome in restoring the river to its present reasonably healthy condition and considers whether present day controls are adequate for future environmental needs.

INTRODUCTION

Prior to 1800 in Britain and, indeed throughout most of western Europe, water pollution was mainly caused by small industries which used water for power and for processing materials. The pollution was generally local in character and only affected waters in the immediate vicinity of the discharges. During the ensuing five decades, rapid industrialization combined with the growth of urban population led to a massive deterioration in water quality and, of course, to outbreaks of waterborne diseases such as typhoid and cholera. In Britain, the problem of river pollution was forcibly brought to the government's attention because the Houses of Parliament were situated on the bank of the Thames River which, by 1850, was the worst polluted river in Europe. It is of interest to note that in the long, hot summers of 1858 and 1859 Parliamentary debates were at times brought to a standstill by the stench of the river despite the use of curtains saturated with chloride of lime at every window!

The problem became so serious that the government set up successive Royal Commissions in 1865 and 1868 to "inquire into the best means of preventing the pollution of rivers." The 1868 Rivers Pollution Commission sat for 6 years and produced six major reports. These were models of Victorian thoroughness and included much scientific data including chemical analyses of effluents and river waters. The distinguished chemist, Edward Frankland, FRS, was a key member who did much to ensure the high scientific quality of the commission's reports.

The Fourth Report, published in 1872, was devoted to Scottish rivers and comprised two volumes. It recorded every source of pollution on the major rivers and, for each factory, listed the number of employees, quantities of raw materials, and the nature of effluents discharged. The Commission also included, in its minutes of evidence, interviews with a wide range of witnesses which included local councillors, landowners, industrialists, medical officers of health and sanitary inspectors.

The report demonstrated that the Clyde River (Figure 1) was the worst polluted river in Scotland and one of the worst in Britain. It revealed that the river through Glasgow and the lower reaches of all the tributaries, i.e., the South Calder, North Calder, Kelvin, Black Cart and White Cart, were

0-56670-138-4/97/$0.00+$.50
© 1997 by CRC Press, Inc.

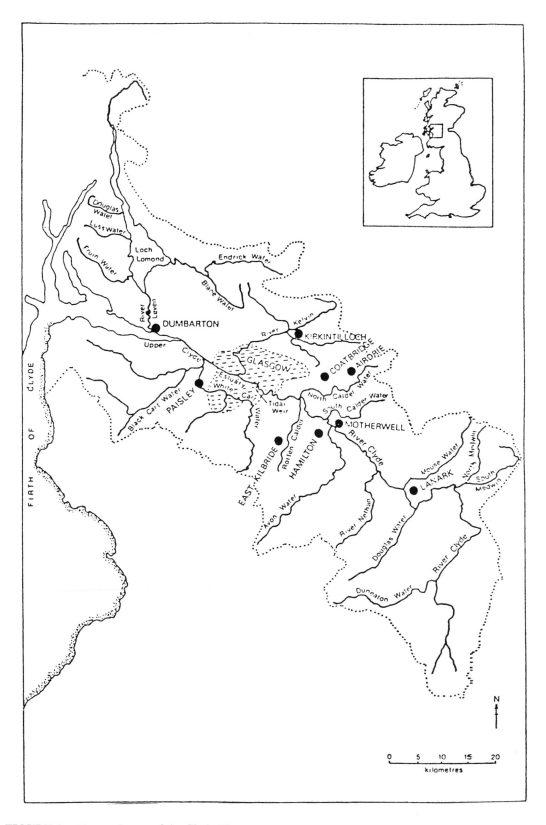

FIGURE 1. The catchment of the Clyde River.

so severely polluted as to be toxic to fish life. It was shown that, in the upper estuary, the stench of hydrogen sulfide had driven the bulk of the passenger traffic from the river to the railways and, furthermore that the water was corrosive to ships' plates!

The devastating impact of pollution on the lower reaches of this river are admirably summed up in the words of the report:

> Description of the Basin of the Clyde...
> The Clyde, which is laden with so much filth before it joins the sea, is, in its upper reaches and for more than two thirds of its course, one of the most beautiful of Scottish rivers — watering the pleasantest of upland pastoral valleys, traversing rich and fertile lowland landscapes, falling through abrupt and rocky wooded defiles and furnishing, in one portion of its course, some of the finest river scenery in the island. Nowhere is there a greater contrast than that which exists between the unpolluted waters which come down to Lanark, or even as far as Hamilton, and the foul and stinking flood to which they have been changed not twenty miles beyond that point. In this short interval, the river has received the Cadzow Burn, which brings the drainage of the town of Hamilton — the comparatively clean South Calder, which however, brings down the drainage of a considerable population in and around Wishaw and Motherwell — the North Calder, draining the thickly populated districts of Airdrie and Coatbridge — the Kelvin, contributing the liquid refuse of Kirkintilloch and of the manufactories near it — the Cart fouled by Johnstone, Barrhead, Pollokshaws and Paisley — and the drainage of the large manufacturing city of Glasgow. Within the space of a few miles the subject of river pollution is thus exhibited in almost all its forms, and may be witnesed in every degree of intensity.

The report goes on to describe the reasons for this dramatic change in the following paragraph:

> And this change which takes place so rapidly, as the river passes the points thus named, has nearly all arisen within living recollection. It is not very long since the Clyde even at Glasgow was comparatively clean. Now, its water there is loaded with sewage mud, foul with sewage gas and poisoned by sewage waste of every kind — from dye works, chemical works, bleach works, paraffin oil works, tanyards, distilleries, privies and water closets. The cause of change, stated shortly, is to be found in the enormous increase of population and of manufacturing industry, which during the past generation has been witnessed in Clydesdale. The population of Lanarkshire has been tripled within the past fifty years and that of Glasgow alone is increasing at the rate of about 10,000 annually. This, together with the great increase of the water supply of towns, by which their filth is more rapidly and perfectly than ever washed into the watercourses, is quite sufficient to account for the condition of the river Clyde, which we now proceed to describe more particularly.

The reports of the 1868 Rivers Pollution Commission show that many of the fundamental principles of water pollution control were well understood at that time. In its fourth report the Commission recognized for the first time the need for legally enforceable standards for effluents and published a list of ten recommended standards which, before publication, were submitted to five Fellows of the Royal Society and to European chemists including Baron von Liebig (Munich), Dr. Hoffmann (Berlin), and M. Dumas, Secretary of the French Institute in Paris. All supported the proposed standards and M. Dumas proposed an additional standard with the following remark: "Even where the cause has not been ascertained by chemical analysis, all water which has become unfit to support the life of fish shall be considered as having received a pollution from which it must be purified." This must be the first ever proposal for a fish toxicity test!

With these proposed standards, the commission also recommended the appointment of government inspectors to monitor discharges and enforce standards and also the setting up of river conservancy boards with limited powers for river maintenance.

RIVERS POLLUTION ACT (1876)

Following publication of the Fourth Report in 1872, the government acted with commendable speed and, in the same year, introduced a Bill before Parliament which incorporated all the recommen-

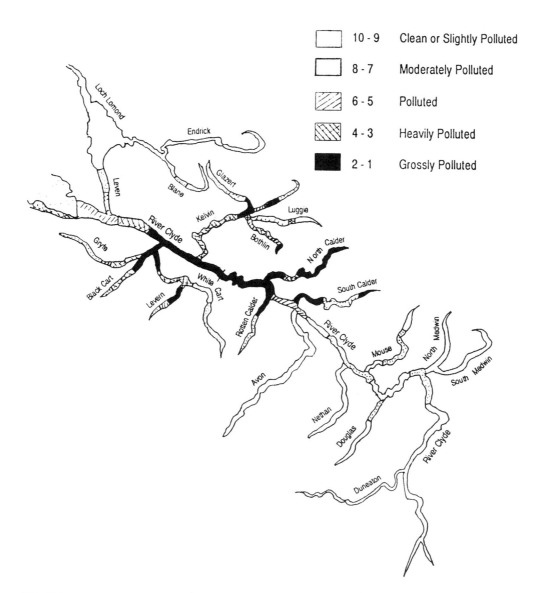

FIGURE 2. Biological quality of the Clyde River in 1968.

amenable to treatment after dilution with domestic sewage and also that most industries did not have the expertise to run effluent treatment works.

During the decade following the 1965 Act some 40 sewage works were either rebuilt or extended to meet more stringent conditions. About 40 factory effluents were connected to public sewers and another 14 trade effluents ceased discharging to rivers because they were converted to closed-circuit systems. Seventeen trade effluent plants were built where these solutions were impracticable.

However, despite the efforts of the Board, progress proved very much slower than desired. A major reason for this was that local authority expenditure was tightly controlled by central government which had to approve all major capital expenditure. Thus there was no point in prosecuting a local authority for failure to meet new consent conditions if the government had refused permission to spend money on a new plant. It was also the case that many industries faced cash-flow problems and the program had to be adapted to take account of these. Moreover, in the Clyde catchment, there were certain dying industries, notably some paper mills which were barely viable or losing money and where the cost of a new treatment plant would result in immediate closure. In such cases, particularly where the life of the factory was obviously limited, the Board took a realistic view and imposed only low cost remedial measures.

PRIVATE LEGISLATION

A serious barrier to progress occurred in the late 1960s when several firms took advantage of a loophole in the law and decided to discharge their liquid wastes underground either to abandoned mine shafts of the former coal mines or through boreholes. In one example a distillery discharged its waste waters to a mine shaft about half a mile from the North Calder Water. After about one year the effluent broke out through a former day level directly into the river at a rate of over 2 million L/day. The liquid waste, many times the strength of raw sewage, turned the river black for a distance of 10 km downstream and also produced poisonous gases in the vicinity of the breakout. The Board found that it was powerless to take action against the firm unless it could prove that the effluent in the river was the same as that discharged to the mineshaft. The firm refused to allow the inspectors to take samples of the effluent. This discharge was only brought under control by the vigorous action of the Clyde RPB in successfully promoting a private act, the Clyde River Purification Board Act (1972), after it failed to convince the Secretary of State for Scotland of the need for national legislation! The Act not only provided powers to control all discharges to underground strata but gave the Board a number of other valuable powers such as control over the diversion, culverting, or piping of rivers, powers to perform stream cleaning operations and to control the direct extraction of sand and gravel from rivers — a major cause of pollution in the Clyde.

MARINE POLLUTION CONTROL

Almost from its inception the Board recognized that the most severely polluted section of the Clyde was the upper estuary where lack of oxygen and high levels of ammonia had provided a complete barrier to the migration of salmon and sea trout for more than a century.

In preparation for administration of the Tidal Waters Order granted in 1968 the Board set up an Estuary Survey Section in 1864 (now the Marine Survey Section) which has subsequently built up an unrivaled body of data on the biological and physicochemical characteristics of the Clyde Estuary and the surrounding coastal waters (Figure 3). Such studies were essential for understanding the complex physical and biochemical processes in the tidal waters and for the setting of scientifically sound consent conditions in the marine environment, a procedure to which the Board has made some pioneering contributions (Hammerton, 1984; Mackay et al., 1986). In the three decades since its inception over 100 papers have been published and the work of the section has become an essential management tool for environmental management in the marine waters.

LOCAL GOVERNMENT REFORM (1975)

Under the Local Government (Scotland) Act (1973) the creation of nine Regional Councils greatly improved the quality of sewerage management and sewage treatment in Scotland. In Strathclyde Region a large number of sewerage authorities were unified into a single department whose area coincided fairly closely with that of the Clyde RPB which was reconstituted under the same Act. This greatly simplified the task of strategic planning and ensured that a high standard of management was available to all the sewage works in the area.

CONTROL OF POLLUTION ACT (1974)

This comprehensive piece of legislation incorporated many of the recommendations of the Royal Commission on Environmental Pollution (RCEP). First, it brought together in a single Act most of the previous legislation relating to solid waste on land, air pollution, water pollution, and noise

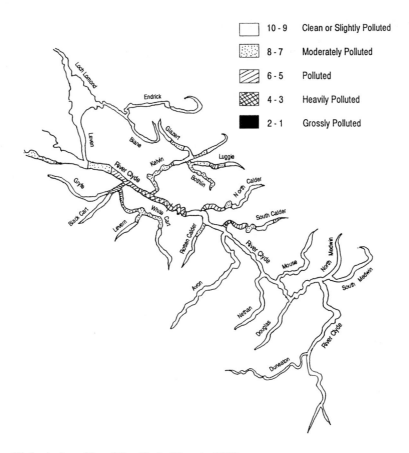

☐	10 - 9	Clean or Slightly Polluted
▦	8 - 7	Moderately Polluted
▧	6 - 5	Polluted
▨	4 - 3	Heavily Polluted
■	2 - 1	Grossly Polluted

FIGURE 3.   Biological quality of the Clyde River in 1993.

pollution. Second, it strengthened many of the existing controls and, in particular, paved the way for implementation of European Community legislation in order to provide uniform controls in accordance with the Community Action Program on the Environment. The aims of this program were to prevent or reduce pollution, to husband natural resources and the balance of ecological systems, and to improve the quality of life and working conditions. The Act also provided for the enforcement of international agreements such as the Oslo and Paris Conventions and North Sea Ministerial Decisions. Third, the Act provided for public participation by means of public registers and the advertising, for public comment, all significant applications for consent. The Act also brought under control all discharges to tidal waters and to underground strata. However, it should be noted here that, although the Act received the Royal Assent in 1974, implementation of the new provisions relating to water was delayed for ten years because of the recession.

PROGRESS SINCE 1975

The Annual Reports of the Clyde RPB and the 5-yearly surveys published by the Scottish Office show that, although financial constraints during the lengthy recession resulted in a reduced rate of progress, improvements continued to be made. In 1975, for example, large numbers of sticklebacks appeared in the stretch of the Clyde upstream of the tidal weir in the heart of Glasgow. This was the first confirmed report of fish life in the Clyde within the City boundary since biological sampling commenced. The routine chemical and biological surveys also showed steady improvements and the return of fish life to the lower reaches of all the main tributaries around Glasgow. A steady improvement in water quality in the upper estuary also led to an increasing number of species of both bottom living fauna and of fish.

RETURN OF SALMON TO THE CLYDE

The return of salmon and sea trout to the Clyde came earlier than expected and resulted in intense media interest, particularly because the first salmon was caught on a rod and line on August 18th, four days before the first salmon was caught in the Thames River in London. In both rivers no salmon had been recorded for more than 100 years. In the case of the Thames, the return of the salmon followed a ten year stocking program with salmon fry from the North of Scotland, whereas in the case of the Clyde the fish returned naturally, a large run occurring in the autumn of 1983 and every year since. However, it should be noted that, while the Clyde estuary through Glasgow is now of reasonably good quality during medium to high flows, it still suffers serious deoxygenation during low flows and further remedial measures now in the pipeline will need to be completed before the estuary returns to full health.

POLLUTION CONTROL PHILOSOPHY

As already indicated, in the early years of the Board, the setting of consent conditions for discharges was, of necessity, based on rule of thumb methods using wholly inadequate data. However, from its inception, the standards were designed, at least theoretically, to meet the needs of the receiving water and thus varied according to dilution in contrast to the system of fixed emission standards widely used in continental Europe.

During the subsequent years, with the accumulation of experience, a rapidly growing data base and the availability of toxicity data from institutions such as the Water Research Center it has become possible to set scientifically based effluent standards with a considerable degree of confidence. In common with British practice these standards are designed to protect the receiving waters under the worst conditions of flow, temperature, or (in coastal waters) the tidal condition. In rivers the aim is to protect the water quality at minimum flow when dilution is at a minimum taking into account the maximum permitted flow of the discharge. Each consent is designed to ensure that, after mixing, the receiving water will meet the Environmental Quality Standard for each component including heavy metals, microorganic pollutants, and pesticides. Moreover, the consent must ensure that the requirements of various European Directives are met. These include the Freshwater Fisheries Directive, the Dangerous Substances Directive, and the Shellfish Directive (for marine waters). Where additive or synergistic effects are anticipated (for example, in the case of highly complex mixtures) or where there is inadequate toxicity data, the Board carries out extensive toxicity tests on the effluent in admixture with varying dilutions of the actual receiving water. As a final safeguard any consent may be reviewed after 2 years if adverse effects are detected. The discharger must inform the Board of any changes that may affect effluent quality.

National Environmental Quality Standards (EQSs) are set by the Department of Environment for a wide range of substances and the Board has a duty to monitor the inland and coastal waters to ensure that these standards are met. The EQSs have a built-in safety factor and further safety factors are often incorporated in the consent conditions. Every effort is made to ensure that possible long-term effects such as bioaccumulation, mutagenicity, and carcinogenicty are taken into account.

CONCLUSIONS AND FUTURE OUTLOOK

The Clyde River, throughout its length, is in better condition today than at any time in the past 150 years. Fish life has returned to all the lower reaches where it had been absent since about 1870 while salmon have been migrating up the river as far as the Falls of Clyde ever since 1983. However, the river through Glasgow above and below the tidal weir remains in poor condition during low flows and will not show much improvement until a major new sewage treatment works to replace the existing one at Dalmuir in Glasgow is completed in 2000 AD at an estimated cost of 120 million. Together with other improvements scheduled for the various tributaries a considerable further improvement is expected within the next ten years. Although eutrophication is not considered to be a serious problem in the Clyde basin, controls on the use of farm fertilizers may be required if the recommendations of the Royal Commission on Environmental Pollution are to be met. The Commission, in its sixteenth report on Freshwater Quality (1992), said inter alia that rivers should be restored as near as possible to their original condition. The Commission also recommended that there should be less reliance on the use of pesticides in agriculture and that there should be a general tightening of Environmental Quality Standards, coupled with an incentive charging scheme to reduce pollution to the lowest possible level.

Unfortunately the United Kingdom government has given a rather negative response to the Commission and is reportedly seeking to repeal or water down recent European Environmental legislation. Nevertheless the author is convinced that, while progress may slow down in the immediate future, public concern for the environment will ensure that more stringent controls will continue to be applied whenever scientific justification can be demonstrated.

INTEGRATED POLLUTION CONTROL

At the time of writing far reaching changes in the way environmental regulation is administered are well under way. On April 1st 1996 a new regulatory body, the Scottish Environment Protection Agency, will come into existence and take over the functions of air, land, and water pollution control presently carried out respectively by Her Majesty's Industrial Pollution Inspectorate, the Environmental Health Departments of District Councils, and the River Purification Authorities. This major reorganization, partly modeled on U.S. practice, follows earlier recommendations of the Royal Commission on Environmental Pollution which criticized the "compartmental" approach to the environment and recommended an integrated approach by a single regulatory body which would have regard to environmental pathways as well as loads. Not only was this seen as environmentally desirable but also as a positive benefit to industries which would make a single application for all the discharges from a factory, whether to air, water, or land or to all media.

As far as the Clyde is concerned the author believes that there is much to commend the new organization. In the past 40 years there have been many problems arising from the use of landfill for the disposal of domestic rubbish and toxic waste which have proved difficult to resolve, involving protracted negotiations with a variety of district councils some of which lacked expertise in solid waste management. There have also been some catchments severely affected by acid rain resulting from industrial emissions over which the Board has no control. The new agency will have a tremendous opportunity to build on the successes of the river boards and usher in a new era in which the technical expertise of a national body will be available for all aspects of environmental management at a local as well as a regional or national level.

To conclude, exactly 40 years after its inception, the Clyde River Purification Board will hand over to the new organization a river which has already been transformed and which, with a few years further dedicated effort, should once again become one of Scotland's finest salmon and trout rivers.

REFERENCES*

Hammerton, D., The history of environmental water quality management in Scotland, *J. Instit. Water Engineers Scientists,* 37, 336, 1983.
Hammerton, D., Assessing the impact of major developments on water resources, in *Planning and Ecology,* Roberts, R. D. and Roberts, T. M., Eds., Chapman and Hall, London, 1984.
Hammerton, D., Cleaning the Clyde — a century of progress?, *J. Operational Res. Soc.,* 37, 911, 1986.

* See also Clyde River Purification Board, Annual Reports 1956–1995. The Clyde RPB recognized that it would be unrealistic to expect all local authority and industrial effluents to be treated to desirable standards overnight and therefore the costs would have to be spread over a considerable number of years. By this time the Board had built up sufficient data to identify the worst "blackspots" in terms of water quality.

Hammerton, D., The impact of environmental legislation, *Water Pollut. Control,* 86, 333, 1987.

Hammerton, D., Domestic and Industrial Pollution, in *The Freshwaters of Scotland,* Maitland, Boon, and McLusky, Eds., John Wiley & Sons, Chichester, 1994.

Klein, *Aspects of River Pollution,* Butterworths, London, 1957.

Mackay, D. W., Haig, A. J. N., and Allcock, R., Licensing a major discharge to coastal waters: the practical application of the EQO/EQS approach, *Water Sci. Technol.,* Vol. 18, BC, Plymouth, 1986.

Richardson, G., Ogus, A., and Burrows, P., *A Study of Regulation and Enforcement,* Oxford Socio-Legal Studies, Clarendon Press, Oxford, 1982.

Rivers Pollution Commission, Fourth Report: Pollution of Rivers of Scotland, HMSO, London, 1872.

Royal Commission on Environmental Pollution, Sixteenth Report, Freshwater Quality, HMSO, London, 1992.

40 Intergovernmental and International Aquatic Ecological Programs: Approaches for Successful Implementation

Stuart W. McKenzie

Programs involving intergovernmental and/or international cooperation must have certain characteristics to be successful. Several tangible ingredients are needed to accomplish ecological assessments that involve multiple governments and agencies:

- knowledge of hydrologic principles,
- knowledge of ecological principles, and
- knowledge of, and agreement on, issues that should be addressed.

For any program there also are intangibles such as trust, recognition, willingness, and commitment. Current developing intergovernmental programs are used to illustrate some of these tangible and intangible ingredients.

ISSUES FACING INTERGOVERNMENTAL WATER PROGRAMS

Intergovernmental and international programs exist that assess, from an ecological perspective, the water resources in many geographic areas of the world. Their continued existence requires that all participating groups agree on issues that warrant the expenditure of resources (money, time, and attention). The following common issues might be thought of as mutual principles:

- Minimum quantities of water are necessary to meet identified beneficial uses. Until these minimum quantities of water are available, considerations of water quality are secondary.
- Water identified for more than one use at a location should ideally be of a quality sufficient for all uses. For example, it is desirable for a drinking-water supply to have minimal taste and odor, whereas a cold-water fishery supported by the same source needs low temperatures and high dissolved oxygen concentrations.
- Uniform water quality standards should be adopted among political units. Such a policy leads to other "good neighbor" policies and promotes interstate activities. This is particularly true when surface and groundwaters move from one political unit to another and are used by both political units. Although the need for cooperation might seem logical, the desire to control our own destiny makes us reluctant to change our standards to be in agreement with our neighbor.
- Local management of river basins and watersheds should be allowed; however, such small- or local-scale management can lead to different water quality standards for adjacent watersheds. Because of the recognized importance of the nonpoint source contri-

bution of many contaminants, the need for watershed, community, and small-area management of land-use activities and waterways is critical to most successful programs.

- Multiple organizations would benefit by sharing data of known quality. Data of unknown or variable quality can lead to multiple conclusions and misunderstandings. Data from all organizations need to be supported by compatible protocols and quality-control data that are readily available and that document the quality of the data. The most common reasons that groups do not use existing water quality data are: (1) their inability to obtain these data in an efficient and timely manner, and (2) deficiencies of the data base(s) that make identification of the quality of the data therein impossible (Intergovernmental Task Force on Monitoring Water Quality, 1992).

ECOLOGICAL PRINCIPLES AND THE MANAGEMENT OF WATER RESOURCES

Water-resource managers have an opportunity to manage and impact our environment in a more ecological manner today than ever before; however, to take advantage of this opportunity, governmental agencies must cooperate to align their water and terrestrial programs if environmental management is to succeed. The opportunity now exists because of a better understanding of the science of ecology and advances in scientific technology in general. Whereas hydrologists and water-resource managers were once concerned only with the source, cause, transport, and fate of chemicals within watersheds and hydrologic cycles, today they need to consider the effects of the resulting water quality on aquatic biota and to infer the quality of water from the presence or absence of those biota.

The evolving science of ecology has identified scientific principles that are important to resource managers:

- Bacteria exists in soils at high concentrations and are effective at breaking down organic material. In the soil, most detrital organic material is broken down into inorganic chemicals that become dissolved in water, de-gas to the atmosphere, or sorb to soil particles that may be transported to water bodies. These bacteria enhance soil fertility which allows us to grow crops year after year in the same field. In the soil, many synthetic chemicals (pesticides) are broken down into nontoxic substances. For example, the herbicide atrazine will break down in a few days in the soil, whereas when dissolved atrazine is transported to a reservoir and held for 6 to 9 months, it is essentially preserved and will leave the reservoir in concentrations similar to those when it entered (Goolsby et al., 1993).
- Water used in the laboratory is purified by the distillation process to rid it of dissolved inorganic constituents. Nature has a similar process called the hydrologic cycle, which is composed of evaporation, transpiration, and precipitation. This natural still, operated over the Earth and powered by the Sun, produces most of our fresh water. We must fully understand this process if we are to manage our freshwater resource efficiently and effectively.

Advances in scientific technology and knowledge of ecological principles have had significant impacts on ecological assessments, resource managers, and water quality regulators in the following ways:

- Advances in analytical chemistry have paralleled developments in ecology. Analytical instrumentation is now capable of producing lower detection levels, cheaper analyses, and a better awareness of new and complex findings in our environment.

- The development of the computer has been critical to our ability to store and analyze the large amounts of data collected in the last two decades. Computers allow us to store and retrieve large amounts of data efficiently and to be responsive to data requests in a timely manner. Computers also have allowed us to develop methods for analyzing data using models and statistics which are effective tools for today's scientists and policymakers. Information is now being shared beyond the dreams of our parents. The Internet gives the public, researchers, data managers, and policymakers access to very large data sets.

- The understanding and misunderstanding of ecology causes the public to demand that managers of our natural resources think, regulate, and manage ecologically so that our children and grandchildren will be able to have water of good quality and experience the same vistas and habitats as we do now. An example of a program born of this concern is the President's Forest Management Conference in Portland, OR, where President Clinton brought resource managers, environmental groups, industrialists, and scientists together to establish a policy of ecological management for our national forest lands. Carefully crafted, such a policy would provide the wood-products industry with their necessary resources, allow the public to enjoy the recreational attributes of forest land, and preserve natural habitats and clean water for future generations.

HYDROLOGIC ISSUES CAN BE BENEFICIALLY ADDRESSED BY INTERGOVERNMENTAL WATER PROGRAMS

Many issues related to hydrology, sources of contaminants, and uses of water are amenable to cooperative efforts between agencies that collect data and those that manage the natural resource. In the U.S., our Congress, State Legislatures, and Tribal Councils are asking the government workforce to be more efficient by using fewer resources to accomplish more difficult tasks over larger geographic areas. This reduction in force is the opportune time to integrate programs confronting common issues that no longer have support and that cannot be achieved by a single agency. First, however, the identification of these issues must come from universities, research organizations, and other groups that collect, analyze, and interpret data. Water issues can be identified by their temporal and spatial variability, downstream transport, airborne transport, and use.

TEMPORAL AND SPATIAL VARIABILITY

There is no such thing as a steady state or homogeneity in nature. The measurement of the variability and interrelation of water quality and ecological parameters can be time consuming and expensive. Therefore, because there are limited resources to identify water quality problems, we must carefully consider both temporal and spatial variability in program design and in the optimization of sampling.

The existence of temporal variability in hydrologic data is well recognized. Examples include 24-hr variability of air and water temperatures, solar radiation, dissolved oxygen concentrations, and pH in water; local variability in flows, turbidity, and rate of water use; seasonal patterns of flow, water temperature, aquatic growth, and water use; long-term cycles (multiple years) of precipitation and drought; and long-term trends resulting from land-use changes, such as conversion of land use from agricultural to urban activities.

Examples of temporal patterns include pesticide application at specific times to control insects and noxious weeds, resulting in detectable pesticide concentrations in waterways; seasonal storms that cause soils to erode, resulting in highly turbid streams; and the presence of benthic invertebrates of sufficient maturity to result in correct species identification. There are also seasons when aquatic organisms are more sensitive to environmental conditions, such as during spawning and rearing of anadromous fish, when high dissolved oxygen concentrations are important.

In addition to varying in time, water quality conditions can vary spatially. Local natural characteristics either directly or indirectly control the land-use and water-use activities of an area. Examples of natural characteristics include geology, topography, groundwater and surface-water availability, soil fertility and drainage, precipitation, and solar radiation. The human activities that are influenced by these natural characteristics determine which contaminants might enter groundwater or surface water in a particular area.

DOWNSTREAM TRANSPORT

An important hydrologic issue that affects downstream water users in a watershed, but that involves upstream water users as well, is downstream transport. For every action that occurs upstream, there is some response downstream. Things placed in a stream will move downstream, and if the rights to using water of acceptable quality and quantity are to be protected for all users in a watershed, responsibility must be accepted by all the water users in the basin. This is one of the primary reasons why management of water resources on a local level, such as a river basin or watershed level, is more likely to be successful.

AIRBORNE-CONTAMINANT TRANSPORT

The transport of airborne contaminants is an issue that is not directly related to watershed and hydrologic principles but is nonetheless a hydrologic issue. The principle is the same as for downstream transport, but the medium is air instead of water. Several examples of impacts on water and vegetation that are the result of the downwind contamination by point and nonpoint sources can be cited:

- Nitrates in Sierra Nevada snow that are downwind of Los Angeles and San Francisco.
- Acid rain and sulfates in water bodies that are downwind of large industrial and urban areas where high sulfur coals are burned, such as in the midwestern and eastern areas of the U.S. and Canada and in areas of Europe.
- Trace elements in snow downwind of a smelter in Tacoma, WA.
- Eradication of natural vegetation on a hillside downwind of an industry at Luke, MD.

WATER USE

Efficient and equitable use of water for agricultural, urban, and industrial activities is important because of the limited availability of high-quality water. Not all uses are equal, and where water is used, the quality often has been altered. Examples include the following:

- Use of water to cool a thermoelectric plant will cause the temperature of a stream or estuary to rise, will adversely affect fish habitat, but will consume very little of the resource.
- Use of water for irrigation consumes from 25 to 90% of the water diverted from streams or pumped from wells; return flows typically carry suspended sediment, pesticides, and nutrients.
- Mining activities can place sediment in streams, disrupt habitat, and place toxic contaminants in surface and groundwater, streambed sediment, and aquatic tissue.

Generally, sufficient funds are not available in Federal, State, Tribal, or County treasuries to police all watershed activities. River basin and watershed or community acceptance of water use and water quality goals will be much greater if program coordination is on a watershed level.

EXAMPLES OF INEFFECITVE (BUT IMPROVING) INTERGOVERNMENTAL WATER PROGRAMS

The U.S. has policies that encourage the development of natural resources in the western states, including the harvest of trees from watersheds managed by the U.S. Forest Service and the Bureau of Land Management. This policy has resulted in endangering several species (e.g., spotted owl, bull trout, and salmon), and has caused public concern. Environmental groups are demanding that federal agencies include in their forest management policy, the management of ecological health and long-term productivity of forest and habitats. This form of management, which is still developing avenues of communication, compatible data sets, and trust, requires cooperation among groups that often have been on opposite sides of policy issues; these groups include forest managers, environmentalists, State and Tribal representatives, industrial representatives, Federal and local regulators of water and wildlife, and the general public.

Canada has problems that are similar to those of the U.S. One of the difficulties for Environment Canada, as an agency that assesses and regulates water resources, is the delegation of responsibility to different independent agencies. The agency responsible for water quantity often does not have to coordinate with the agency responsible for water quality. While the management responsibility for surface-water quantity and quality is in the federal sector (Environment Canada), Provincial governments are responsible for groundwater. Coordination and collaboration are still possible but require goodwill by all participants. With no overriding agency or policy to see that progress will be made when conflict occurs, changes are often slow to nonexistent. Recent reorganization of Environment Canada is trying to correct some of these problems.

Historical discharge records for the Jordan River in Jordan are available for only up to when Israel took control of the West Bank of Jordan. Since that time, no discharge records for the Jordan River have been available to the Jordanian government. The need for placing the collection of acceptable scientific information within the policies of both countries has been missing. Recently, a greater acceptance of common goals within policies of both governments has emerged that could lead to a change of environmental goals.

Russia and countries of the former Soviet Union have a history of strong central governments, including those during the rule of the Czars, and later, the Communists. In the past political environment, fragmentation of responsibility to collect and assess water-resource data was not a problem because there was an overriding regulatory responsibility by a higher authority. Today, however, the strong central government is absent, and individual agencies from different countries are having to learn how to work in coordination and collaboration with each other. Because of the limited funding in support of these public agencies, progress is often very slow. Natural resources will likely be the loser because of the lack of good data and coordinated management. Recently, assistance by the World Bank has provided an opportunity for several agencies responsible for the drinking water supply for Moscow to work together to revise the water-quality monitoring program.

DESIRABLE COMPONENTS OF ENVIRONMENT-BASED INTERGOVERNMENTAL WATER PROGRAMS

WORKING AGREEMENTS

To be most effective, intergovernmental working agreements must include Federal, State, Tribal, and local public agencies as well as private and voluntary organizations. This is particularly true when managing a river basin or watershed. All the "stakeholders" that use or are affected by the use of the resource must be willing to agree on the management of the resources and support the effort (new taxes, new regulations and restrictions, promotions of goals of preserving and improving the resources).

MONITORING PROGRAMS WITH IDENTIFIED GOALS

In order for monitoring programs to have the data necessary to assess environmental conditions, there must be a recognition of the desired goals and specific questions to be answered. Often times a data-collection program is started, the data are never interpreted, the program is never revised, and when questions are asked, the data needed to answer the questions are not available. There are very logical and rational ways of developing monitoring programs. An American Society for Testing and Materials method for designing monitoring programs has been developed. To use water-quality monitoring programs that are not well planned wastes public money and breaks down confidence in the public management of our resources.

QUALITY ASSURANCE PLANS

We must be able to use each other's data with confidence and to accomplish this the data must be collected according to (1) protocols that are accepted by all, (2) techniques and methods that are compatible, and (3) quality-control data guidelines that provide an acceptable measure of the quality of the data. These three preceding items describe a quality-assurance plan. The National Water-Quality Monitoring Council is currently working to identify compatible sample collection methods and performance-based analytical methods that will make up a usable quality-assurance plan (Intergovernmental Task Force on Monitoring Water Quality, Technical Appendixes I, N, and O, 1995a).

ENVIRONMENTAL INDICATORS AND DATA SETS

Cooperating public agencies and private organizations need to identify the required environmental indicators and specify the minimal amount of data needed to answer their agreed on questions. This is particularly true of ecological programs where different agencies have specific responsibilities. It is critical that all agencies work together and share their data for a complete assessment to be successful. A minimal data set includes date, time, location, and the measured constituent value. Ideally, the data should be located in a system that can easily be accessed and geographically portrayed. New geographic information systems (GIS) are capable of handling data in this manner.

DATA STORAGE AND TRANSFER

Data storage is a critical part of any water program. Characteristics of a modern data-collection program ideally should include the reliable storage of relational data (including the listing of protocols, analytical methods, and quality-control data), the efficient and timely transfer of data when requested, and the frequent update and maintenance of the data by the agency or organizations collecting the data. The transfer of data among agencies would be made easier if the agencies that collect and store data would use software that is mutually compatible and would develop programs to facilitate the transfer of data between computers (Intergovernmental Task Force on Monitoring Water Quality, Technical Appendix M, 1995b).

INTERPRETATION AND COMMUNICATION

Interpretive tools (models and statistical programs) and communication skills are needed to turn data into information and make it understood by managers and the public. Interpretation is often done by universities, research institutes, consulting firms, and public agencies. The computer has significantly enhanced our ability to interpret data through the development of mathematical models, nonparametric statistical packages, time-trend analyses, and GIS software. We must still inform the managers and the public of the results in an understandable and timely manner if the results are to be meaningful. The day of writing a 50 to 100 page technical report to display on a library shelf is over. We need layreader reports to facilitate communication among technical experts, lay managers, and the public. Examples of lay publications and their characteristics are listed in the Intergovernmental Task Force on Monitoring Water Quality report (Technical Appendix K, 1995c).

INTANGIBLE INGREDIENTS FOR SUCCESSFUL INTERGOVERNMENTAL WATER PROGRAMS

Finally, intangible ingredients must be present for intergovernmental water programs to succeed. These include:

- The establishment of trust among the participants. This trust often comes from working together, and in time, the credibility of an organization would be recognized by other organizations.
- The recognition that the goal can be accomplished only by working together. Indeed, the economic savings may be so great that the advantage of cooperation would be self evident. Or, a higher organization can recognize the need and direct the cooperation.
- A willingness of each agency and organization to accept the responsibility for its actions and to risk making decisions; some of which almost surely will be contested by the public and (or) some of the participants.
- A commitment that the scientific integrity of the program will not be compromised to accommodate political objectives.

These intangibles are basic to successful intergovernmental assessments and management programs. Decisions need to be made by the heads of agencies before any movement toward cooperation can begin. Even then, there is a tremendous bureaucratic weight to shift before cooperation and collaboration will occur in a meaningful way.

The alternative to heads of agencies causing change is for the emphasis to come from the grassroots level. Certainly, the most likely route to success (causing a change) is from a watershed area perspective; stream basins and watershed boards need more encouragement. However, large river basins will be left out because of their size; the Columbia River Basin and perhaps the Willamette River Basin are in this category. Water-resource agencies in a river basin must have sufficient monetary resources, often related to population and (or) industrial activity, to fund the needed data collection, assessments, and structural and nonstructural changes necessary to improve existing environmental conditions. Watershed boards could, however, cause bureaucracy to react if they presented a unified front.

EXAMPLES OF EFFECTIVE WATER PROGRAMS ON THE WATERSHED AND RIVER BASIN LEVELS

The Nooksack Watershed Project was started in 1994 and has stakeholder participation in the form of agriculture, businesses, environmental groups, fisheries, recreational interests, water utilities, and representatives of local, state, tribal, and federal agencies. The Nooksack Basin, located in Northwest Washington, covers 1250 mi^2, has over 1000 miles of rivers and streams, and has a population of 143,000. The project deals with the issue of limited streamflow throughout the basin and water-quality problems in the forested, rural lowlands, and urban areas. For example, 33 water bodies do not meet Federal water quality standards at this time. The effort to change these conditions has started with the inception of the project (Richard Grout, Washington Department of Ecology, written commun., 1996).

The McKenzie Watershed Study in Oregon was started in 1991 by two local agencies seeking ways to resolve existing issues. They have collected all the existing water quantity and water quality data and have begun monitoring key locations. There are 20 members on their council representing federal, state, city, county, utility, parks and recreation, agricultural industry, timber industry, environmental groups, and local neighborhoods. The four issues that they currently deal with, of 50 that were identified, include (1) water quality conditions, (2) fish and wildlife habitat, (3) recreation, and (4) human habitat. One unique aspect of the McKenzie Study is that the effort has been carried out by volunteers (Laurie Power, McKenzie Watershed Council, oral commun., February 1996).

The St. Croix River Basin is 7700 mi^2 in drainage area in Minnesota and Wisconsin, has 134 dams on 1770 tributary streams, and has 624 open lakes. One hundred ten species of fish have been identified in the basin. In 1992, the heads of the St. Croix Riverways Office, the Minnesota and Wisconsin Departments of Natural Resources, and the Minnesota Pollution Control Agency agreed "to provide for the coordination of the planning and implementation of measures to protect and improve the water quality in the St. Croix River Basin" (Minnesota-Wisconsin Boundary Area Commission, 1996). The plan of study was completed in 1993 by the St. Croix River Basin Team, composed of member representatives and representatives from the Minnesota-Wisconsin Boundary Area Commission, University of Minnesota, and the U.S. Geological Survey. Two issues being addressed by the team in the basin include (1) How much phosphorus reduction is needed to protect Lake St. Croix? and (2) How much mercury reduction is needed (and over what timeline) to eliminate fish consumption advisories in the St. Croix River?

The Delaware River Basin Commission, formed in 1961, oversees a compact among Pennsylvania, New York, New Jersey, and Delaware, and the Federal Government. Funding for the compact is from these five members. The commission has strong management powers for resolving problems between states, including issues related to drought flows, storage and release of water from reservoirs, and maintaining sufficient flow to keep the salt wedge downstream of the recharge area in the lower Delaware Basin. The Commission has been active in cleaning up and controlling municipal and industrial sources of contaminants, starting in 1968, using guidelines now known as Total Maximum Daily Loads. Recently, the Commission has been working to establish uniform criteria and acceptance of common waste load allocation methods for the river. The Commission has also been active in securing special classification for drinking water supplies and for some waters to prevent further degradation (David Pollison, Delaware River Basin Commission, oral commun., February 1996).

REFERENCES

Goolsby, D. A., Battaglin, W. A., Fallon, J. D., Aga, D. S., Koplin, D. W., and Therman, E. M., Persistence of herbicides in selected reservoirs in the Midwestern United States — Some preliminary results, in *Selected Papers on Agricultural Chemicals in Water Resources of the Midcontinental United States,* Goolsby, D. A., Boyer, L. L., and Mallard, G. E., Eds., U.S. Geological Survey Open-File Report 93-418, 1993, 51.

Intergovernmental Task Force on Monitoring Water Quality, Ambient water-quality monitoring in the United States: First year review, evaluation, and recommendations: Reston, VA, U.S. Geological Survey, 1992, 51.

Intergovernmental Task Force on Monitoring Water Quality, The national strategy for improving water-quality monitoring in the United States: Technical appendixes I, N, and O, Reston, VA, U.S. Geological Survey, 1995a, 59.

Intergovernmental Task Force on Monitoring Water Quality, The national strategy for improving water-quality monitoring in the United States: Technical appendix M, Reston, VA, U.S. Geological Survey, 1995b, 83.

Intergovernmental Task Force on Monitoring Water Quality, The national strategy for improving water-quality monitoring in the United States: Technical Appendix K, Reston, VA, U.S. Geological Survey, 1995c, 69.

Minnesota-Wisconsin Boundary Area Commission, St. Croix Stewards Journal, Fall 1995–Winter 1996, Hudson, WI, 1996, 14.

41 A Review of the Federal Water Pollution Control Act, Amendments of 1972, 1977, 1981, and 1987: How These Acts Improved the Waters in Oregon

William J. Sobolewski

The year 1994 marks the 22nd anniversary of the Federal Water Pollution Control Act, Amendments of 1972. Many sections in this act and subsequent amendments in 1977, 1981, and 1987 were instrumental in restoring and maintaining the "chemical, physical, and biological integrity" of the waters in Oregon. A section of the Act that was largely "ignored" for 14 years by the U.S. Environmental Protection Agency (EPA) and Oregon Department of Environmental Quality (DEQ) was Section 303(d). A lawsuit filed in 1986 by a local environmental group and citizen established a "new" direction of water-quality management in the state. Though the initial lawsuit filed in December 1986 focused mainly on the Tualatin River, a tributary to the Willamette River, it was amended in March 1987 to include other bodies of water in the state. An "out-of-court" settlement was reached in June 1987. The settlement required EPA and DEQ to establish total maximum daily loads for pollutants and subsequent effluent limitations based on water-quality standards rather than technology-based standards for bodies of water that were classified as "water quality limited" or waters in which various pollutants exceed water-quality standards.

INTRODUCTION

The Federal Water Pollution Control Act (FWPCA) of 1972 set up the most comprehensive federal regulatory program for controlling water pollution discharges.

The Act established the National Pollution Discharge Elimination System (NPDES) permit program. The 1972 amendments also authorized EPA to award grants for municipal wastewater treatment works in the amount of 75% of the cost of planning, design, and construction.

The Amendments in 1977 (referred as the Clean Water Act) made other significant changes in the federal legislation of water pollution control. It also established the pretreatment program, which regulates the discharge of industrial wastes into publicly owned treatment works (POTW).

In the early 1980s, concern over the increasing federal deficit resulted in President Ronald Reagan "freezing" the federal construction grant funds until Congress passed amendments to the Clean Water Act. These amendments were passed in 1981. The amendments dealt primarily with revisions to the construction grants program. The major provision of this amendment was the reduction of the federal share to 55% of the cost of construction after October 1984.

In 1987, the Clean Water Act was amended by the Water Quality Act of 1987. This Act added the development and implementation of the nonpoint source management program as a national goal and the national estuary program. Title VI or the state water pollution control revolving funds was added, which resulted in the termination of the construction grants program in 1990 and added a state revolving loan program for 1989 through 1994.

NATIONAL POLLUTANT DISCHARGE ELIMINATION SYSTEM (NPDES)

Section 402, the FWPCA, Amendments of 1972 established the first federal discharge permitting system to control the discharge of pollutants into navigable waters. The NPDES permit program was probably the most significant water pollution control programs enacted in 1972.

According to the Act, all point sources of industrial and municipal wastewater were required to obtain a permit that regulates the facility's discharge of pollutants into "waters of the United States." NPDES permits were issued to individual dischargers (individual NPDES permits) or to a group of similarly situated dischargers subject to the same terms and conditions (general NPDES permits).

The more stringent of either technology-based effluent limits (e.g., BPT and BAT for industrial dischargers and secondary sewage treatment for municipal dischargers) or water quality-based effluent limits are incorporated in the NPDES permit.

Technology-based effluent limits are industry specific and are either based on nationally promulgated effluent guidelines, state treatment requirements, or best professional judgment of the permit writer (BPJ). Examples of technology-based requirements are best conventional technology (BCT) for conventional pollutants (e.g., suspended solids), best available technology (BAT) for nonconventional pollutants (e.g., chlorine and ammonia), and toxic pollutants (e.g., pentachlorophenol), and new source performance standards (NSPS) for new sources.

Water-quality-based effluent limits are based on the state's water quality standards, including both numeric criteria and, a narrative "free from" criterion. The effluent limits are developed to ensure that the discharge will not exceed state water quality standards at the edge of the designated mixing zone. Along with the more stringent effluent limits, a compliance schedule, monitoring, and reporting requirement are incorporated in the permit.

The CWA provides for civil, criminal, and administrative penalties for enforcing compliance schedules in permits. All of these levels of enforcement response can be applied, in varying degrees to all CWA statutory programs, i.e., Sections 301, 302, 306, 307, 402, and 404.

Other sources of pollution that are regulated under the NPDES program are stormwater and combined sewer overflows (CSOs). Stormwater is a leading source of water pollution in the U.S., causing an estimated one third of the impaired water quality nationwide.

Oregon cities such as Portland, Astoria, and North Bend are now embarking on programs to control CSOs. Portland signed a stipulated order with the Environmental Quality Commission to eliminate CSOs in the Willamette River and Columbia Slough.

The Act authorizes the U.S. EPA to approve or delegate the administration of the NPDES program to the states. In Oregon, the Oregon DEQ was granted this authority on September 26, 1973. Authority to issue permits to federal facilities was granted on March 2, 1979.

The DEQ regulates 471 industrial and agricultural sources of pollution and about 219 municipal sources of pollution under the NPDES permit program. The industrial and agricultural sources are regulated either by an "individual" NPDES permit or a "general" NPDES permit which covers specific categories of waste disposal. There are 10 general NPDES permits in Oregon, which regulate about 1,767 sources of pollution.

Oregon has developed 14 stormwater general permits. The main component of these general permits is the development of Best Management Practices.

In March 1989, DEQ developed their own enforcement procedures and civil penalties which closely relates to the federal procedures. The goal of DEQ's enforcement is to:

- obtain and maintain compliance with DEQ's statutes, rules, permits, and orders
- protect public health and the environment
- deter future violators and violations
- ensure an appropriate and consistent statewide enforcement program

From 1989 to 1993, DEQ assessed over $666,000 in civil penalties associated with water quality violations.

Besides the issuance of permits to point source dischargers, another major component of the NPDES permit program is the National Pretreatment Program. Industrial wastes entering municipal wastewater treatment plants (e.g., via sewers) are regulated under the CWA by the pretreatment program.

The pretreatment program is designed to regulate industrial wastewater discharges into a POTW to protect the treatment works, sludge, and water quality. Actual limits on the amount of industrial pollutants discharged to a POTW are set by the EPA and municipal authorities that have been delegated pretreatment authority.

Since DEQ's authority to approve municipal pretreatment programs was granted on March 12, 1981, there has been measurable improvements in EPA's pretreatment program. POTWs managing local pretreatment programs are seeing lower levels of toxic pollutants in their influent. (This then results in lower toxic metals in the sludges and allows for beneficial use of the sludge.) There are also less treatment plant upsets and lower level of toxics passing through the treatment plant. There is also an increase in enforcement efforts by local pretreatment POTWs against noncomplying facilities.

GRANTS FOR CONSTRUCTION OF TREATMENT WORKS

Section 201, the FWPCA, Amendments of 1972 established a federal grant program for the construction of POTWs. The federal share was initially 75% of the cost of construction of the treatment works including engineering planning and design. Treatment works included the municipal sewage treatment plant, interceptor, and collection sewers; land that is used for the treatment process is used for the ultimate disposal of sludges, etc. The federal share was eventually reduced to 55% in 1981 and the entire construction grants program was phased-out in 1990 and replaced by the state revolving loan fund program.

Since 1972, an enormous investment of federal dollars has been made to construct sewerage facilities in Oregon. More than $548.9 million has been awarded to POTWs from 1972 to 1991. This was matched by local funds of $204.3 million.

WATER-QUALITY STANDARDS

When the FWPCA was passed in 1972, EPA embarked on a long-term program aimed at restoring and maintaining the chemical, physical, and biological integrity of U.S. bodies of water.

Section 303(c)(2)(B) of the Water Quality Act of 1987 required states to adopt criteria for all toxic pollutants as part of their water quality standards reviews. The EPA National Toxics Rule was promulgated on December 22, 1992. This rule promulgated the chemical-specific, numeric criteria for priority pollutants necessary to bring a state into compliance with the CWA.

Oregon adopted the toxic criteria as part of their water-quality standards. The 1992 water-quality assessment report (305b) identified 12 bodies of water throughout the state that had levels of toxic pollutants above guidance values for fish tissue and sediment. In the Willamette Basin, arsenic, chromium, copper, lead, zinc, DDT, and dioxin (2,3,7,8-TCDD) are the main toxic pollutants.

NONPOINT SOURCE ASSESSMENTS AND MANAGEMENT PROGRAMS

Oregon's effort to control nonpoint source (NPS) pollution began under Section 208 of the FWPCA, Amendments of 1972. The initial effort was aimed at several critical needs. DEQ coordinated statewide studies on silviculture (forestry) confined animal feeding operations (dairies and feedlots) primarily in Tillamook County, stream habitat restoration, and on-site (septic) system waste disposal.

The Water Quality Act of 1987 shifted the approach of NPS water pollution control toward a new national NPS Action Program by creating Section 319. NPS has been identified as the dominant fraction of the remaining surface-water pollution problem in the U.S. In addition, the role of NPS in the pollution of groundwater resources has also become increasingly evident.

Section 319 requires that states assess their waters and develop management programs to control nonpoint sources, and authorizes federal grants to help implement these programs. The Water Quality Act targeted the development of the NPS programs that integrated surface and groundwater control efforts.

The primary purpose of the Section 319 assessment-planning effort is to provide the states and tribes with a new blueprint for implementing integrated programs to address high priority NPS water-quality problems. This focus is needed in order to identify innovative funding opportunities and to effectively direct limited resources toward the highest priority issues and bodies of water. The secondary purpose involves the fulfillment of the CWA requirements in order for states and tribes to compete for Section 319 grants for implementing NPS controls.

In Oregon, DEQ has completed their comprehensive NPS assessment-planning effort. Future Section 319 funds in Oregon will be directed to the implementation phase of controlling NPS based on priority needs.

SLUDGE

Section 405 of the Water Quality Act of 1987 requires NPDES permits (or other permits affording equivalent restrictions) to regulate the use and disposal of sewage sludge. The primary purpose of the sludge program is to ensure that sewage sludge is disposed in a manner that protects human health and the environment.

On February 19, 1993, the U.S. EPA promulgated technical standards establishing acceptable sludge quality levels for land application, marketing, and for municipal sludge incinerators. The regulations set national standards for pathogens and ten heavy metals. The document also defines management practices for the safe handling and use of sewage sludge.

The sludge regulations require sewage treatment plants to develop a sludge management plan and sample the sludge for toxic metals. Sewage sludge permit applications are to be submitted in phases. Municipalities with sludge incinerators receive the most attention initially and have to comply with the air toxics rule under the Clean Water Act. Incinerators were required to submit an application in mid-August 1993.

Though the Oregon DEQ has not adopted EPA's regulations pertaining to the application of sewage sludge, DEQ does regulate the land application and disposal of sewage treatment plant sludge and sludge-derived products, including septage through a NPDES or Water Pollution Control Facility (WPCF) permit, which is a state permit issued to facilities that do not discharge to a receiving stream. Their regulations require that a sludge management plan must be submitted and approved by the Department by all persons engaged in sludge disposal or application activity.

The sludge management plan must include an identification of the site, a determination of sludge stability, sludge analysis including heavy metals, and application rates.

DREDGE AND FILL PROGRAM

EPA and the Corps of Engineers (COE) currently use the methods contained in the COE's Wetland Delineation Manual published in 1987. This manual requires that positive indication of all **three** parameters (i.e., plant communities, soils, and hydrology that are characteristic of wetlands) must be present to identify an area as a wetland.

The protection of wetlands in Oregon is primarily the responsibility of the Oregon Division of State Lands (DSL) and DEQ. The DSL's and DEQ's wetland protection activities are currently based on administration of Oregon Removal and Fill Law and Section 401 of the Clean Water Act.

Since June 1990, DEQ has been in the process of developing a wetlands program consistent with its responsibilities under Section 401. The DEQ wetland program will include:

- wetlands in the definition of "waters of the state"
- developing of designated uses specifically for wetlands
- developing narrative criteria specific to wetlands, including those in maintaining hydrology
- applying narrative criteria and appropriate existing numeric criteria through a classification system
- developing narrative biological criteria specific to wetlands
- developing numeric biological criteria specific to wetlands

Once implemented, the DEQ wetland program will greatly clarify the state's role in the protection of wetlands.

1986 AND 1987 LAWSUITS AGAINST EPA

Due to the magnitude of the water pollution problems and the program priorities in Oregon in the 1970s and early 1980s, both the U.S. EPA and the state of Oregon administered a water pollution control program that was oriented toward point source controls. The issuance and elimination of industrial NPDES permits with technology-based controls, assuring compliance of municipal dischargers with secondary treatment requirements, and the certification and awarding of construction grants within specified time frames were the major program priorities during this period. The effort came to an end in 1986 as a result of a lawsuit against the EPA.

On December 12, 1986, the Northwest Environmental Defense Center and John R. Churchill filed a lawsuit against the EPA for failure to comply with and enforce the federal Clean Water Act, as it applies to waters of the United States within the state of Oregon, specifically the Tualatin River Basin and Lake Oswego in Oregon.

The complaint stated that the EPA failed to establish and enforce Total Maximum Daily Loads (TMDLs) to reduce nutrient discharges to meet water-quality standards in and on the Tualatin River and Lake Oswego. On March 20, 1987, the complaint was amended to 22 other waterbodies in Oregon where TMDLs were not established.

After several meetings with the plaintiffs, the EPA signed a consent decree on June 3, 1987 whereby TMDLs would be established on 12 waterbodies in 1987 and 1988 and that future TMDLs would be set on all other waterbodies identified as "water quality limited" in the state of Oregon Section 305(b) water-quality reports at a rate of 20% annually, but in no event less than 2 annually.

CONCLUSIONS

The Oregon DEQ routinely monitors 3500 miles of streams in its ambient monitoring program. These streams receive approximately 90% of the point source loads for the state. In addition to the

TABLE 1
Estimates of Categorized Beneficial Uses in the 1972, 1982, 1986, and 1992 Water-Quality Assessment Reports

Year	Fully supported	Partially supported	Not supported	Unknown	Total miles assessed
1972	2687 (60%)	1077 (24%)	353 (8%)	362 (8%)	4479 (100%)
1982	3309 (74%)	897 (20%)	273 (6%)	—	4479 (100%)
1986	9665 (82%)	1915 (16%)	265 (2%)	—	11,855[a] (100%)
1992	322	1321	2829	—	4472
	12,330	7381	4926		24,637[b]
	12,652	8702	7755		29,109

[a] Includes nonpoint source related to on-site septic tank/drainfield systems, agricultural, and other nonpoint sources.

[b] Includes information from DEQ nonpoint sources assessment compiled in 1988.

routine sampling at established river and estuary stations, DEQ also conducts intensive studies on water-quality-limited waterbodies where water-quality standards are being violated, conducts compliance monitoring for waste discharge permits, performs bioassessments, and directs special studies. Approximately 24,000 miles have been evaluated in the 1988 Statewide Nonpoint Source Assessment as shown in Table 1.

River miles are categorized as "fully," "partially," or "not supporting" beneficial uses. The estimates in the 1992 water quality assessment report identified 12,652 miles as fully supported, 8702 miles as partially supporting, and 7755 as not supporting beneficial uses. These figures include both point and nonpoint source contribution.

When a comparison of water-quality assessment reports was made, it appeared there was an increasing percentage of streams that fully supported beneficial uses during the period of 1972 to 1982. The 1986 and 1992 305(b) report factored nonpoint sources of pollution into the beneficial use category, and as a result, the improving quality from a point-source viewpoint in Oregon's streams was not readily reflected.

One can conclude that the Clean Water Act and its implementation in Oregon has improved the quality of rivers in many parts of the state. However, in other parts of the state, the rivers continue to be stressed due to various contaminates from point and nonpoint sources of pollution.

REFERENCES

Oregon Department of Environmental Quality, Water Quality Control in Oregon, December 1970.

Oregon Department of Environmental Quality, Water Quality Oregon 1980, October 1981.

Oregon Department of Environmental Quality, Oregon's 1984 Water Quality Status Assessment Report (305b) Report, July 1984.

Oregon Department of Environmental Quality, Oregon's 1986 Water Quality Status Assessment Report (305b) Report, June 1986.

Oregon Department of Environmental Quality, Oregon's 1992 Water Quality Status Assessment Report (305b) Report, April 1992.

Thomson, E., Oregon Department of Environmental Quality, Personal discussion, January 27, 1994.

U.S. Congress, Federal Water Pollution Control Acts, Amendments of 1972, 1977, 1981, and 1987.

U.S. Environmental Protection Agency, *Federal Register*, Vol. 39, No. 9, Rules and Regulations, February 11, 1974.

U.S. Environmental Protection Agency — The Clean Water Act, Compliance/Enforcement Guidance Manual, 1985.

U.S. General Accounting Office, Water Pollution — More EPA Action Needed to Improve the Quality of Heavily Polluted Waters, GAO/RCED-89-38, January 1989.

U.S. House of Representatives, Committee on Public Works, Laws of the United States Relating to Water Pollution Control and Environmental Quality, 93-1, March 1973.

William, D. C., The Guardian EPA's Formative Years 1970–1973, September 1993.

Index

urban distribution system, 394
water quality, 395
watershed. see watersheds

C

Cadmium. see metals
Carbonaceous biochemical oxygen demand,
    158, 162
  Tualatin River Basin, 155–156
Carpathian Mountain Range, 5
Cascade Range, 3, 7, 132, 133, 134, 187, 199
  drainage density comparison, 110, 112
  ecoregions, 275
CBOD. see carbonaceous biochemical oxygen
    demand
CE-QUAL-W2, 155, 156
Channelization, 40
Channel loss, Willamette River
  channel constraints, 35
  channel cut-offs, 36
  channel width, 28–29
  dams, 27
  deposition, 25
  downcutting, 25
  history, 24–25
  land-river interface, effect on, 27
  large woody debris, 33, 34, 35
    snags, 33, 35
    wood jams, 33, 35
  large woody debris, role, 25, 27
  scour, 25
  secondary channels, 24
  sediment, 25, 33, 35
  side-channel elimination, 27
  simplification, 25, 26, 37
  survey plats, 24
Channel morphology, 153
Chlopyrifos. see pesticides
Chromium. see metals
  in water. see trace elements
Cleanup, Willamette River. see Willamette
    River
Clean Water Act, 69, 443, 444
Clyde River
  Control of Pollution Act (1974), 429–430
  effluents, 428
  future outlook, 431–431
  history, 423, 424
  integrated pollution control, 432
  local government reform, 429
  pollution
    control philosophy, 431
    marine, control, 429
    progress, 430
    severity, 427
  private legislation, 429
  Purification Board, 426, 427, 428
  river catchments, 426
  salmon return, 431
  standards for effluents, 425
Coast Range, 132, 152, 229, 300

Colloid
  adsorption experiments, 207, 208
  -associated elements, 205
  coprecipitation experiments, 207, 208–209
  definition, 205
  filtration artifacts. see filtration artifacts
  iron oxides, 205
    average concentrations, Tualatin River
      Basin, 208
    effect on phosphorus, 207
    fractions in suspended sediment, Tualatin
      River Basin, 209–210
  phosphorus, 207–208
    average concentrations, Tualatin River
      Basin, 208
    fractions in suspended sediment, Tualatin
      River Basin, 209–210
  pools, 207
  silicate experiments, 207, 209, 210
Colorado Discharge Permit System, 339
Colorado water-quality assessment
  history, 341
  Standley Lake evaluations, 343
    control regulation proposals, 344
    settlement agreement, 344
    streamflow-quality modeling, 341–343
    water-supply description, 339–340
Columbia River, 69, 131, 133, 134, 163, 165,
    170, 213
Columbia River Basalt
  aquifer, 131, 133, 135, 136, 138
  Group, 131, 133, 138, 194
Columbia Slough, 213, 219
Conservative constituent transport, 128
Control of Pollution Act (1974), 429–430
Copper. see metals
Coprecipitation experiments. see colloid
Cracow, 12
Cross-sectional numerical flow model. see
    modeling
Culture, effect on perception of rivers, 42–44

D

Dams
  Congressionally authorized, Willamette
    River Basin, 71
  effect on Willamette River floods, 41
  LOC diversion, 156
  pre-dam vs. regulated flows, Willamette
    River, 39, 40
  pre-flow vs. post flow, Willamette River
    Basin, 113
  proposed locations, Vistula River, 372
  types
    closing, 27
    cut-off, 27
    water-contracting low, 27
    water-storage, 37
    wing, 30
  Willamette River Basin flood control system,
    70